Control of Uncertain Systems
A Linear Programming Approach

Munther A. Dahleh
Ignacio J. Diaz-Bobillo

Department of Electrical Engineering and Computer Science
Massachusetts Institute of Technology

PRENTICE HALL, Englewood Cliffs, New Jersey 07632

Library of Congress Cataloging-in-Publication Data

Dahleh, Munther A.
    Control of uncertain systems: a linear programming approach /
Munther A. Dahleh, Ignacio J. Diaz-Bobillo.
        p.   cm.
    Includes bibliographical references and index.
    ISBN 0-13-280645-2
    1. Automatic control.   2. Linear systems.   I. Diaz-Bobillo, Ignacio J.
II. Title.
TJ213.D24   1995
629.8'312—dc20

                                        94-8256
                                        CIP

Acquisitions editor: **LINDA RATTS**
Editorial/production supervision
    and interior design: **RICHARD DeLORENZO**
Copy editor: **JAMES TULLY**
Cover design: **DeLUCA DESIGN**
Cover illustration: The picture on the cover is our own visualization of the oldest known feedback system; namely, a
    water clock. This clock was built around 300 B.C. by the Greeks. The main purpose of this mechanism shown is to
    regulate the rate by which the water enters the main vessel. A detailed description of the clock and its function can be
    found in: Otto Mayr, *The origins of feedback control*, M.I.T. Press, Cambridge, 1970. The drawing was made by Nasser
    Rabbat.
Production coordinators: **DAVID DICKEY & PHIL ZOLIT**
Editorial assistant: **JENNIFER KLEIN**

The author and publisher of this book have used their best efforts in preparing this book. These efforts include the
development, research, and testing of the theories and programs to determine their effectiveness. The author and
publisher make no warranty of any kind, expressed or implied, with regard to these programs or the documentation
contained in this book. The author and publisher shall not be liable in any event for incidental or consequential damages
in connection with, or arising out of, the furnishing, performance, or use of these programs.

Printed in the United States of America

10 9 8 7 6 5 4 3 2

ISBN 0-13-280645-2

Prentice-Hall International (UK) Limited, London
Prentice-Hall of Australia Pty. Limited, Sydney
Prentice-Hall Canada, Inc., Toronto
Prentice-Hall Hispanoamericana, S.A., Mexico
Prentice-Hall of India Private Limited, New Delhi
Prentice-Hall of Japan, Inc., Tokyo
Simon & Schuster Asia Pte. Ltd., Singapore
Editora Prentice-Hall do Brasil, Ltda., Rio de Janeiro

Contents

Preface

The field of control theory has been steadily evolving in the past several decades. It has made a giant leap in the past fifteen years with the introduction of the *small gain paradigm* as a framework for designing robust multivariable control systems. Since then substantial advancements have been made to bring this paradigm to a level that is usable by practicing engineers. While the story is still not complete, it has already resulted in methodologies within the above paradigm that can be utilized for design. Although some very good books have been written on specialized aspects of the subject of robust control, we have not come across a book that treats this subject in a unified fashion. In this text we made an attempt in that direction.

This volume has evolved as a consequence of a graduate course in robust control that the first author has taught at MIT for the past five years. It is heavily biased by our view of the field as well as the research we are conducting; however, it is not exclusively limited to that. The title *Control of Uncertain Systems* has specific connotations: *Control* means designing a feedback control system that delivers some given performance specifications, and *Uncertain Systems* refers to a description of a real process in terms of a class of plants in which the process is believed to lie. The subtitle: *A Linear Programming Approach* indicates that the book emphasizes computations and attempts at showing the power of linear programming in controller design. The book is written to be suitable for self-reading as well as teaching in a first-year or second-year graduate course in control theory.

The main purpose of this book is to present a computational theory for robust control that highlights the fundamental limitations and capabilities of linear controller design for plants with uncertainty. It is unique in that it gives a detailed discussion of the general stability and performance robustness problems in the presence of various classes of structured perturbations, and then it covers in detail synthesis problems using the celebrated $\mathcal{H}_2$, $\mathcal{H}_\infty$, and ℓ_1 methods. In particular, it emphasizes the role of infinite dimensional linear programming in the synthesis of controllers that meet a variety of performance specification. The book presents various explicit algorithms for solving such problems and offers a variety of examples.

This book can be read in many different ways and accordingly has different prerequisites. For the user who is not interested in the derivations, the final results are quite accessible with standard linear algebra knowledge. Sufficient examples are presented that show the utility of the results. For the reader interested in the theory, some basic knowledge from functional analysis is required. The essential mathematical facts required are all summarized in Appendix A. The necessary tools from optimization theory, particularly duality theory and linear programming, are discussed in detail within the text. Because most of the discussion requires discrete-time function spaces, details on continuous-time function spaces are not needed.

ACKNOWLEDGMENTS

Many people have helped in getting this book to its present form, and to them we are quite indebted. In particular, we would like to thank J. Boyd Pearson: Boyd has been an inspiration to us all through our careers and was one of the first people who encouraged us to write this book. He continued to play a very active role in the development of the book through his very detailed and valuable comments while testing the book in his classes. Also, we would like to thank Bassam Bamieh, Mohammed Dahleh, Mustafa Khammash, Petros Voulgaris, and Brett Ridgely for testing earlier drafts of the book in their classes and giving valuable comments and suggestions about the contents and organization of the book, Stephen Boyd for detailed comments on the first draft, Pablo Iglesias for various comments on Chapter 13, John Doyle, Sanjoy Mitter and George Zames for the many discussions we had that influenced our view of the field, Jinane Abounadi and Bonny Peirce for copyediting and many stylistic suggestions, and the reviewers of the first draft (Pramod Khargonekar and Tryphon Georgiou) for their helpful comments and suggestions, which considerably affected the outcome of this project. We also would like to thank the students from MIT and Caltech who relayed comments and suggestions to us at different stages of the book, in particular Raffaello D'Andrea, Nicola Elia, Marcos Escobar, Mitch Livston, Steve Patek, and Venkatesh Saligrama.

We would like to acknowledge the support of ARO, NSF, AFOSR, and Draper Laboratory during the period in which this manuscript was written. Also, we would like to acknowledge the EECS Department at MIT for providing support for book development, and the EE Department at Caltech for providing the opportunity to test the book while the first author was on leave at Caltech.

Writing this volume required a lot of time, much of which was taken from time that belonged to our families. We are very blessed to have such patient, supporting, and loving spouses, Jinane and Ivana, without whom this book would not have come to completion. They were certainly instrumental in this project and to them we are very grateful. Also while working on the book, we have been blessed with the addition of Deema, Marcos, and Victoria to our families. Finally, we would like to thank our parents, brothers and sisters, and the rest of our families and friends for their continuous support.

To my parents, Wisam and Abdullah

——*Munther A. Dahleh*

To the memory of my father, Victor Diaz-Bobillo

——*Ignacio J. Diaz-Bobillo*

CHAPTER 1 ─────────────

Introduction to Robust Control

With the growing complexity of dynamic systems and the need for automatic control with high-quality performance, it is necessary to develop a deeper understanding of the fundamental limitations and capabilities of control system design. This problem can be translated into two basic questions: Can a target performance be achieved? If so, what is the design that achieves it? The first question deals with the fundamental limitations of achievable performance in the presence of uncertainty, whether it is in the inputs or in the process itself. For some single-input single-output (SISO) systems, these limitations may be captured in a precise analytical way. For more complicated systems, however, no analytical solutions are available, and it is necessary to resort to numerical techniques. The second question is concerned with developing a systematic procedure for designing control systems that optimizes performance in the presence of uncertainty. The design issue is, of course, not at all separate from the limitations issue, and the two questions are strongly tied together.

The theory of control systems has evolved considerably in the past fifty years. Earlier, design techniques focused on SISO systems and frequency domain methods such as those based on Bode plots and Nyquist plots. Feedback control was used to desensitize control systems to changes in the process as well as to stabilize unstable systems. The need for more sophisticated control systems pushed researchers to move toward time-domain descriptions, which resulted in a state-space theory. This theory deals with basic system-theoretic questions and offers a design methodology based on Linear Quadratic Regulator, and Kalman Filter theory (LQG/$\mathcal{H}_2$). While the state-space LQG theory remains popular up to the present time, the direct incorporation of plant uncertainty in synthesis leads to computationally difficult problems. Hence robustness of the closed-loop system is guaranteed only a posteriori (to some level of uncertainty). Motivated by the interest to address uncertainty in a systematic and tractable way, Zames formulated what is now called the $\mathcal{H}_\infty$ problem. This problem, later solved through the work of many researchers, is strongly rooted in classical techniques and has motivated a powerful paradigm for control system design under uncertainty. In this paradigm, uncertainty is incorporated a priori in the design, and the question of fundamental limitations is directly addressed. The structured perturbation problem, formulated by Doyle, has highlighted the differences between MIMO systems and SISO systems in an elegant way and made it possible to address multivariable systems through the structured singular value analysis and synthesis. Nevertheless, the theory developed is purely a frequency domain theory and time-domain specifications such as tracking or saturation constraints are not considered explicitly. This led to the development of the ℓ_1 theory. This new theory, which fits under the same paradigm as the $\mathcal{H}_\infty$ theory, addresses a different class of problems and captures the fundamental limitations of achievable time-domain performance under uncertainty. Both the $\mathcal{H}_\infty$ and the ℓ_1 problems have motivated a class of numerical formulations that deal with combined objectives and various uncertain environments. Although these are not fully developed yet, they do provide usable computational methods for designing controllers and for highlighting the limits of performance.

1.1 APPLICATIONS

There are a variety of complex systems that require sophisticated control strategies to achieve acceptable performance within the uncertain environment in which they operate. This section lists a few of such examples to highlight some of the main issues addressed in this book.

Robotics. Robots are used in many applications ranging from unmanned automated systems to robotic arms used to aid human operators. Typically, the various parameters of the robot are not known precisely; however, a crude estimate is available. A design strategy based on linear control may also require linearizing the model of the robot around an equilibrium point, which suggests that the linear model describes the process only approximately. Even if the control strategy is based on feedback linearization, the resulting system is never linear, and the ignored dynamics have to be accounted for. Despite the lack of an accurate model, typically the robot is intended to have good tracking performance, while observing constraints on the allowable control effort.

Space structures. These are examples of very complex systems with many inputs and outputs, and only crude models are available. Designing controllers for such systems is a major challenge even when stability is the only objective. Such applications highlight the power of the current machinery in optimizing performance when the uncertainty is structured.

Aircraft. The controller design for a highly maneuverable aircraft is a good example of a problem in which the performance requirements are of prime importance.

The control action is limited in both magnitude and bandwidth to prevent exciting the high frequency bending modes. The model of the process is valid only in a region around the equilibrium point assuming that the aircraft is a rigid body. Not only should the controller account for the unmodeled dynamics, unexpected changes in the dynamics that are not captured in the model, and unexpected disturbances, but it should also deliver high performance in terms of tracking trajectories for a variety of commands.

High purity distillation columns.   High purity distillation columns are good illustrations of the differences between multi-input multi-output (MIMO) systems and single-input single-output systems (SISO). Depending on the particular configuration of a distillation column, certain directions of the inputs have much higher gain than other directions, resulting in a plant that is ill conditioned. Both classical designs (where a MIMO system is treated as a collection of SISO systems) as well as LQG do not address this problem effectively. When applied to distillation columns, the resulting designs may turn out to be quite sensitive to certain directional uncertainties.

1.2 GENERAL DESCRIPTION

Figure 1.1 depicts a general description of a control system. The process is viewed as an element of a larger set of plants that reflects the modeling errors. The system is subjected to *exogenous inputs* such as disturbance, noise, or commands. These inputs represent, in general, the input uncertainty. The system has a set of *control inputs* and a set of *measured outputs*, both accessible for feedback. Finally, a set of fictitious outputs is defined that captures performance objectives in terms of error variables. Such outputs are known as *regulated outputs*.

Plant uncertainty.   In general, plant uncertainty is described using certain classes of perturbations. This description is an outcome of modeling and prior knowledge of the physical process.

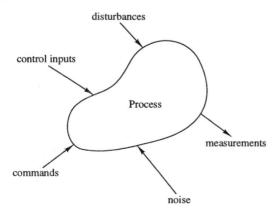

Figure 1.1   General uncertain system.

1. **Nonparametric Perturbations:** These are perturbations of the input-output description of the process. In general, they are assumed to be bounded in some precise sense. Such a class can describe various aspects of plant uncertainty. For example, it is quite straightforward to represent unmodeled high-frequency dynamics as an *additive* perturbation with bounded frequency response. The effect of linearization or time variation of the system can be described by a nonlinear time-varying perturbation of a linear time-invariant model. In general, we will refer to such perturbations as nonparametric norm-bounded perturbations. The selection of the norm in this context depends on the particular application.

   Nonparametric perturbations may be structured. This means that independent perturbations occur at different locations of the closed loop system. For instance, we might want to consider sensor unmodeled dynamics and actuator unmodeled dynamics as independent blocks of perturbations. In general, some blocks are linear time invariant and others may be linear time varying or nonlinear.

2. **Parametric Perturbations:** These are perturbations of specific parameters of the process. In many situations these parameters have physical meaning such as a spring constant in a mechanical system or a resistance in a circuit. In general, however, parametric perturbations are described in a state-space setting, and may not correspond to actual physical parameters.

A general description of uncertainty should have both of these classes of perturbations. Although parametric perturbations are quite natural, they are in effect much harder to deal with than nonparametric perturbations. The theory for analysis and synthesis in the presence of mixed perturbations is not fully developed.

Exogenous inputs. There are two major directions in modeling disturbances.

1. **Stochastic Models:** This has been the traditional method for describing disturbance and noise. In this setup, the signal is assumed to be a sequence of random variables with a known joint probability distribution. Because of the difficulty of modeling such signals in the full generality, only second-order statistics (mean and covariance) are typically discussed or modeled.

2. **Deterministic Models:** Such models describe the class of signals as unknown but bounded in some precise sense. Examples of these classes are bounded-energy signals, bounded-magnitude signals, bounded-power signals, etc. More generally, signals bounded by a set of convex constraints are good models for exogenous inputs.

Regulated outputs. A regulated output contains the error signals that capture the performance specifications for the given system. Typical performance specifications involve good disturbance rejection, good tracking, limits on the control authority and bandwidth, overshoot, undershoot and settling time specifications for the response to a fixed command such as a step input, etc. Such specifications are in general stated in terms of "a norm" of the error. Examples of such norms are the maximum magnitude of the error over time, or the energy of the error. In a more general setting, such specifications are described in terms of a set of convex constraints such as a time template in which the response (possibly to a specific input) is forced to lie.

Finally, we note that if the exogenous inputs are modeled as stochastic signals, then the regulated output is stochastic, and the measures above will be taken with expected values.

Control inputs and measured outputs.   These are the signals available for feedback. The outputs are in general assumed to be noisy, and the noise input is included in the exogenous inputs. Both the inputs and outputs are generally multivariable.

1.3 STABILITY AND PERFORMANCE ROBUSTNESS

The major problem in controller design is finding a controller that can deliver good performance in the presence of uncertainty, both in the model and the inputs. To address this problem, it is convenient to study an intermediate problem in which stability is the only objective. We define these two problems below.

Stability robustness problem.   Find a feedback controller such that the system is stable for all plants in the set of plant uncertainty.

Performance robustness problem.   Find a feedback controller such that the system is stable and meets the desired performance objectives for all plants in the set of plant uncertainty, in the presence of all possible exogenous signals.

Worst-case paradigm.   When no probability distribution is assigned to the uncertainty, the above robust performance problem is cast under the worst-case paradigm. This is because the performance specifications are to be met for all exogenous signals and for all plants in the set of uncertainty. An alternate view of such a problem is game-theoretic. We think of the control input as a signal trying to satisfy the performance objectives. Plant perturbations and exogenous inputs are regarded as adversaries that are trying to maximize the errors. Both of these adversaries have access to the control input. It may seem initially that this methodology will yield a conservative design since the final closed-loop system guards against exogenous inputs and plant perturbations that may not be very realistic. It turns out, however, that such a design has more desirable robustness properties than a design based on very specific descriptions of uncertainty, or on stochastic averaging techniques.

Two-phase solution.   The solution to robust stability and performance problems are carried out in two stages: the analysis and synthesis stages.

1. **Analysis:** This entails the development of computable methods for verifying whether or not a given closed-loop system meets certain design specifications in the presence of uncertainty. Such methods are quite desirable from a practical point of view.

2. **Synthesis:** This entails developing techniques for designing controllers that meet given performance specifications in the presence of uncertainty. Synthesis methods are usually different for different classes of uncertainty as well as for different specifications. Finding such controllers involves solving optimization problems, which may or may not have exact solutions in closed form. Closed-form solutions can be derived for simplified scenarios, and are usually quite helpful in understanding the limits of performance and some general properties of the optimal controllers.

1.4 MOTIVATION

We give some reasons behind the development of such a mathematically advanced methodology.

1. **Complex Systems.** Many of today's systems, ranging from space structures to high purity distillation columns, are quite complex. The complexity comes from the very high order of the system as well as the large number of inputs and outputs. Modeling such systems accurately is a difficult task and may not be possible. A powerful methodology that deals systematically with multiple inputs and outputs and with various classes of structured uncertainty is essential.

2. **High Performance Requirement.** Systems are built to perform very specific jobs with very high accuracy. Robots are already used to perform very accurate jobs such as placing components on integrated circuits boards. Aircraft are built with high maneuverability and are designed specifically for such tasks. Classical SISO design techniques cannot accommodate these problems and usually result in designs that have conservative performance.

3. **Limits of Performance.** In complex systems, it is time-consuming to establish by trial and error whether a system can meet certain performance objectives (even without uncertainty). Thus, it is necessary to develop systematic methods to quantify the fundamental limitations of systems and to highlight the trade-offs of a given design.

4. **A Systematic Design Process.** It is inevitable that designing a controller for a system will involve iterations. Unless this procedure is made systematic, the design process can become very cumbersome and slow. The design procedure should not only exhibit a controller. It should also provide the designer with indicators for the next iteration by showing which of the constraints are the limiting ones and which part of the uncertainty is causing the most difficulty. Note also that a general procedure should be able to accommodate a variety of constraints, both in the time and in the frequency domain.

5. **Computable Methods.** It is quite straightforward to formulate important control problems, but it is not so easy to formulate solvable problems that can provide computable methods for analysis and synthesis. Much of the current research invokes high-level mathematics to ultimately provide simple computable results. This will be demonstrated in almost every chapter of this book. In a sense, the computability of a methodology is the test of its success.

6. **Technological Advancement.** Many aspects of technological development will affect the design of control systems. The availability of "cheaper" sensors and actuators that are well modeled allows for designing control systems with a large number of inputs and outputs. The availability of fast microprocessors, as well as memory, makes it possible to implement more complex feedback systems with high order. The limiting factor in controller implementation is no longer the order of the controller. Instead, it is the computational power and the memory availability.

7. **Impact on Technology.** While technology advancement helps to implement complex control systems, a robust control theory impacts the design of the actual processes. By understanding the limitations of certain dynamic systems, it is possible to design a process that can exhibit high performance in the face of uncertainty. In the future, it is hoped that the design of the process will be done concurrently with the analysis of the control capabilities, a philosophy that is not widely practiced at the present time.

1.5 STEPS FOR ADDRESSING THE PERFORMANCE ROBUSTNESS PROBLEM

This book, mostly based on the worst-case paradigm, can be divided into three major parts.

1. The first part addresses the development of nonconservative conditions that guarantee the performance robustness of the closed loop system. These conditions are different depending on the classes of perturbations, the exogenous inputs, and the performance requirements.

2. The second part gives a complete characterization of achievable input-output maps for a given process using arbitrary stabilizing controllers. This is in general an algebraic characterization and is analyzed from both a theoretical and computational point of view.

3. The third part poses the controller design as an optimization problem using the robust performance conditions and the characterization of the closed-loop maps. Most solutions of such problems can be obtained only by numerical techniques. There are, however, special cases where solutions can be obtained exactly and these will be studied in detail.

Finally, examples and case studies are presented to show how these three steps are used in the design of a system.

1.6 PHILOSOPHY OF THE BOOK

This work is better described as one on the foundations of robust control. By this we mean that we present the theory and show its potential value for applications. We do not take the approach of presenting the experience gained in applying the different methods of design to real-life problems.

The presentation is entirely for discrete-time systems. The main motivation behind this is that many controllers are implemented in a digital fashion. Of course, this makes the system a *hybrid* one, i.e., governed by both discrete-time and continuous-time dynamics. Such systems are known as *sampled-data* systems. In this book, we have chosen not to address the general hybrid problem, although an interesting theory is emerging to accommodate it. Instead, we show that such a system can be dealt with via approximation with a multirate discrete-time system, which provides a systematic way for addressing the general problem. However, basic ideas involved in digital control, such as sampling and the consequences of choosing a sampling rate, are not covered. Such material can be found in any basic digital control textbook. Finally, we note that many of the results for discrete-time systems that are discussed in this book can be extended to continuous-time systems in a straightforward fashion. References for such results will be given.

The book puts a major emphasis on time-domain specifications. Such problems in general give rise to objectives and constraints in terms of the peak values of signals. Typical controller design methodologies focus on energy descriptions and avoid working with magnitude descriptions—possibly since the mathematics do not seem as elegant. However, we study in full detail the problem of stability and performance robustness when the inputs are bounded in magnitude and the objective is to keep the maximum amplitude of the regulated output bounded, in the presence of bounded-input bounded-output perturbations. We also give a complete account of the ℓ_1 design methodology—a methodology based on minimizing the Peak-to-Peak gain of the system. Although the emphasis is on ℓ_1, we present the $\mathcal{H}_\infty$ and $\mathcal{H}_2$ counterparts to the ℓ_1 results.

For several reasons, our emphasis is on time-domain problems. First, most of the relevant specifications in control systems are in the time domain. For instance, tracking problems are quite common in control systems and can be quite difficult if the system is subjected to persistent disturbances with limits on the actuator's authority. This is entirely a time-domain problem, and frequency domain methods deal with it only indirectly. Second, time-domain specifications, in general, give rise to linear constraints on the set of all achievable input-output maps. This enables us to utilize the power of linear programming algorithms for designing control systems. If some of the specifications are in the frequency-domain, resulting from bandwidth constraints, they can be well approximated by linear constraints and are easily incorporated in the design. Finally, as we mentioned earlier, problems with direct time-domain specifications have not been addressed by many textbooks on robust control, possibly due to their mathematical difficulty. We hope to show in this book that it is possible to extract from the unfamiliar mathematical tools a very simple and elegant theory for designing controllers for such problems.

Our objective is to motivate a research direction that focuses on computations. Indeed, a good computational algorithm for synthesizing robust controllers should provide three different pieces of information: upper and lower approximations of the objective functions, feasible solutions, and indicators to the hardest objectives to meet. By numerical solutions, we do not mean that optimization problems are formulated and then solved in a brute force way using optimization algorithms. Instead, we mean deriving a computer-aided design environment that provides qualitative (as well as quantitative) information about the solutions.

Complete solutions for current methodologies ($\mathcal{H}_2$, $\mathcal{H}_\infty$ and ℓ_1) are presented for SISO systems and the results are compared. We believe that many of the properties of the optimal solutions are apparent in SISO systems, and comparisons between these techniques can give much insight into these methodologies.

The development in the book follows the philosophy: *Think in terms of input-output maps but compute in state-space*. The material presented is based on input-output theory. With this thinking, plant uncertainty is represented as input-output perturbation and is independent of the representation of the model itself. The performance objectives are stated in terms of input-output behavior, where the inputs are the exogenous inputs and the outputs are fictitious signals that contain important signals from the closed loop. On the other hand, much of the computations necessary to synthesize the controller are performed using matrix computations, and thus a state-space representation of the system becomes quite useful. Matrix computations are inherently more numerically robust than symbolic manipulations or manipulations of transfer functions, and are generally preferred for controller design.

Finally, it is important to note that any theory dealing with uncertainty is developed within some "paradigm" that defines acceptable descriptions of uncertainty in the process and in the inputs. In reality, the physical process is not uncertain. However, our lack of knowledge of its exact description makes it necessary for us to describe it with uncertainty. In that sense, any description of uncertainty is unverifiable. On the other hand, the uncertainty description is reasonable if it captures some of the information lost in the modeling process. A theory will prove to be effective if it is flexible enough to accommodate a variety of design specifications in the presence of different classes of uncertainty and to provide computational methods for design.

1.7 ORGANIZATION OF THE BOOK

The material presented in the book is organized in a chronological order, each chapter depending primarily on past chapters and the appendices. Nevertheless, depending on the objectives of the reader, a certain degree of flexibility is possible. Below are possible sequences for reading the book.

1. **Basic Results and SISO Synthesis.** This sequence is shown in Figure 1.2a. Of course, not all the sections in the chapters have to be covered. With such a sequence, the reader learns the basic definitions of signal and system norms, a systematic approach for posing general design specifications in this framework, the ideas behind the parametrization of all stabilizing controllers, the solution to the

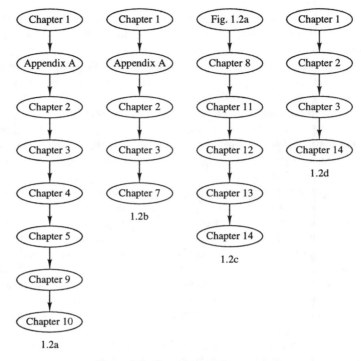

Figure 1.2   General uncertain system.

ℓ_1 model matching problem and its relations to linear programming, the solution to the $\mathcal{H}_2$ model matching problem using the projection theorem, and the solution to the $\mathcal{H}_\infty$ model matching problem using Hankel operator theory.

2. **Stability and Performance Robustness.** This topic is discussed in full detail in Chapter 7. The necessary sequence is shown in Figure 1.2b. All of the subsequent chapters except for Chapter 14 are independent of Chapter 7.

   The combined sequences of Figures 1.2a and 1.2b can make a one semester course. Of course, not all the results in all chapters need to be proved. We have also included a selection of exercises at the end of each chapter. Some of these exercises are not straightforward; however, detailed steps are given to direct the reader to the solution. We believe that these problems enhance the reader's understanding of the underlying theory. We have included references where the solutions to many of these problems can be found.

3. **MIMO Synthesis.** The required sequence is shown in Figure 1.2c. This can be covered in a second semester class. Since the book emphasizes numerical techniques using linear programs, we have explained the ℓ_1 theory in complete detail. The $\mathcal{H}_2$ problem is a well known problem and is explained in the book both for reference purposes as well as to outline the geometry of inner product spaces underlying the standard separation structure in the controller. The $\mathcal{H}_\infty$ problem,

is stated for reference purposes only and only the ideas behind the derivations are mentioned. If this book is used for a second semester class, some auxiliary material may be needed to cover the proofs behind the $\mathcal{H}_\infty$ problem.

4. **Applications.** Chapter 14 is written in such a way that an application-oriented user of the theory can start at the end, and work backward through the rest of the chapters to find out what she/he needs in order to use the theory. Some basic background is of course necessary. The required sequence is shown in Figure 1.2d.

CHAPTER 2 —————————————————————

Signals and Systems

Our view of systems is that they are operators acting on certain classes of signals. Although this view ignores the internal physical states of the system, it provides a useful tool for characterizing the system's uncertainty in a mathematically tractable way. It also makes it possible to provide a general framework for posing performance specifications in the presence of input and plant uncertainty.

In this chapter we introduce various signal spaces with emphasis on bounded signals. We give a complete characterization of operators on such signals, and we study basic notions associated with these operators such as linearity, causality, time invariance, and stability.

One important concept discussed is the notion of the *gain* of a system. By this we mean the amount of amplification that a system exerts on normalized bounded signals. This gain is a function of the norms of both the input and output signals and provides a very good measure of size for the system. Comparisons of different gains are conducted.

2.1 SIGNAL SPACES

In this section we study classes of signals that arise frequently in control systems. These classes consist of bounded energy signals, bounded magnitude signals, and bounded power signals. Other classes are also discussed.

12

2.1.1 The Classical ℓ_p Spaces

Let $\ell_p^n(\mathbf{Z})$ denote the space of all vector-valued real sequences on integers, of dimension n, i.e., $x = (\ldots, x(-1), x(0), x(1), \ldots)$ with $x(k) \in \mathbb{R}^n$, such that

$$\|x\|_p := \left(\sum_{k=-\infty}^{\infty} \sum_{i=1}^{n} |x_i(k)|^p \right)^{\frac{1}{p}} = \left(\sum_{k=-\infty}^{\infty} |x(k)|_p^p \right)^{\frac{1}{p}} < \infty. \qquad (2.1)$$

Here, we have adopted the notation $|\cdot|_p$ to denote the p-norm on $\mathbb{R}^n$ (See Example A.1.1). In the future, we will be particularly interested in the cases where $p = 1, 2, \infty$. In the case where $p = \infty$, the norm is defined as

$$\|x\|_\infty = \sup_k \max_i |x_i(k)|. \qquad (2.2)$$

Notice that the norm on the finite dimensional components is chosen consistently with the infinite dimensional norm. For the case $p = 2$, the norm $\|x\|_2$ is simply the amount of energy contained in the signal. If $p = \infty$, then $\|x\|_\infty$ is the maximum magnitude that the signal attains over all time. It is evident that every finite energy signal is also finite magnitude, but not vice-versa. In fact, the $\ell_p(\mathbf{Z})$ spaces are nested with $\ell_\infty(\mathbf{Z})$ as the largest. This is depicted in Figure 2.1.

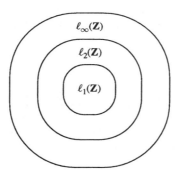

Figure 2.1   Spaces of sequences as nested sets.

Example 2.1.1

The one-sided sequence $\{1/k\}_{k=1}^{\infty}$ is in $\ell_\infty(\mathbf{Z})$ and $\ell_2(\mathbf{Z})$ but is not in $\ell_1(\mathbf{Z})$.

Example 2.1.2

All exponentially decaying one-sided scalar sequences of the form $x = \{a^k\}_{k=0}^{\infty}$, where $|a| < 1$, are in $\ell_p(\mathbf{Z})$ for $1 \le p \le \infty$. In fact,

$$\|x\|_p = \left(\frac{1}{1 - |a|^p} \right)^{\frac{1}{p}}.$$

If in the above sequence $a = 1$, then $x = \{1, 1, , \ldots\}$ which belongs to $\ell_\infty(\mathbf{Z})$ but not to $\ell_p(\mathbf{Z})$ for $p < \infty$.

The space $\ell_2^n(\mathbf{Z})$ is a Hilbert space with inner product defined as

$$\langle x, y \rangle = \sum_{k=-\infty}^{\infty} y^T(k)x(k). \tag{2.3}$$

Define the Fourier Transform of a sequence x as

$$\hat{x}(e^{i\theta}) = \sum_{k=-\infty}^{\infty} x(k)e^{ik\theta}. \tag{2.4}$$

The space of all Fourier Transforms of $\ell_2^n(\mathbf{Z})$ is also an inner product space with product defined as

$$\langle \hat{x}, \hat{y} \rangle = \frac{1}{2\pi} \int_0^{2\pi} \hat{y}^*(e^{i\theta})\hat{x}(e^{i\theta})d\theta, \tag{2.5}$$

where $*$ denotes conjugate-transpose. The Fourier Transform preserves the norm in $\ell_2^n(\mathbf{Z})$ in the following sense:

$$\langle x, y \rangle = \langle \hat{x}, \hat{y} \rangle.$$

In the next example, complex valued functions are discussed.

Example 2.1.3

It is quite common in the case of bounded energy signals to define complex-valued spaces. Consider the space $\ell_2^n(\mathbf{Z})$ of complex-valued sequences, with inner product defined as

$$\langle x, y \rangle = \sum_{k=-\infty}^{\infty} y^*(k)x(k). \tag{2.6}$$

Let $\mathcal{L}_2^n[0, 2\pi]$ denote the space of all complex vector-valued Lebesque-integrable functions on $[0, 2\pi]$ such that $\langle \hat{x}, \hat{x} \rangle < \infty$. If $x \in \ell_2^n(\mathbf{Z})$, then $\hat{x} \in \mathcal{L}_2^n[0, 2\pi]$. The converse of this statement is also true, since the inverse Fourier Transform is defined for every element in $\mathcal{L}_2^n[0, 2\pi]$. The space $\ell_2^n(\mathbf{Z})$ can be written as the direct sum of two spaces of one-sided sequences, namely $\ell_2^n(\mathbf{Z}_+)$ and $\ell_2^n(\mathbf{Z}_-)$, where $0 \in \mathbf{Z}_+$, i.e.,

$$\ell_2^n(\mathbf{Z}) = \ell_2^n(\mathbf{Z}_+) \oplus \ell_2^n(\mathbf{Z}_-). \tag{2.7}$$

Fourier Transforms of elements of $\ell_2^n(\mathbf{Z}_+)$ have the property that they admit analytic continuation in the open unit disc. The space of all such functions is denoted by $\mathcal{H}_2^n$, i.e., if $\hat{g}(\lambda) \in \mathcal{H}_2^n$; then $\hat{g}(e^{-i\theta}) \in \mathcal{L}_2^n[0, 2\pi]$, and $\hat{g}(\lambda)$ is analytic in the open unit disc. Similarly, $\ell_2^n(\mathbf{Z}_-)$ is transformed to functions that are analytic in the complement of the disc. Such a space is denoted by $\mathcal{H}_2^{n\perp}$. It follows that

$$\mathcal{L}_2^n[0, 2\pi] = \mathcal{H}_2^n \oplus \mathcal{H}_2^{n\perp}. \tag{2.8}$$

Notation.   Most of the signals that we deal with in real systems are one-sided, i.e., supported on the nonnegative integers. However, it is sometimes convenient to consider two-sided signals. For notational convenience we will drop the argument of $\ell_p^n(\mathbf{Z}_+)$ and simply denote the space by ℓ_p^n. Also, we will denote by ℓ^n the space of all possible vector-valued sequences supported on the positive integers.

2.1.2 Finite-Power Signals

An important class of signals arising frequently in applications is finite-power signals. The square-root of the average power of a signal is defined as:

$$P_x = \lim_{N \to \infty} \left(\frac{1}{N} \sum_{t=0}^{N-1} x^T(t)x(t) \right)^{\frac{1}{2}}. \tag{2.9}$$

The signal x is a power signal if the above limit exists. The quantity P_x is also known as the RMS value (root mean square value) of x. Every signal in ℓ_2 has finite power; however, the converse is not true. For instance, all nonzero periodic signals have finite power, but not finite energy. Notice that the power of a signal is not a norm, since a signal of zero power is not necessarily equal to zero.

Example 2.1.4

How do finite-power signals relate to ℓ_p signals? First note that, if $x \in \ell_2$, then

$$P_x^2 \le \lim_{N \to \infty} \frac{1}{N} \|x\|_2^2 = 0.$$

Thus, all sequences in ℓ_p, for $1 \le p \le 2$, have zero average power (i.e., such signals decay too fast to have nonzero average power over all time). If a scalar signal is a power signal, then its power is bounded by the ℓ_∞-norm:

$$P_x^2 \le \lim_{N \to \infty} \frac{1}{N} \sum_{k=0}^{N-1} \|x\|_\infty^2 = \|x\|_\infty^2 < \infty.$$

For instance, the sequence $x = \{1, \ 1, \ldots\} \in \ell_\infty(\mathbf{Z})$ has $P_x = 1$.

2.1.3 Signals Satisfying Linear Constraints

Sometimes, it is difficult to describe certain classes of signals as norm-bounded signals. A more general approach is to characterize the class through linear constraints, with possibly an infinite number of conditions. For instance, suppose that a class of signals contains sequences of bounded magnitude for some time, after which they all decay faster than some fixed rate. To describe this class consider the following. Let the signal g be fixed, for example:

$$g(t) = \begin{cases} M & 0 \le t \le T \\ Ma^{t-T} & t \ge T \end{cases}. \tag{2.10}$$

A class of signals can be characterized as:

$$\{x \ | -g(t) \le x(t) \le g(t) \ \text{ for all } t\}.$$

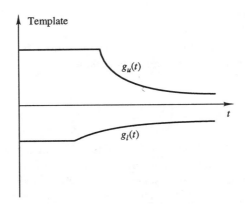

Figure 2.2 Template constraints on signals.

The set of constraints does not have to be symmetric (see Figure 2.2). For any choice of g_u and g_l as the upper and lower bound functions, the set of signals can be defined as:

$$\{x \mid g_l(t) \le x(t) \le g_u(t) \quad \text{for all } t\}. \tag{2.11}$$

It should be noted that the set defined by Equation (2.11) is a convex one. Indeed, if any two signals x and y satisfy the linear constraints, then a convex combination of them will also satisfy the linear constraints, i.e.,

$$g_l(t) \le \alpha x(t) + (1 - \alpha)y(t) \le g_u(t)$$

for any real $\alpha \in [0, 1]$ and $t \in \mathbf{Z}_+$.

2.1.4 White Signals

We say a signal is white (and of unit intensity) if

$$\lim_{N \to \infty} \frac{1}{N} \sum_{i=0}^{N-1} x(i)x^T(t + i) = \begin{cases} I & \text{If } t = 0 \\ 0 & \text{otherwise} \end{cases}.$$

This definition arises naturally in describing a stochastic process in which different events are uncorrelated.

2.2 SYSTEMS

The abstract definition of a system is simply an operator between two signal spaces. Such operators may be described in terms of a state-space realization, or an input-output map. In this section, concepts of causality, linearity, time-invariance, and stability are studied and some examples are provided.

In general, we study systems operating on one-sided signals. Of course, the same systems can be viewed as operators on two-sided signals, and in some situations, it is advantageous to do that. This issue arises later on in this chapter.

2.2.1 Linearity

T is a linear operator from X to Y if it satisfies

$$T(\alpha x_1 + \beta x_2) = \alpha T(x_1) + \beta T(x_2), \qquad \text{for all } \alpha, \beta \in \mathbb{R}. \tag{2.12}$$

Examples of linear operators are abundant. A few are listed below.

Example 2.2.1

A sequence $h \in \ell_1$ defines an operator h on ℓ_p as follows:

$$hx(t) = h * x(t) := \sum_{i=0}^{\infty} h(i)x(t-i). \tag{2.13}$$

The above operation is called convolution.

Example 2.2.2

A function f defined on $\mathbb{Z}_+ \times \mathbb{Z}_+$ defines a linear operator on ℓ_p as follows

$$y(t) = (fx)(t) = \sum_{j=0}^{\infty} f(t, j)x(j). \tag{2.14}$$

This operator can be represented as a multiplication operator with an associated infinite matrix:

$$\begin{pmatrix} y(0) \\ y(1) \\ \vdots \end{pmatrix} = \begin{pmatrix} f(0,0) & f(0,1) & f(0,2) & \cdots \\ f(1,0) & f(1,1) & f(1,2) & \cdots \\ \vdots & \vdots & \vdots & \ddots \end{pmatrix} \begin{pmatrix} x(0) \\ x(1) \\ \vdots \end{pmatrix}. \tag{2.15}$$

2.2.2 Causality and Time-Invariance

Denote by P_k, $k \in \mathbb{Z}_+$, the standard truncation operator on ℓ^n, i.e.,

$$P_k(x(0), x(1) \ldots) = (x(0), x(1), \ldots, x(k), 0, 0 \ldots). \tag{2.16}$$

Definition 2.2.1.   An operator T is causal (proper) if

$$P_t T = P_t T P_t \quad \text{for all } t, \tag{2.17}$$

and is strictly causal (strictly proper) if

$$P_t T = P_t T P_{t-1} \quad \text{for all } t. \tag{2.18}$$

                                                                                     ■

In words, an operator T is *causal* if the current output does not depend on future inputs. It is strictly causal, if the current output depends on past inputs, not including the current input.

Denote by S the standard unit shift operator, i.e.,

$$S(x(0), x(1), \ldots) = (0, x(0), x(1), \ldots). \tag{2.19}$$

Definition 2.2.2.   An operator T is *time-invariant* if it commutes with the unit shift operator, i.e.,

$$ST = TS \tag{2.20}$$

■

In words, a system is time-invariant if its action is independent of the starting time.

Example 2.2.3

The operator defined in Example 2.2.1 is a linear, time-invariant, and causal operator. The operator defined in Example 2.2.2 is in general time varying and noncausal. It is time-invariant if and only if $f(t, j) = f(t + k, j + k)$, for all $k \geq 0$. It is causal if and only if $f(t, j) = 0$ for all $j > t$. Thus, if f is linear time-invariant causal, then it has a representation similar to Example 2.2.1. The matrix representation in this case has a lower-triangular Toeplitz structure:

$$\begin{pmatrix} y(0) \\ y(1) \\ \vdots \end{pmatrix} = \begin{pmatrix} f(0) & 0 & 0 & \dots \\ f(1) & f(0) & 0 & \dots \\ \vdots & \vdots & \vdots & \vdots \end{pmatrix} \begin{pmatrix} x(0) \\ x(1) \\ \vdots \end{pmatrix}. \tag{2.21}$$

2.2.3 Stability and Maximum Amplification

Let T be an operator from X to Y, two normed linear spaces. From Theorem A.7.1 it follows that T is a bounded operator from X to Y (i.e., continuous) if and only if its induced norm is finite, i.e.,

$$\|T\| := \sup_{x \neq 0} \frac{\|Tx\|}{\|x\|} < \infty.$$

The induced norm of T indicates the amount of amplification the operator T exerts on the space X.

Definition 2.2.3.   A linear system T is stable with respect to some input/output space X if it is bounded as a linear operator on X. In general, the input space has a different dimension from the output space.   ■

Bounded operators on a space X have a nice property known as the sub-multiplicative property. Suppose T_1 and T_2 are bounded operators on the space X to itself. The new operator defined from the composition of T_1 and T_2 is also a bounded operator and its norm satisfies:

$$\|T_2 T_1\| \leq \|T_1\| \, \|T_2\|. \tag{2.22}$$

The same property holds if T_1 is a bounded operator from X to Y, and T_2 is a bounded operator from Y to Z. In this case $T_2 T_1$ is a bounded operator from X to Z with the property $\|T_2 T_1\| \leq \|T_1\| \, \|T_2\|$.

Example 2.2.4

Given any matrix $A = (a_{ij}) \in \mathbb{R}^{m \times n}$. The induced norm of A as an operator from $(\mathbb{R}^n, |\cdot|_p)$ to $(\mathbb{R}^m, |\cdot|_p)$ is given by:

$$\|A\| := |A|_q := \sup_{|x|_p \leq 1} |Ax|_p \quad \text{where} \quad \frac{1}{p} + \frac{1}{q} = 1. \tag{2.23}$$

The induced norm can be computed exactly for $p = 1, 2, \infty$:

$$\begin{aligned}
|A|_1 &= \max_{1 \leq i \leq m} \sum_{j=1}^{n} |a_{ij}|, \\
|A|_\infty &= \max_{1 \leq j \leq n} \sum_{i=1}^{m} |a_{ij}|, \\
|A|_2 &= \sigma_{\max}(A).
\end{aligned} \tag{2.24}$$

These induced norms are related in the following sense:

$$\frac{1}{\sqrt{n}} |A|_1 \leq |A|_2 \leq \sqrt{m} |A|_1. \tag{2.25}$$

It is interesting to notice that the notion of stability is tied to the particular space under consideration. One particular operator can be stable with respect to one space but not the other.

2.3 LINEAR OPERATORS ON ℓ_p^n

In this section, classes of linear, causal operators on ℓ_∞^n and ℓ_2^n are characterized.

2.3.1 General Time-Varying: $\mathcal{L}$

The space of all linear, causal operators from ℓ^n to ℓ^m is characterized by infinite block lower triangular matrices of the form

$$\begin{pmatrix} R(0, 0) & 0 & 0 \\ R(1, 0) & R(1, 1) & \\ \vdots & \vdots & \ddots \end{pmatrix} \tag{2.26}$$

where $R(i, j) \in \mathbb{R}^{m \times n}$. This infinite matrix representation of R acts on elements of ℓ^n by multiplication, i.e., if $u \in \ell^n$, then $y := Ru \in \ell^m$ where $y(t) = \sum_{j=0}^{t} R(t, j)u(j) \in \mathbb{R}^m$. The space of such operators is denoted by $\mathcal{L}$ (to indicate the dimensions, it is sometimes denoted by $\mathcal{L}^{m \times n}$). Such operators may or may not be stable (bounded).

2.3.2 Bounded Operators on ℓ_∞^n

Time-varying operators on ℓ_∞^n. The following theorem characterizes all bounded linear, causal operators, from ℓ_∞^n to ℓ_∞^m.

Theorem 2.3.1.   The space of all bounded, linear, causal operators from ℓ_∞^n to ℓ_∞^m is given by the space of all infinite block lower triangular matrices, with a finite induced norm, i.e.,

$$\sup_k |(R_k)|_1 < \infty, \tag{2.27}$$

where (R_k) is the kth dimensional block of the matrix R:

$$\begin{pmatrix} R(0,0) & 0 & \cdots & 0 \\ R(1,0) & R(1,1) & 0 & \cdots \\ \vdots & \vdots & \vdots & \ddots \\ R(k,0) & R(k,1) & \cdots & R(k,k) \end{pmatrix}.$$

Equivalently,

$$\sup_i |(\, R(i,0) \quad \cdots \quad R(i,i)\,)|_1 < \infty. \tag{2.28}$$

■

Time-invariant operators on ℓ_∞^n.   From Definition 2.2.2, R is time-invariant if $SR = RS$. This implies that

$$R(i,j) = R(i+1,j+1) \qquad \text{for all } i, j. \tag{2.29}$$

Equivalently, the associated block lower triangular matrix representation has a block Toeplitz structure, and can be completely characterized by one sequence $R(k)$, i.e.,

$$\begin{pmatrix} R(0) & 0 & 0 \\ R(1) & R(0) & \\ \vdots & \vdots & \ddots \end{pmatrix}. \tag{2.30}$$

The induced operator norm in Equation (2.28) simplifies to

$$\|R\| = \lim_{i\to\infty} |(\, R(0) \quad \cdots \quad R(i)\,)|_1. \tag{2.31}$$

In fact, an easier way to compute the above norm is given in the formula

$$\|R\| = \max_{1\le i\le m} \sum_{j=1}^{n} \sum_{t=0}^{\infty} |r_{ij}(t)| = \max_{1\le i\le m} \sum_{j=1}^{n} \|r_{ij}\|_1 \tag{2.32}$$

where $R(t) = \{r_{ij}(t)\}$. In the SISO case, this is nothing but the ℓ_1 norm of the pulse response sequence. Thus ℓ_∞-stability is guaranteed if and only if the pulse response is in ℓ_1. Notice also that for the MIMO case, the induced norm is the composition of the finite-dimensional matrix norm with the ℓ_1-norm. For this reason, the norm defined in Equation (2.32) will be denoted as $\|\cdot\|_1$.

λ-Transforms.   For every element $R \in \ell_1^{m\times n}$, the λ-Transform $\hat R$ is defined as

$$\hat R(\lambda) = \sum_{i=0}^{\infty} R(i)\lambda^i. \tag{2.33}$$

Then, $\hat{R}(\lambda)$ is analytic on the open unit disc, and continuous on the boundary. The collection of all such elements equipped with the norm defined in Equation (2.32) is traditionally denoted by $\mathbf{A}$ (or $\mathbf{A}^{m \times n}$). From this definition, it is clear that the spaces $\mathbf{A}^{m \times n}$ and $\ell_1^{m \times n}$ are different representations of the same space.

2.3.3 Bounded Operators on ℓ_2^n

Time-varying operators on ℓ_2^n. All time-varying operators are given in $\mathcal{L}$. An operator $R \in \mathcal{L}$ is bounded on ℓ_2 if and only if

$$\sup_k \sigma_{\max}(R_k) < \infty, \tag{2.34}$$

where (R_k) is the kth dimensional block of the matrix R:

$$\begin{pmatrix} R(0,0) & 0 & \ldots & 0 \\ R(1,0) & R(1,1) & 0 & \ldots \\ \vdots & \vdots & \vdots & \ddots \\ R(k,0) & R(k,1) & \ldots & R(k,k) \end{pmatrix}. \tag{2.35}$$

Time-invariant operators on ℓ_2^n ($\mathcal{H}_2^n$). Recall that the space ℓ_2^n is identified with the space $\mathcal{H}_2^n$, the space of Fourier Transforms. It turns out that it is more convenient to identify the class of linear, time-invariant, bounded, causal operators on ℓ_2^n through multiplication operators on $\mathcal{H}_2^n$.

Let $\mathcal{L}_\infty^{m \times n}$ denote the space of all complex-valued matrix functions on the unit circle that are bounded, i.e., if $\hat{R} \in \mathcal{L}_\infty^{m \times n}$, then

$$\|\hat{R}\|_\infty = \operatorname{ess\,sup}_\theta \sigma_{\max}[\hat{R}(e^{i\theta})] < \infty. \tag{2.36}$$

The subspace of $\mathcal{L}_\infty^{m \times n}$ of all the elements that admit analytic continuations in the unit disc is denoted by $\mathcal{H}_\infty^{m \times n}$, i.e., if $\hat{R} \in \mathcal{H}_\infty^{m \times n}$, then $\hat{R}$ is analytic in the open unit disc and bounded on the unit circle.

Given $\hat{R} \in \mathcal{L}_\infty^{m \times n}$, define a multiplication operator on the space $\mathcal{L}_2^n[0, 2\pi]$ as follows

$$\begin{aligned} \hat{R} &: \mathcal{L}_2^n[0, 2\pi] \to \mathcal{L}_2^m[0, 2\pi], \\ \hat{g} &\quad\quad \to \hat{R}\hat{g}. \end{aligned} \tag{2.37}$$

Notice that:

$$\|\hat{R}\hat{g}\|_2 = \left(\frac{1}{2\pi} \int_0^{2\pi} |\hat{R}(e^{i\theta})\hat{g}(e^{i\theta})|_2^2 \, d\theta \right)^{\frac{1}{2}} \le \operatorname{ess\,sup}_\theta \sigma_{\max}[\hat{R}(e^{i\theta})] \, \|\hat{g}\|_2.$$

Hence, the induced norm of this multiplication operator satisfies

$$\|\hat{R}\| \le \|\hat{R}\|_\infty.$$

In fact, it can be shown that equality holds. This is captured in the following theorem.

Theorem 2.3.2

1. Every bounded linear time-invariant operator on $\ell_2(\mathbf{Z})$ can be identified by a multiplication operator on $\mathcal{L}_2^n[0, 2\pi]$ with an element $\hat{R} \in \mathcal{L}_\infty^{m \times n}$. The induced norm of this operator is equal to $\|\hat{R}\|_\infty$.
2. Every bounded linear-time invariant, causal operator on $\ell_2(\mathbf{Z}_+)$ can be identified by a multiplication operator on $\mathcal{H}_2^n$ with an element $\hat{R} \in \mathcal{H}_\infty^{m \times n}$. The induced norm of this operator is equal to $\|\hat{R}\|_\infty$. ∎

The reader should not confuse this norm with the ℓ_∞-norm defined on signals. Even though the above notation is self-explanatory (note that the infinity norm is taken on a complex-valued function indicated with a hat) we will sometimes use the notation $\| \cdot \|_{\mathcal{H}_\infty}$ as a reminder.

Example 2.3.1

Let a function $\hat{R}$ be given by

$$\hat{R}(\lambda) = \frac{-2.5}{(\lambda - .5)(\lambda - 3)} = \frac{1}{\lambda - .5} - \frac{1}{\lambda - 3}.$$

It is straightforward to verify that $\hat{R}(e^{i\theta})$ is bounded for all θ, i.e., $\hat{R}(e^{i\theta}) \in \mathcal{L}_\infty$. We observe that:

1. $\hat{R}$ is a bounded operator from $\mathcal{L}_2[0, 2\pi]$ to $\mathcal{L}_2[0, 2\pi]$.
2. $\hat{R}$ is not an operator from $\mathcal{H}_2$ to $\mathcal{H}_2$. To verify, notice that if $\hat{g} \in \mathcal{H}_2$ then $\hat{R}\hat{g}$ is not analytic inside the unit disc.
3. Similarly, $\hat{R}$ is not an operator from $\mathcal{H}_2^\perp$ to $\mathcal{H}_2^\perp$.

Let us decompose $\hat{R}$ in the following way,

$$\hat{R}_1 = \frac{1}{\lambda - .5}, \qquad \hat{R}_2 = \frac{-1}{\lambda - 3}.$$

Then $\hat{R} = \hat{R}_1 + \hat{R}_2$. We observe:

1. $\hat{R}_2$ is a bounded operator from $\mathcal{H}_2$ to $\mathcal{H}_2$.
2. $\hat{R}_1$ is a bounded operator from $\mathcal{H}_2^\perp$ to $\mathcal{H}_2^\perp$.
3. $\hat{R}_2$ is not an operator from $\mathcal{H}_2^\perp$ to $\mathcal{H}_2^\perp$.

To understand the time-domain interpretation of these operators, define the two-sided λ-Transform of a two-sided sequence $R(k)$ as:

$$\hat{R}(\lambda) = \sum_{k=-\infty}^{\infty} R(k)\lambda^k.$$

For the above example, R is given by

$$R(k) = \begin{cases} (1/3)^{k+1} & k \geq 0 \\ 2^{k+1} & k \leq -1 \end{cases}.$$

Also, $R_2(k) = (1/3)^{k+1}$, $k \geq 0$, and $R_1(k) = 2^{k+1}$, $k \leq -1$. From the definition of convolution, it is easy to see that

$$R_2 : \ell_2(\mathbf{Z}_+) \to \ell_2(\mathbf{Z}_+),$$

$$R_1 : \ell_2(\mathbf{Z}_-) \to \ell_2(\mathbf{Z}_-),$$

and

$$R_2 : \ell_2(\mathbf{Z}_-) \to \ell_2(\mathbf{Z}),$$

$$R_1 : \ell_2(\mathbf{Z}_+) \to \ell_2(\mathbf{Z}).$$

2.3.4 Relations Between Induced Norms

Relationship between $\mathcal{H}_\infty$ and A.   The space $\mathcal{H}_\infty$ is the space of linear time-invariant ℓ_2-stable operators, and the space **A** is the space of linear time-invariant ℓ_∞-stable operators. Both spaces contain functions that are analytic in the open unit disc. Let $\hat{R} \in \mathbf{A}$ (SISO), then $\hat{R}$ is the λ-transform of a sequence $R \in \ell_1$. This guarantees that $\hat{R}(e^{i\theta})$ is continuous as a function of θ. With this observation we have:

$$\sup_\theta \left| \sum_{k=0}^\infty R(k)e^{i\theta k} \right| \leq \|R\|_1 = \|\hat{R}\|_{\mathbf{A}}.$$

Equivalently

$$\|\hat{R}\|_{\mathcal{H}_\infty} \leq \|\hat{R}\|_{\mathbf{A}}. \tag{2.38}$$

For MIMO systems, combining the above inequality with the results in Example 2.2.4 we have

$$\|\hat{R}\|_{\mathcal{H}_\infty} \leq \sqrt{m}\|\hat{R}\|_{\mathbf{A}}. \tag{2.39}$$

In other words, an ℓ_∞-stable linear time-invariant operator is also ℓ_2-stable or equivalently $\mathbf{A}^{m \times n} \subset \mathcal{H}_\infty^{m \times n}$. The converse, however, is not true.

Example 2.3.2

The function

$$\hat{R}(\lambda) = e^{\frac{1}{\lambda - 1}} \tag{2.40}$$

is in $\mathcal{H}_\infty$ but not in **A**. The pulse response of $\hat{R}$ is $R(k) = \frac{1}{k!}\hat{R}^{(k)}(0)$. These coefficients are exactly the Taylor series expansion of $\hat{R}$ around zero. The pulse response is not absolutely summable since $\hat{R}(\lambda)$ is not continuous on the unit circle.

Example 2.3.3

Equation (2.39) asserts that ℓ_∞ stability of linear time-invariant operators implies ℓ_2-stability. This is not true for time-varying operators. In fact, the two classes of operators are not comparable. For instance, let R be an LTV operator given by

$$Rf(t) = f(0). \tag{2.41}$$

This is not an ℓ_2-stable operator, but is clearly ℓ_∞-stable, with norm equal to one.

It turns out that if $\hat{R} \in \mathcal{H}_\infty$, then

$$\sum_{k=1}^{\infty} \frac{1}{k} |R(k)| \leq 2\pi \|\hat{R}\|_\infty, \tag{2.42}$$

i.e., the sequence $\{\frac{1}{k} R(k)\} \in \ell_1$.

Relations between ℓ_p-induced norms.   Consider a SISO system $h \in \ell_1$. We have already established that h acts as a bounded operator on the spaces ℓ_∞ and ℓ_2. It turns out that h acts as an operator on all ℓ_p-spaces for $p \geq 1$, and satisfies

$$\|h * u\|_p \leq \|h\|_1 \|u\|_p. \tag{2.43}$$

This implies that the induced norm of h on ℓ_p is bounded by its ℓ_1 norm. Furthermore, a stronger result can be shown for all $p \geq 1$

$$\|\hat{h}\|_{\mathcal{H}_\infty} \leq \|h\|_{\ell_p-\text{ind}} \leq \|h\|_1. \tag{2.44}$$

2.3.5 Stable Inversion

For a given system $h \in \ell_1$, it is desirable to find out when h^{-1} is also in ℓ_1. Intuitively, this should be true if $\hat{h}$ has no zeros in the disc. This turns out to be precisely the condition and is given in the following theorem known as Wiener's theorem.

Theorem 2.3.3.   Let h be in ℓ_1. Then h^{-1} is in ℓ_1 if and only if

$$\inf_{|\lambda| \leq 1} |\hat{h}(\lambda)| > 0. \qquad \blacksquare$$

It turns out that the same condition holds for the inversion of functions in $\mathcal{H}_\infty$.

2.3.6 Other System Measures

$\mathcal{H}_2$ norm.   There are other measures of size for linear time-invariant systems that are not necessarily induced norms on some space. One example is measuring the energy contained in the pulse response, otherwise termed the $\mathcal{H}_2$ norm, and is given by

$$\|\hat{R}\|_2^2 = \frac{1}{2\pi} \int_{-\pi}^{\pi} \text{Trace}\, \hat{R}(e^{i\theta}) \hat{R}^T(e^{-i\theta}) d\theta$$

$$= \sum_{k=0}^{\infty} \text{Trace}\, R(k) R^T(k). \tag{2.45}$$

Let $y = Ru$ and assume that u is a white signal. The power (RMS value) of y is equal to $\|\hat{R}\|_2$. This gives another interpretation of the $\mathcal{H}_2$-norm. A third interpretation of this norm, as will be seen shortly for SISO systems, is that it is equal to the induced norm from ℓ_2 to ℓ_∞.

Weighted norms.   Other measures can be obtained by introducing weighting functions, both in time and frequency domains. Let $C(t)$ be a given matrix valued time function. The weighted ℓ_1 norm is defined as

$$\|R\|_C = \max_{1 \le i \le m} \sum_{j=1}^{n} \sum_{t=0}^{\infty} |c_{ij}(t) r_{ij}(t)|. \tag{2.46}$$

If $c_{ij}(t) = t$, then this measure is known as the ITAE (Integral Time Absolute Error). If $c_{ij}(t) = a^t$, then the measure is known as the exponentially weighted ℓ_1 norm.

Summary.   We have seen so far that a causal linear time-invariant operator R is ℓ_p-stable if $R \in \ell_1^{m \times n}$. The induced norm of this operator depends on the space; for instance,

$$\|R\|_{\ell_2-\text{ind}} = \|\hat{R}\|_{\infty},$$

and

$$\|R\|_{\ell_{\infty}-\text{ind}} = \|R\|_1.$$

Table 2.1 contains a summary of the induced norms of $R \in \ell_1$ on different input/output signal spaces. It is assumed for simplicity that R is SISO. Notice that the $\mathcal{H}_2$ norm arises as the induced norm from ℓ_2 to ℓ_{∞} and thus has the interpretation of measuring the worst-case maximum amplitude of the output in the presence of bounded energy signals. The $\mathcal{H}_{\infty}$ norm has two interpretations: the ℓ_2/ℓ_2 gain and the *Power/Power* gain. Also it is an upper bound on the $\ell_{\infty}/Power$ gain (assuming that the input is restricted to a smaller set such that the output is a power signal). Notice that the table shows that bounded power signals can produce unbounded outputs, thus both the *Power/ℓ_2* and *Power/ℓ_{∞}* gains are infinite.

TABLE 2.1 COMPARISON OF INDUCED NORMS

Output\Input	ℓ_{∞}	ℓ_2	Power
ℓ_{∞}	ℓ_1	$\mathcal{H}_2$	∞
ℓ_2	∞	$\mathcal{H}_{\infty}$	∞
Power	$\le \mathcal{H}_{\infty}$	0	$\mathcal{H}_{\infty}$

2.3.7 Finite Pulse Response Systems

Such systems are generally known as FIR systems (finite impulse response). If R is FIR, then there exists an N such that $R(k) = 0$ for all $k \ge N$. It follows immediately that $R \in \ell_1$.

Finite pulse response systems play an important role in solving optimization problems, as will be demonstrated in future chapters. The main reason behind this is that we can approximate any stable system in ℓ_1 by a FIR system arbitrarily closely. This follows since if $R \in \ell_1$, then for every $\epsilon \geq 0$, there exists an N such that

$$\sum_{k \geq N} |R(k)| < \epsilon.$$

We note that FIR systems do not approximate every element in $\mathcal{H}_\infty$. The closure of these systems in $\mathcal{H}_\infty$ is a proper subspace whose elements are continuous on the unit circle.

2.4 NONLINEAR OPERATORS

Most of the nonlinear operators considered in this book are bounded gain operators. The gain of an operator over the space ℓ_p is given by the same induced-norm formula

$$\|T\| = \sup_{x \neq 0} \frac{\|Tx\|}{\|x\|}. \tag{2.47}$$

In fact, the same notation is used to indicate the gain of the operator. A nonlinear operator is stable if it has finite gain. It follows that if T has finite gain, then $T0 = 0$.

Example 2.4.1

There are many examples of bounded gain nonlinear operators, e.g.,

$$T(x)(k) = \sin(x(k))$$

with a gain $\|T\| = 1$ over any ℓ_p space. Another example is the saturation operator:

$$T(x)(k) = \begin{cases} x(k) & if \ \ |x(k)| \leq 1 \\ \mathrm{sgn}(x(k)) & if \ \ |x(k)| > 1 \end{cases}.$$

On the other hand, the operator

$$T(x)(k) = x^2(k)$$

is not bounded gain, for any ℓ_p.

Example 2.4.2

A Class of Fading Memory Operators. A system has fading memory if the current outputs do not depend heavily on very far away past inputs. For example, all systems in ℓ_1 have this property. Nonlinear fading memory operators on ℓ_∞ can be approximated by systems of the form

$$T(x) = g(h * x),$$

where h is an element in ℓ_1 and g is a continuous function on the interval $[-\|h\|_1, \|h\|_1]$.

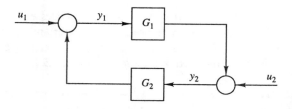

Figure 2.3 Feedback connection.

2.5 FEEDBACK CONNECTIONS

Consider the system shown in Figure 2.3. The systems G_1 and G_2 can be arbitrary non-linear time-varying systems. The relations between the inputs and outputs are given by

$$y_1 = G_2 y_2 + u_1,$$

$$y_2 = G_1 y_1 + u_2.$$

Well posedness. The feedback system in Figure 2.3 is said to be well posed if for any choice of u_1, $u_2 \in \ell$, there corresponds a unique solution y_1, $y_2 \in \ell$. Necessarily, the operator $(I - G_1 G_2)$ is one-to-one (i.e., different elements have different values) and onto (i.e., the range is all ℓ).

Example 2.5.1

Let $\hat{G}_1 = 1 - \lambda$ and $\hat{G}_2 = 1$. It follows from the above equations that

$$y_1(k - 1) = u_1(k) + G_2 u_2(k),$$

since $(I - \hat{G}_1 \hat{G}_2) = \lambda$. But the range of λ is not equal to ℓ. Hence, this system is not well posed.

The following theorem gives a sufficient condition that guarantees well posedness.

Theorem 2.5.1. The feedback system in Figure 2.3 is well posed if the operator $G_1 G_2$ is strictly causal.

Proof. If $G_1 G_2$ is strictly causal, then for any given u_1, $u_2 \in \ell$, a unique solution y_1, $y_2 \in \ell$ can be computed recursively.

If both G_1 and G_2 are LTI, then strict causality implies that $\hat{G}_1 \hat{G}_2$ has the form $\lambda \hat{R}$. Well posedness is equivalent to the existence of a causal inverse of $(I - \lambda \hat{R})$, which in turn is equivalent to $det(I - \lambda \hat{R}) \neq 0$ at $\lambda = 0$. But this is immediately true. ∎

Stability. Let $H(G_1, G_2)$ be the closed-loop operator defined as

$$\begin{pmatrix} y_1 \\ y_2 \end{pmatrix} = H(G_1, G_2) \begin{pmatrix} u_1 \\ u_2 \end{pmatrix}.$$

The closed-loop feedback system is ℓ_p-stable if

$$\| H(G_1, G_2) \|_{\ell_p - \text{ind}} < \infty.$$

In the future, stability of interconnected systems will be addressed extensively, particularly systems connected in feedback as in Figure 2.3. An important stability result is based on what is called the *Small Gain Theorem*. Simply, this theorem provides a sufficient condition for the stability of the feedback system in Figure 2.3.

Theorem 2.5.2.   Let $G_1 : \ell_p^n \to \ell_p^m$ and $G_2 : \ell_p^m \to \ell_p^n$ be two stable operators of bounded ℓ_p-gain and assume that the closed loop system is well posed. Then the closed loop system is stable if $||G_1||||G_2|| < 1$.

Proof. Well posedness guarantees that for any u_1, $u_2 \in \ell$, there exists a unique solution y_1, $y_2 \in \ell$. The norms are bounded by:

$$||y_1|| \leq ||G_2||||y_2|| + ||u_1||$$

and

$$||y_2|| \leq ||G_1||||y_1|| + ||u_2||.$$

Since $||G_1||||G_2|| < 1$, it follows that

$$||y_1|| \leq \frac{1}{1 - ||G_1||||G_2||}(||G_2||||u_2|| + ||u_1||).$$

Similarly for y_2.   ∎

If both G_1 and G_2 are linear operators, then the small gain condition guarantees well posedness. This, however, is not sufficient if either operator is nonlinear.

2.6 SUMMARY

We have presented an overview of important classes of signals and of classes of operators that arise in the analysis of control systems. The ideas presented here will be extensively used in future chapters. In particular, signal norms and system norms will be used to state performance specifications in a unified mathematical framework. The small gain theorem will play an instrumental role in stability robustness and performance robustness problems.

EXERCISES

2.1. Define the Fourier Transform of a sequence $x = \{x(n)\} \in \ell_2(\mathbf{Z})$ as:

$$\hat{x}(e^{i\theta}) = \sum_{n=-\infty}^{\infty} x(n)(e^{i\theta})^n.$$

Show that

$$\langle \hat{x}, \hat{y} \rangle := \frac{1}{2\pi} \int_0^{2\pi} \hat{y}^*(e^{i\theta})\hat{x}(e^{i\theta})d\theta = \langle x, y \rangle.$$

2.2. Show that

$$\ell_2(\mathbf{Z}) = \oplus_i M_i$$

where $dim M_i = 1$ and $M_i \perp M_j$ for all $i \neq j$

2.3. Use the projection theorem to solve the problem:

$$\min_{y \in \mathbb{R}^n} ||x - Ay||_2$$

where $x \in \mathbb{R}^m$ is fixed, A is an $m \times n$ real matrix, $(m > n)$, with rank $A = n$. Is the solution unique?

2.4. Use the projection theorem to solve the problem:

$$\min_{x \in \mathbb{R}^n} \{x^T Q x \mid Ax = b, \ Q \text{ is positive definite}\}$$

where A is an $m \times n$ real matrix, $(m < n)$, with rank $A = m$. Is the solution unique?

2.5. Let S be a closed subspace in a Hilbert space H. Let Π_S denote the projection operator from H to S. Show that:

(a) Π_S is well defined, i.e, maps every element $x \in H$ to a unique element in S.
(b) Π_S is a bounded linear operator with induced norm equal to 1.
(c) $(\Pi_S)^2 = \Pi_S$.
(d) Let $S = \mathcal{H}_2$, $H = \mathcal{L}_2$, compute

$$\Pi_S \left(\frac{(\lambda + 3)(\lambda + 2)}{(\lambda + .5)(\lambda + .2)(\lambda + 5)} \right).$$

2.6. Let $\hat{F} \in \mathcal{L}_\infty$, real rational. Define an operator Γ_F as follows:

$$\Gamma_F : \mathcal{H}_2^\perp \to \mathcal{H}_2,$$

$$\Gamma_F(\hat{h}) = \Pi_{\mathcal{H}_2} \hat{F} \hat{h}.$$

(a) Show that $\Gamma_{F+Q} = \Gamma_F$ for any anti-causal $\hat{Q}$.
(b) Compute Γ_F for $\hat{F} = \frac{1}{\lambda+2}$.
(c) Compute the induced norm of Γ_F for the above F.
(d) Show that $||\hat{F} - \hat{Q}||_{\mathcal{L}_\infty} \geq ||\Gamma_F||$.

2.7. Verify the entries of Table 2.1.

2.8. Show that R is a bounded operator on ℓ_1 if and only if $R \in \ell_1$. Compute the induced norm and verify that it is equal to the ℓ_1 norm of R. Is this true in the matrix case? Compute the induced norm for the latter.

2.9. Given two non-zero vectors $v, w \in \mathbb{R}^n$. Show that there exists a matrix A such that $v = Aw$ and $\sigma_{max}(A) = \sqrt{v^T v / w^T w}$.

2.10. Given two non-zero vectors $v, w \in \mathbb{R}^n$. Does there exist a matrix A such that $v = Aw$ and $|A|_1 = ||v||_\infty / ||w||_\infty$? Prove or disprove by a counter example.

2.11. Given two signals $v, w \in \ell_\infty$ such that

$$||P_n v||_\infty < ||P_n w||_\infty \quad \text{for all } n \geq 0.$$

Show that there exists a causal time-varying operator G with $||G||_{\ell_\infty\text{-ind}} \leq 1$ such that $v = Gw$. Does there exist a time-invariant G?

2.12. A causal linear operator G is periodic with a period N if it commutes with S_N, i.e., $GS_N = S_N G$.

(a) For $N = 2$, show that there exist two infinite sequences, g_1, g_2, such that

$$
G = \begin{pmatrix}
g_1(0) & 0 & \cdots & & \\
g_2(1) & g_2(0) & 0 & \cdots & \\
g_1(2) & g_1(1) & g_1(0) & 0 & \cdots \\
\vdots & \vdots & \vdots & \vdots & \vdots
\end{pmatrix}.
$$

(b) Generalize for arbitrary N.

(c) Define the average-delayed operator as

$$
\overline{G} = \frac{1}{N} \sum_{k=0}^{N-1} S_{-k} G S_k
$$

Show that $\overline{G}$ is time invariant.

(d) Show that

$$
\|\overline{G}\|_{\ell_p\text{-ind}} \leq \|G\|_{\ell_p\text{-ind}}.
$$

2.13. Let $f(t) = f_1(t) + a\delta(t)$, with $f_1 \in \mathcal{L}_1(\mathbb{R}_+)$. Show that f defines a bounded operator on $\mathcal{L}_\infty(\mathbb{R}_+)$ defined as:

$$
Fg = f * g.
$$

Show that the induced operator norm is given by:

$$
\|F\| = |a| + \|f_1\|_1.
$$

2.14. Let $f(t) = f_1(t) + a\delta(t)$ with $f_1 \in \mathcal{L}_1(\mathbb{R}_+)$. Show that f defines a bounded operator on $\mathcal{L}_2(\mathbb{R}_+)$ with an operator induced norm

$$
\|\hat{f}\|_\infty := \sup_w |\hat{f}(iw)|
$$

where $\hat{f}$ is the Laplace Transform of f. This is the continuous-time $\mathcal{H}_\infty$ norm.

2.15. Define the Sampling operator $\mathbf{S}_T$ from continuous continuous-time functions to discrete-time sequences as:

$$
\begin{aligned}
\mathbf{S}_T : C(\mathbb{R}_+) \cap \mathcal{L}_p &\to \quad \ell_p, \\
g(t) &\to g(nT).
\end{aligned}
$$

Show that $\mathbf{S}_T$ is a bounded operator for $p = \infty$ and compute its norm. Is this true for $p < \infty$? Prove or show a counter example.

2.16. Define the Hold operator $\mathbf{H}_T$ from discrete-time signals to continuous-time signals as:

$$
\begin{aligned}
\mathbf{H}_T : \ell_p &\to \qquad\qquad \mathcal{L}_p, \\
g(n) &\to \tilde{g}(t) = g(n) \quad nT \leq t < (n+1)T.
\end{aligned}
$$

Show that $\mathbf{H}_T$ is a bounded operator and compute its induced norm.

2.17. Let $G \in \mathcal{L}_1(\mathbb{R}_+)$. Let $\tilde{G}_T$ denote the zero order hold equivalence of G, i.e., $\tilde{G}_T = \mathbf{S}_T G \mathbf{H}_T$. Show that

$$
\|G\|_1 = \lim_{T \to 0} \|\tilde{G}_T\|_1.
$$

NOTES AND REFERENCES

This view of systems as input-output maps follows [DV75], in which much of the material can be found, see also [Wil71]. A good discussion on complex-valued spaces, $\mathcal{H}_2$ and $\mathcal{H}_\infty$ can be found in [Dur70, Hof62, Rud73a]. The characterization of the space $\mathcal{H}_\infty$ as the space of causal, bounded LTI operators on $\mathcal{H}_2$ can be found in [Vid85]. Similarly for the space $\mathcal{L}_\infty$.

The small-gain theorem and its use for control applications was first introduced in [Zam66].

Some comparisons between the ℓ_1 and $\mathcal{H}_\infty$ norms can be found in [BD87]. The fact that the norm of LTI stable systems is an upper bound of all other induced norms, Equation (2.43), is known as Young's Inequality and is proved in [CD82]. Equation (2.42) is known as Hardy's inequality and is proved in [Hof62], and Equation (2.44) is proved in [BD92]. Wiener's theorem can be found in several functional analysis books; see, for example, [Con85, Rud73a].

For recent books discussing norms arising in control systems for LTI plants, see [BB91, DFT92]. Table 2.1 is proved in detail in [DFT92] for continuous-time systems.

CHAPTER 3 ———————————

Performance Constraints

Feedback controllers are designed for two main reasons: to stabilize a given system and to meet certain performance requirements. Indeed, these objectives need to be met in the presence of plant and input uncertainty. Many practical performance requirements are difficult to state in a precise mathematical formulation owing to the fact that mathematical descriptions rarely represent real environments. A simple example of this is modeling noise or disturbance inputs. Nevertheless, mathematical descriptions provide tools for dealing with the real situations and are effective only if they are not too far from the real environment.

In this chapter a general setup is introduced in which both stability and general performance requirements can be formulated. We will describe various performance specifications that lead to linear constraints on the closed loop function. This is done in a general setting allowing multiple objectives for different input/output pairs. An example of such specifications is the ℓ_1 problem. Also, we will consider performance specifications that result in convex constraints, and can be well approximated by linear constraints. The $\mathcal{H}_\infty$ problem is an example of such specifications.

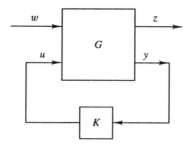

Figure 3.1   General setup.

3.1 PROBLEM SETUP

Figure 3.1 shows a general setup for posing performance specifications. The variables as defined in Figure 3.1 are:

$$u = \text{Control Inputs}$$
$$y = \text{Measured Outputs}$$

$$w = \text{Exogenous Inputs} = \begin{cases} \text{Fixed commands} \\ \text{Unknown commands} \\ \text{Disturbances} \\ \text{Noise} \\ \vdots \end{cases}$$

$$z = \text{Regulated outputs} = \begin{cases} \text{Tracking Errors} \\ \text{Control Inputs} \\ \text{Measured Outputs} \\ \text{States} \\ \vdots \end{cases}$$

The operator G is a 2×2 block matrix mapping the inputs w and u to the outputs z and y:

$$\begin{bmatrix} z \\ y \end{bmatrix} = \begin{bmatrix} G_{11} & G_{12} \\ G_{21} & G_{22} \end{bmatrix} \begin{bmatrix} w \\ u \end{bmatrix}.$$

The actual process or *plant* is the submatrix G_{22}. Both the exogenous inputs and the regulated outputs are auxiliary signals that need not be part of the closed-loop system. The feedback controller is denoted by K. The dimensions of the signal spaces will be denoted by n_u, n_y, n_w and n_z, where the association is explicitly given by the subindex.

3.1.1 Well Posedness and Stability

In Figure 3.2 the inputs and outputs are related as follows:

$$\begin{bmatrix} I & -G_{12} & 0 \\ 0 & I & -K \\ 0 & -G_{22} & I \end{bmatrix} \begin{bmatrix} z \\ u \\ y \end{bmatrix} = \begin{bmatrix} G_{11} & 0 & 0 \\ 0 & I & K \\ G_{21} & 0 & 0 \end{bmatrix} \begin{bmatrix} w \\ v_1 \\ v_2 \end{bmatrix}. \tag{3.1}$$

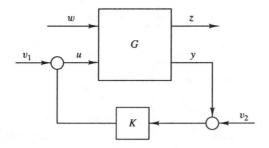

Figure 3.2   Closed-loop system.

Let $H(G, K)$ denote the map

$$\begin{pmatrix} z \\ u \\ y \end{pmatrix} = H(G, K) \begin{pmatrix} w \\ v_1 \\ v_2 \end{pmatrix}.$$

We will assume that the closed-loop system is well posed, i.e., the LHS operator in Equation (3.1) has a causal inverse. One way to guarantee this property is by assuming that the plant G_{22} is strictly causal, i.e., has a delay. Then the λ-transform of the matrix is invertible at $\lambda = 0$ and, by continuity, it will have an inverse defined in a neighborhood of zero.

Definition 3.1.1.   The closed-loop system is ℓ_p-stable if the ℓ_p-induced norm of $H(G, K)$ is finite. In such a case, K is said to be *stabilizing* in the ℓ_p sense. ■

The map of interest is the map between w to z, denoted by Φ:

$$\Phi = G_{11} + G_{12} K (I - G_{22} K)^{-1} G_{21}. \tag{3.2}$$

This particular mapping represents the *performance objectives*. Whenever it is necessary, we will denote this map by T_{zw} to explicitly state the inputs and outputs of the map. For a given map Φ, we will talk about two kinds of constraints:

- *feasibility constraints*, i.e., whether Φ can be written as in Equation 3.2 for some stabilizing controller, and
- *performance constraints* representing the performance objectives.

The discussion on the first will be postponed until Chapters 5 and 6. A map Φ satisfying the feasibility constraints will be referred to as a feasible closed-loop map.

3.1.2 Linear Fractional Transformation (LFT)

The discussion above falls under a special case of a more general description of feedback interconnections. This description will be used in all subsequent chapters.

Given a 2×2 block matrix transfer function G such that G_{22} has n_u inputs and n_y outputs, and a $n_u \times n_y$ transfer function matrix K, we can define a *lower* Linear Fractional Transformation as the following map:

$$F_\ell(G, K) = G_{11} + G_{12} K (I - G_{22} K)^{-1} G_{21}.$$

Similarly, given a transfer function Δ with the appropriate dimensions (i.e., consistent with G_{11}) an *upper* LFT is defined as

$$F_u(G, \Delta) = G_{22} + G_{21}\Delta(I - G_{11}\Delta)^{-1}G_{12}.$$

These definitions are possible provided $(I - G_{22}K)^{-1}$ and $(I - G_{11}\Delta)^{-1}$ exist, i.e., the closed loops are well posed. The block diagram representation of upper and lower LFT's is shown in Figure 3.3. In both diagrams, the upper and lower LFT's are given by the mapping from w to z.

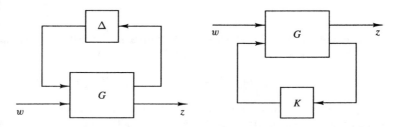

Figure 3.3   Upper and lower LFT's.

Star product.   The star product is simply the process of obtaining the operator representations for two 2-input 2-output systems connected in a feedback fashion. Consider the diagram in Figure 3.4. The diagram shows two systems M, N, connected in the obvious way. The star product is simply the resulting 2×2 composite system (assuming compatible dimensions and well-posedness):

$$\begin{pmatrix} z_1 \\ z_2 \end{pmatrix} = \begin{pmatrix} F_\ell(M, N_{11}) & M_{12}(I - N_{11}M_{22})^{-1}N_{12} \\ N_{21}(I - M_{22}N_{11})^{-1}M_{12} & F_u(N, M_{22}) \end{pmatrix} \begin{pmatrix} w_1 \\ w_2 \end{pmatrix}.$$

This operation is defined only if the two systems have compatible dimensions. The lower and upper LFT's are the diagonal elements of the star product.

In the subsequent chapters a general robust controller design problem is formulated using the framework of LFT's. In the sequel, however, we will use only the lower fractional transformation to represent the operator T_{zw}.

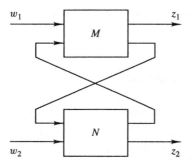

Figure 3.4   Star product.

3.2 WORST-CASE PERFORMANCE

If the exogenous signal is not known exactly but is known to lie in a set, then a reasonable measure for performance is one that looks at the worst possible output. In particular, assume that the set of exogenous inputs is given by

$$\left\{ w \in \ell_p \mid \|w\|_p \le 1 \right\}.$$

A good measure of performance is given by

$$\sup_w \|z\|_p,$$

which is the norm of the worst possible output as the exogenous signal ranges over the allowable set. The controller design problem is given by

$$\inf_{K \text{ stabilizing}} (\sup_w \|\Phi w\|_p) = \inf_{K \text{ stabilizing}} \|\Phi\|_{\ell_p - ind}.$$

This performance objective is known as a *minimax* objective. The controller is designed to guard against all exogenous signals in the allowable set. Hence, any minimization problem involving an induced norm of the closed loop operator will be considered a worst-case design method.

Notice that this formulation does not require any probabilistic assumptions on the exogenous signals. Classes of signals modeled in terms of a norm are known as *unknown but bounded signals*.

As a comparison let us consider an example of a probabilistic formulation. Assume that w is a white signal with unit intensity. A good measure of performance is given by

$$E(\|z\|_2^2),$$

where E denotes the expected value. This turns out to be precisely the $\mathcal{H}_2$ norm of Φ (assume Φ is SISO). We can always arrive to this measure through a deterministic interpretation; namely, it is equal to $\|z\|_2$ for a pulse input, or the induced norm from ℓ_2 to ℓ_∞. So for performance objectives involving the energy of the output, the $\mathcal{H}_2$ norm is not a worst-case objective. However, for performance objectives involving the maximum amplitude of the output, the $\mathcal{H}_2$ norm is a worst-case objective over bounded energy disturbances. As mentioned above, it also has a pure stochastic interpretation. In the case where Φ is MIMO, the $\mathcal{H}_2$-norm does not have an interpretation as an induced norm (see Chapter 13). Both the ℓ_1 and the $\mathcal{H}_\infty$ norms are worst-case methods and are discussed next.

3.3 PERFORMANCE: LINEAR CONSTRAINTS

In this section we discuss design specifications that result in linear constraints on the closed-loop map, Φ. Let $\mathcal{P}$ be a positive cone in $\ell_p^{r \times s}$ defined as

$$\mathcal{P} = \{H \in \ell_p^{r \times s} \mid h_{ij}(k) \ge 0 \text{ for all } i, j, k\}.$$

Let $\mathcal{A}$ be a linear operator from $\ell_1^{n_z \times n_w}$ to $\ell_p^{r \times s}$ for some p, and $b \in \ell_p^{r \times s}$ be a fixed element. Then Φ satisfies the set of linear constraints given by $\mathcal{A}$ and b if

$$b - \mathcal{A}\Phi \in \mathcal{P}.$$

With a more familiar notation, the above condition takes the form

$$\mathcal{A}\Phi \leq b \tag{3.3}$$

where the inequality is with respect to the cone $\mathcal{P}$, i.e., pointwise. It turns out that many performance specifications can be posed in terms of linear constraints, as in Equation (3.3).

3.3.1 ℓ_1-Norm Constraints

Suppose the exogenous inputs satisfy $w \in \ell_\infty$ with $\|w\|_\infty \leq 1$ but are otherwise arbitrary. If the objective is to minimize the maximum amplitude of the regulated output, then the nominal ℓ_1 performance problem is defined as

$$\nu^o := \inf_{K \text{ stabilizing}} (\sup_w \|\Phi w\|_\infty) = \inf_{K \text{ stabilizing}} \|\Phi\|_1. \tag{3.4}$$

Before we present prototypes of such performance specifications, it is interesting to show that the corresponding performance constraints are in fact linear. The complete verification of this fact depends on the following result (which will be shown later): *The feasible set of closed-loop maps is characterized by a set of linear equations.*

To bring the objective function $\|\Phi\|_1$ into linear form and to avoid the nonlinearity built into the norm (i.e., the absolute value function), we use a standard change of variables from linear programming. Let $\Phi = \Phi^+ - \Phi^-$, where Φ^+ and Φ^- are sequences of $n_z \times n_w$ matrices with nonnegative entries. That is, Φ^+, $\Phi^- \in \mathcal{P}$. Then, replace the ℓ_1 norm of Φ by

$$\max_i \sum_{j=1}^{n_w} \sum_{t=0}^{\infty} (\phi_{ij}^+(t) + \phi_{ij}^-(t))$$

which is linear in (Φ^+, Φ^-). This expression is equal to the norm only if, for every (i, j, t), either $\phi_{ij}^+(t)$ or $\phi_{ij}^-(t)$ is zero. It turns out that this is a guaranteed property of the optimal solution to Problem (3.4). Indeed, if a feasible solution is such that $\phi_{ij}^+(t)$ and $\phi_{ij}^-(t)$ are strictly positive and $m := \min[\phi_{ij}^+(t), \phi_{ij}^-(t)]$, then reducing both variables by m does not violate feasibility since the difference remains the same (i.e., Φ remains the same), and further, one of the two variables becomes zero. Clearly, this adjustment decreases the value of the cost, when $\phi_{ij}^+(t)$ and $\phi_{ij}^-(t)$ are added up. Therefore, the optimal solution will always be such that either $\phi_{ij}^+(t)$ or $\phi_{ij}^-(t)$ is zero. Note that this transformation doubles the number of variables representing the closed-loop response.

Consequently, the ℓ_1 minimization problem (3.4) can be restated as follows:

$$\nu^o = \inf_{\nu, \Phi^+, \Phi^-} \nu$$

subject to

$$\sum_{j=1}^{n_w} \sum_{t=0}^{\infty} (\phi_{ij}^+(t) + \phi_{ij}^-(t)) \le \nu \quad \text{for} \quad i = 1, \ldots, n_z \tag{3.5}$$

$$\Phi = \Phi^+ - \Phi^- \text{ is feasible.}$$

Finally, a compact representation of the ℓ_1 norm constraints can be obtained by defining an operator $\mathcal{A}_{\ell_1} : \ell_1^{n_z \times n_w} \to \mathbb{R}^{n_z}$ such that

$$(\mathcal{A}_{\ell_1} \Phi)_i = \sum_{j=1}^{n_w} \sum_{t=0}^{\infty} \phi_{ij}(t) \quad \text{for} \quad i = 1, \ldots, n_z,$$

and a vector with all the elements equal to one, $\mathbf{1} \in \mathbb{R}^{n_w}$. It follows that

$$\sum_{j=1}^{n_w} \sum_{t=0}^{\infty} (\phi_{ij}^+(t) + \phi_{ij}^-(t)) \le \nu \quad \text{for} \quad i = 1, \ldots, n_z \Longleftrightarrow \mathcal{A}_{\ell_1}(\Phi^+ + \Phi^-) \le \mathbf{1}\nu. \tag{3.6}$$

We will say that $\mathcal{A}_{\ell_1}$ is a norm operator since it replaces the ℓ_1 norm constraints.

Why the ℓ_∞ signal norm?   In many real-world applications, exogenous disturbance and/or noise is persistent, i.e., continues acting on the system as long as the system is in operation. This implies that such inputs have infinite energy, and thus cannot be modeled as "finite energy signals." Quite often, however, it is possible to get good estimates on the maximum amplitude of such inputs. Examples where bounded disturbances arise in practical situations are abundant. Wind gusts facing an aircraft in flight can be viewed as bounded and persistent disturbances. Without a correcting control action, such disturbances will cause the aircraft to deviate from its set path. An automobile driven over an unpaved road experiences disturbances due to the irregularity of the course. Such disturbances, although persistent, are clearly bounded in magnitude. In process control, level measurements of a boiling liquid are corrupted by bounded disturbances due to the constant level fluctuations of the liquid. Because such disturbances are so frequent, a mathematical model describing them is essential. The ℓ_∞ norm is clearly a natural choice for measuring the size of such disturbances. In general, we will assume that the disturbance is the output of a linear-time invariant filter subjected to signals of magnitude less than or equal to one, i.e.,

$$d = Ww, \quad \|w\|_\infty \le 1.$$

Not only is the ℓ_∞ norm useful for measuring input disturbance size, but it can also be very useful as a measure for the size of output (regulated) signals. For example, in many applications it is crucial that the tracking error never exceeds a certain level at *any* time. While this requirement cannot be captured by using the ℓ_2 norm, it can be stated explicitly as a condition on the ℓ_∞ norm of the error signal. Another situation where the ℓ_∞ norm is useful is when the plant, or any other device in the control loop, has a maximum input rating which should not be exceeded. This translates directly to a requirement on the ℓ_∞ norm of that input. An example of such a requirement appears in the next section. In addition, the ℓ_∞ norm plays an important role in designing controllers for nonlinear systems. Because most of the nonlinear controller designs are based on linearization, the linear model gives a faithful representation of the system only if the states remain close to the equilibrium point, a requirement captured directly in terms of the ℓ_∞ norm.

While the ℓ_∞ norm is used as a measure of signal size, the ℓ_1 norm is used to measure a system's amplification of ℓ_∞ input signals. Let H be a linear time-invariant system given by

$$z(t) = (Hw)(t) = \sum_{k=0}^{t} H(t-k)w(k).$$

The inputs and outputs of the system are measured by their maximum amplitude over all time; that is, by their ℓ_∞ norm:

$$\|w\|_\infty = \max_j \sup_k |w_j(k)|.$$

The ℓ_1 norm of the system H is precisely equal to the maximum amplification the system exerts on bounded signals. This measure, as defined in the previous chapter, is given by

$$\|H\|_1 = \max_{1 \le i \le n_z} \sum_{j=1}^{n_w} \sum_{k=0}^{\infty} |h_{ij}(k)|.$$

Disturbance rejection.   In the context of ℓ_∞ signals, the disturbance rejection problem is defined as follows: Find a feedback controller that minimizes the maximum amplitude of the regulated output over all possible disturbances of bounded magnitude. The 2-input 2-output system shown in Figure 3.5 depicts the particular case where the disturbance enters the system at the plant output. Its mathematical representation is given by

$$z = P_0 u + W w,$$

$$y = P_0 u + W w.$$

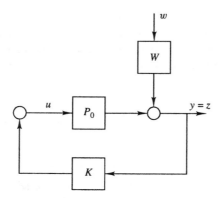

Figure 3.5   Disturbance rejection problem.

Notice that the disturbance rejection problem provides a general enough structure to represent a broad class of interesting control problems.

Command following with saturation. The command following problem, equivalent to a disturbance rejection problem, is shown in Figure 3.6. We will show how to pose this problem in the presence of saturation nonlinearities at the input of the plant, as an ℓ_1-optimal control problem. Define the function

$$\text{Sat}(u) = \begin{cases} u & |u| \leq U_{\max} \\ U_{\max} sgn(u) & |u| > U_{\max} \end{cases}.$$

Let the plant be described as

$$Pu = P_0 \text{Sat}(u)$$

where P_0 is LTI. Let the commands be modeled as

$$r = Ww \quad \text{where} \quad \|w\|_\infty \leq 1.$$

The objective is to find a controller K such that y follows r uniformly in time. Keeping in mind the saturation function, and in order to stay in the linear region of operation, it is clear that the allowable control inputs have to have $\|u\|_\infty \leq U_{\max}$. Let γ be the (tracking) performance level desired, and define

$$z = \begin{bmatrix} (y - r)/\gamma \\ u/U_{\max} \end{bmatrix}$$

with

$$y = P_0 u.$$

The problem is equivalent to finding a controller such that

$$\sup_w \|z\|_\infty < 1,$$

which is an ℓ_1-optimal control problem.

It is interesting to note that the above closed-loop system may remain stable even if the input saturates, as long as it does so infrequently. The solution to the above problem will determine the limits of performance when the system is required to operate in the linear region. Also, stability for such a system will mean local stability of the nonlinear system.

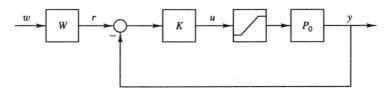

Figure 3.6 Command following with input saturation.

Saturation and rate limits.   In the previous example, actuator limitations may require that the rate of change of the control input be bounded. This is captured in the condition

$$\left| \frac{u(k) - u(k-1)}{T_s} \right| \le U_{\text{der}}$$

where T_s is the sampling period. Let

$$W_{\text{der}} = \frac{1 - \lambda}{T_s U_{\text{der}}}.$$

This condition can be easily incorporated in the objective function by defining z as:

$$z = \begin{bmatrix} (y - r)/\gamma \\ u/U_{\text{max}} \\ W_{\text{der}} u \end{bmatrix}.$$

The result is a standard ℓ_1-optimal control problem.

3.3.2 Performance with Fixed Inputs

So far we have expressed performance specifications in terms of a class of input signals, namely the class of bounded magnitude signals. In many control applications, however, it is desirable to track a specific trajectory with, for example, the least overshoot or undershoot possible. In a more general setting, the controller is designed so that the response to a fixed input is within a time-domain template:

$$g_l(t) \le z(t) \le g_u(t)$$

with both g_l and g_u specified (see Figure 2.2). We will say z is feasible if $z = \Phi w_f$ for some feasible Φ, where w_f is the specific fixed input. Following are some problems of traditional importance in control design.

Minimum overshoot.   For a step input, w_{step}, the tracking error is defined as $z = w_{\text{step}} - y$. Consequently, the no-overshoot problem can be formulated as

$$z \ge 0,$$
$$z \text{ is feasible.} \tag{3.7}$$

Similarly, the minimum overshoot problem is

$$\min \nu$$
$$z \ge -\nu$$
$$\nu \ge 0 \tag{3.8}$$
$$z \text{ is feasible.}$$

The minimum undershoot problem can be posed in an analogous fashion.

Settling time. Specifications on the settling time of the response to a fixed input are also linear. For instance, the tracking error can be required to decay exponentially after some positive time, T. An augmented set of constraints, with the overshoot specifications included, is given by

$$-a^{T-t} \leq z \leq a^{T-t} \qquad \text{for all } t \geq T, \ |a| < 1$$
$$z \geq -\nu \tag{3.9}$$
$$z \text{ is feasible.}$$

Maximum deviation. In certain applications, it may be desirable to keep the response to a fixed input within a small neighborhood of the input trajectory. Such specifications are reflected in norm bounds, i.e.,

$$\|z\|_\infty \leq \nu,$$
$$z \text{ is feasible.} \tag{3.10}$$

These constraints are linear constraints. This is immediately seen from

$$\|z\|_\infty \leq \nu \iff -\nu \leq z(t) \leq \nu \quad \text{for all } t \geq 0.$$

Summary. All of the specifications above result in linear constraints on the closed loop map. By combining all of these constraints, a linear operator $\mathcal{A}_{\text{temp}}$ can be constructed such that the exact input specifications are equivalent to

$$\mathcal{A}_{\text{temp}} \Phi \leq b_{\text{temp}}$$

for some fixed b_{temp}. From the above, it is clear that the set of closed-loop maps that satisfy the performance constraints is a convex one. Also, notice that the range of $\mathcal{A}_{\text{temp}}$ may be infinite dimensional.

3.4 PERFORMANCE: APPROXIMATE LINEAR CONSTRAINTS

Not all design specifications yield linear constraints on the closed-loop map. We will present an example where the constraints are convex (they specify a convex subset of the set of feasible closed-loop maps) but nonlinear, and can be approximated quite well with linear constraints.

3.4.1 $\mathcal{H}_\infty$-Norm Constraints

Suppose the exogenous inputs are such that $\|w\|_2 \leq 1$ but are otherwise arbitrary. If the objective is to minimize the energy of the regulated output, then the nominal performance problem is defined as

$$\inf_{K \text{ stabilizing}} (\sup_w \|\Phi w\|_2) = \inf_{K \text{ stabilizing}} \|\hat{\Phi}\|_\infty.$$

Why the ℓ_2 signal norm? Suppose that in a given control problem, the class of exogenous disturbances is best described as all w's such that the energy of Ww is bounded, where the filter W reflects the spectral content of the signals. Then, if one is concerned with the (weighted) energy of the error signal z, the problem is most naturally posed in ℓ_2. However, this situation is not very typical of control applications. In fact, performance specifications are mostly concerned with amplitudes of signals, which are in general not reflected in an energy bound (see Figure 3.7).

There is, however, an alternative interpretation of the $\mathcal{H}_\infty$ norm as seen in Table 2.1. The $\mathcal{H}_\infty$ norm is not only the gain over ℓ_2 signals, but is also an upper bound on the amplitude gain over persistent sinusoidal signals (i.e., bounded power signals). Such a viewpoint has deep historical roots and is intimately related to traditional frequency Bode plots and loopshaping ideas. In this way, the ℓ_2 signal space is highly relevant to control system design.

Approximate $\mathcal{H}_\infty$ constraints. Given a closed loop-map Φ, the maximum singular value of $\hat{\Phi}(e^{i\omega_0})$ is given by

$$\sigma_{\max}[\hat{\Phi}(e^{i\omega_0})] = \max_{u,v}\{\Re[u^*\hat{\Phi}(e^{i\omega_0})v] \mid |u|_2 = |v|_2 = 1\}. \qquad (3.11)$$

Say that the design specification is such that the $\mathcal{H}_\infty$ norm of Φ needs to be bounded from above by $\gamma > 0$, i.e.,

$$\sigma_{\max}[\hat{\Phi}(e^{i\omega_0})] \leq \gamma \quad \text{for all} \quad \omega_0 \in [0, 2\pi).$$

To approximate these constraints, first N samples from the unit circle are obtained. Then, for each sample, Equation (3.11) is approximated by a polytope. Of course, this kind of approximation can yield a large set of linear inequalities. However, we will see later that only a small number of constraints is generally sufficient to alter a given design. This approximation is particularly simple in the SISO case.

SISO $\mathcal{H}_\infty$-norm constraints. For a scalar complex number $H = H_R + iH_I$,

$$|H| \leq \gamma \Longleftrightarrow H_R \cos\theta + H_I \sin\theta \leq \gamma \quad \text{for all} \quad \theta \in [0, 2\pi).$$

Let $\hat{\Phi}(\lambda) = \sum_{k=0}^{\infty} \phi(k)\lambda^k$ be a SISO transfer function with impulse response $\{\phi(k)\}$. Define $\Phi_R(\omega_n) = \Re[\hat{\Phi}(e^{i\omega_n})]$ and $\Phi_I(\omega_n) = \Im[\hat{\Phi}(e^{i\omega_n})]$, where ω_n are samples of the

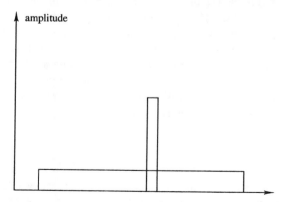

amplitude

time   **Figure 3.7** Two signals with equal energy.

unit circle. A set of linear constraints that approximates the $\mathcal{H}_\infty$-norm constraints is:

$$\Phi_R(\omega_n) \cos \theta_m + \Phi_I(\omega_n) \sin \theta_m \leq \gamma \quad \text{where} \quad \begin{cases} \omega_n \in [0, 2\pi), \ n = 1, \ldots, N \\ \theta_m \in [0, 2\pi), \ m = 1 \ldots, M \end{cases} . \quad (3.12)$$

Note that the evaluation of the λ-transform at some frequency is a linear operation on Φ:

$$\Phi_R(\omega_n) = \sum_{k=0}^{\infty} \phi(k) \cos(k\omega_n), \quad \Phi_I(\omega_n) = \sum_{k=0}^{\infty} \phi(k) \sin(k\omega_n).$$

Combining this with Equation (3.12) we get

$$\sum_{k=0}^{\infty} \phi(k) \cos(k\omega_n - \theta_m) \leq \gamma \quad \text{where} \quad \begin{cases} \omega_n \in [0, 2\pi), \ n = 1, \ldots, N \\ \theta_m \in [0, 2\pi), \ m = 1 \ldots, M \end{cases} . \quad (3.13)$$

The linear constraints in Equation (3.13) can be arranged in a linear operator (infinite matrix), $\mathcal{A}_{\mathcal{H}_\infty} : \ell_1^{n_z \times n_w} \rightarrow \mathbb{R}^{NM}$ such that

$$\|\hat{\Phi}\|_\infty \leq \gamma \Longrightarrow \mathcal{A}_{\mathcal{H}_\infty} \Phi \leq \gamma \mathbf{1}. \quad (3.14)$$

If the samples ω_n and θ_m are dense, then the right hand side of Equation (3.13) approximates the left-hand side.

Phase information.   It is interesting to point out that phase information can also be incorporated to some degree in such a setup, simply by specifying different magnitude constraints on different sectors of the complex plane.

Partition the complex plane into L sectors, that is, define a set of angles, Θ_l, $l = 0, 1, \ldots, L$, such that

$$0 = \Theta_0 < \Theta_1 < \cdots < \Theta_{L-1} < \Theta_L = 2\pi.$$

Let γ_l be the magnitude bound in each sector, then Equation (3.13) can be rewritten as follows: for $l = 1, \ldots L,$

$$\sum_{k=0}^{\infty} \phi(k) \cos(k\omega_n - \theta_m^l) \leq \gamma_l \quad \text{where} \quad \begin{cases} \omega_n \in [0, 2\pi), \ n = 1, \ldots, N \\ \theta_m^l \in [\Theta_{l-1}, \Theta_l), \ m = 1 \ldots, M_l \end{cases} \quad (3.15)$$

and θ_m^l is the mth sample in the lth sector. If the lth sector is unconstrained (i.e., $\gamma_l \rightarrow \infty$), then the corresponding subset of constraints can be ignored.

Note that the above set of constraints is more restrictive than bounding the Nyquist plot on sectors of the complex plane (a nonconvex problem). To see this, consider the case where $\Phi(e^{i\omega})$ is constrained to be of magnitude less than one at all frequencies where $\angle\Phi(e^{i\omega})$ is equal to π. In the context of Equation (3.15) this condition corresponds to $\gamma = 1$, $\theta = \pi$ and all $\omega \in [0, 2\pi)$. But this is equivalent to $\Re[\Phi(e^{i\omega})] > -1$ for all $\omega \in [0, 2\pi)$, a sufficient condition that is certainly not necessary.

3.5 ROBUST STABILITY

In the previous sections we discussed nominal performance given that an exact description of the plant is available. In general, only an approximate model of the process is available, with possibly a description of a set to which the actual process belongs. In these situations the controller is required to stabilize every single plant in the set. We will show shortly that this problem is similar to a nominal performance problem with norm constraints.

Underlying most of the stability robustness results is the Small Gain Theorem as presented in Chapter 2. Basically, the theorem guarantees the stability of a feedback system consisting of an interconnection of two stable systems (as in Figure 3.8) if the gain of their composition is less than unity. This theorem is quite general and applies to nonlinear time-varying systems with any notion of ℓ_p-stability. The small gain condition is in general not necessary for stability; however, it can be necessary if one of the systems in the feedback is arbitrary within a sufficiently large class. We will not analyze the conservatism of the small gain theorem here; a full discussion is left for Chapter 7.

In the context of ℓ_∞ signals, the small gain theorem takes the following form:

Theorem 3.5.1.   Let M be a linear time-invariant system and Δ be a strictly proper ℓ_∞-stable perturbation. The closed-loop system shown in Figure 3.8 is ℓ_∞-stable for all Δ with $\|\Delta\|_{\ell_\infty-\text{ind}} < 1$ if $\|M\|_1 \leq 1$.

Proof. It follows directly from the small gain theorem and the sub-multiplicative property of the norm, i.e.,

$$\|M\Delta\|_{\ell_\infty-\text{ind}} \leq \|M\|_1 \|\Delta\|_{\ell_\infty-\text{ind}} < 1.$$

Strict properness guarantees the well posedness of the closed loop system.   ∎

A similar statement can be made if ℓ_2-stability is required, and the perturbations are ℓ_2-stable. The sufficient condition in such a case is $\|\hat{M}\|_\infty \leq 1$.

Example 3.5.1

Consider a class of linear time-invariant SISO plants, modeled as follows:

$$\Omega = \{P \mid P = M\Delta \text{ and } \|\hat{M}\hat{\Delta}\|_\infty < 1\},$$

and say that the output is connected to the input as in Figure 3.8. A direct application of the small gain theorem guarantees the ℓ_2-stability of such a closed-loop system. Nevertheless, in order to gain more insight, we will apply the standard Nyquist stability criterion. Notice that both elements in the loop (M and Δ) are stable, so the Nyquist stability test calls for no encirclements of the minus one point. But this is exactly the case, since all elements in the class Ω have Nyquist plots that are contained in the open unit disc (as a

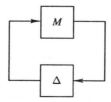

Figure 3.8   Stability robustness problem.

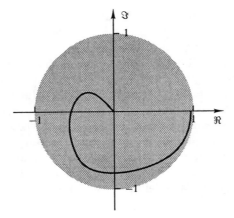

Figure 3.9   Nyquist plot for some $P \in \mathcal{P}$.

consequence of the $\mathcal{H}_\infty$-norm bound, see Figure 3.9). Hence, we conclude that the system is robustly stable.

As a consequence of the small gain theorem, it is possible to provide (sufficient) stability robustness conditions for some classes of perturbed systems.

3.5.1 Unstructured Multiplicative Perturbations

Consider the case where the system has input uncertainty in a multiplicative form as in Figure 3.10, i.e.,

$$\Omega = \{P \mid P = P_o(I + W_1 \Delta W_2) \text{ and } ||\Delta||_{\ell_\infty - \text{ind}} < 1\}.$$

If a controller is designed to stabilize P_0, under what conditions will it stabilize the whole set Ω? By simple manipulations of the closed-loop system, the problem is equivalent to the stability robustness of the feedback system in Figure 3.8, with $M = W_2(I - KP_o)^{-1}KP_oW_1$. In general, this manipulation is done in a systematic way: Cut the loop at the inputs and outputs of Δ and then calculate the map from the output of Δ, w, to the input of Δ, z. A sufficient condition for robust stability is then given by $||M||_1 \leq 1$.

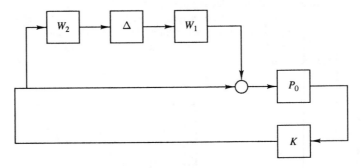

Figure 3.10   Multiplicative perturbations.

The resulting 2-input 2-output description is given by

$$y = P_0 u + P_0 W_1 w,$$

$$z = W_2 u.$$

This is a standard ℓ_1 minimization problem.

3.5.2 Stable Coprime Factor Perturbations

The class of perturbations suggested in the above example allows only stable perturbations of the plant. A more general way of introducing (possibly unstable) perturbations is by perturbing the coprime factors of the plant. A detailed discussion of coprime factorization will be presented in Chapter 5.

    Definition 3.5.1.   Given a SISO LTI plant P_0, a pair of LTI stable operators M, N define a coprime factorization of P_0 if $P_0 = N M^{-1}$ and $\hat{N}$ and $\hat{M}$ have no common zeros inside the closed unit disc.   ■

Let P_0 be a linear time-invariant, finite dimensional plant. As usual, $P_0 = G_{22}$. The graph of P_0 over the space ℓ_q is defined as

$$G^q(P_0) = G_{P_0} \ell_q \text{ where } G_{P_0} = \begin{bmatrix} M \\ N \end{bmatrix}.$$

Define the following class of plants as in Figure 3.11:

$$\Omega_q = \{ P \mid G_P = \begin{bmatrix} M - \Delta_1 \\ N + \Delta_2 \end{bmatrix} \text{ and } \left\| \begin{bmatrix} \Delta_1 \\ \Delta_2 \end{bmatrix} \right\|_{\ell_q - \text{ind}} \le 1 \}.$$

This class corresponds to perturbing the graph of the plant P_0 and depends on the particular coprime factors used. In the sequel, we consider the class Ω_∞.

    To apply the small gain theorem for this problem, we manipulate it to look like Figure 3.8. The 2-input 2-output system is given by

$$y = P_0 w_1 + w_2 + P_0 u,$$

$$z = M^{-1} w_1 + M^{-1} u.$$

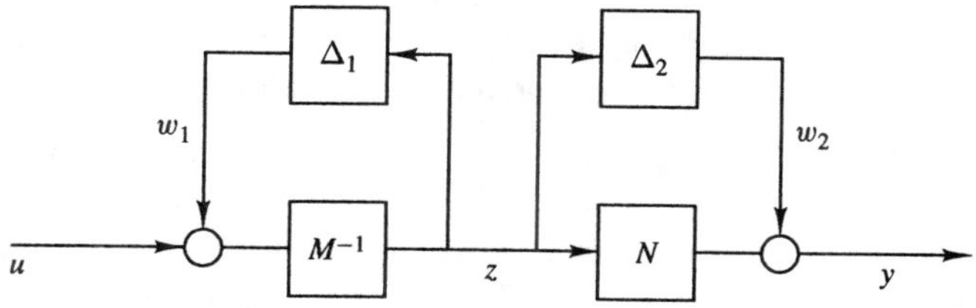

Figure 3.11 Coprime factor perturbations.

It is interesting to note that this class of perturbations is a more natural one in the case of unstable plants. It amounts to perturbing the graph of the operator, rather than the operator directly. This class of perturbations allows unstable perturbations and can result in changing the locations and number of unstable poles of the system. A description of the coprime perturbations of a system can be derived in a natural way from parametric identification techniques in which a fixed-order polynomial is identified for both M and N.

Finally, we note that when the controller is connected in the feedback with the plant, the small gain condition gives an easy test for robust stability in terms of the coprime factors of the controller. This condition can be shown to be

$$\| [X \quad Y] \|_1 \leq 1 \tag{3.16}$$

where $K = YX^{-1}$ with X and Y coprime, and $MX - NY = I$. This problem is discussed in more detail in Chapter 7.

Observations.   In the previous examples, each of the robust stabilization problems was shown to be equivalent to some performance problem where a fictitious disturbance is injected at the output of the perturbation, and an error is measured at the input of the perturbation. So the transfer function to be minimized is simply the function *seen* by the perturbations. This indicates that robust stability is equivalent to some nominal performance problem. In Chapter 7, the dual of this idea will be presented: Performance will be shown to be equivalent to a robust stability problem in the presence of some fictitious perturbation. Such dual equivalence will make the derivation of robust performance conditions a tractable problem.

Notice also that depending on how the perturbations are modeled, a problem involving either the ℓ_1 norm or the $\mathcal{H}_\infty$ norm may arise. The $\mathcal{H}_2$ norm, however, does not arise from robust stability problems even though it has an interpretation as an induced norm from ℓ_2 to ℓ_∞, which are different spaces. This reduces the utility of the $\mathcal{H}_2$ norm in the analysis of perturbed systems.

3.6 MIXED PERFORMANCE OBJECTIVES

To guarantee that a closed-loop map satisfies multiple constraints, we augment all the linear operators in one operator constraint. Notice that different linear constraints can be defined for different closed-loop maps $\Phi = T_{z_i w_j}$, i.e., on the map between the ith input and the jth output. Of course, the augmented set of conditions may not have a feasible solution, indicating that there does not exist a controller that can meet all the stated specifications. A typical augmented operator will have the form:

$$\begin{pmatrix} \mathcal{A}_{\ell_1} & \mathcal{A}_{\ell_1} \\ \mathcal{A}_{\mathcal{H}_\infty} & -\mathcal{A}_{\mathcal{H}_\infty} \\ \mathcal{A}_{\text{temp}} & -\mathcal{A}_{\text{temp}} \end{pmatrix} \begin{pmatrix} \Phi^+ \\ \Phi^- \end{pmatrix} \leq \begin{pmatrix} \nu \mathbf{1} \\ \gamma \mathbf{1} \\ b_{\text{temp}} \end{pmatrix}. \tag{3.17}$$

If the plant is known to lie in a set, then part of the objectives is to guarantee robust stability. This is of course given by some norm constraint.

3.7 ROBUST PERFORMANCE

In general, the controller is required to satisfy robust performance, i.e., to attain a certain level of performance (as discussed before) for all possible plant perturbations. It turns out that the constraints corresponding to such problems are hard to write, and, in fact, the general problem is still an open one. However, when the performance is in terms of norm bounds, we will show in Chapter 7 how such problems can be solved in a nonconservative way. The resulting constraints will not be linear nor convex as functions of the closed-loop map. A few prototype examples in which robust performance is desired are presented next.

Robust disturbance rejection.   In the previously discussed disturbance rejection problems, the plant was assumed to be known exactly. This is rarely the case due to unmodeled dynamics, parameter variations, etc. When the controller designed for a nominal plant model is implemented on the real system, there are no guarantees on the resulting performance of the system. Even requirements as basic as stability may not be met. The deviation from the expected behavior of the system clearly depends on the accuracy of the model. Since modeling uncertainty is inevitable, it is imperative to include stability and performance robustness to model uncertainty as a design objective.

We now take a second look at the disturbance rejection problem discussed earlier. Instead of considering a single nominal time-invariant plant, P_0, we shall consider a collection of plants (see Figure 3.12). Any plant in this collection could be the real plant, but since it is not known which one, the design should consider the entire collection. Take, for instance, the following class of plants:

$$\Omega = \left\{ P \mid P = (I + W_3 \Delta W_2) P_0 \text{ and } \|\Delta\|_{\ell_\infty - \text{ind}} < 1 \right\}$$

where W_2 and W_3 are time-invariant stable weighting functions. From this definition, the plant perturbation, Δ, may be time-varying and/or nonlinear. Any plant belonging to this class is said to be admissible. Note that when $\Delta = 0$, we recover the nominal linear time-invariant plant. Consequently, the collection of admissible plants, Ω, may be

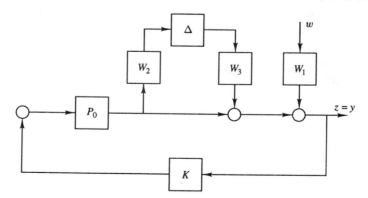

Figure 3.12   Robust disturbance rejection problem.

viewed as a ball of plants centered around the nominal time-invariant plant model. If a closed-loop system property such as stability holds for all admissible plants, then the system is said to be robust. We now add to our original disturbance rejection problem a new objective: robustness. In other words, the controller K is now required to perform the following tasks:

1. to internally stabilize all admissible plants, i.e., all plants in the class Ω.
2. to minimize the worst-case effect of the disturbance w on the magnitude of the output produced by the worst possible admissible plant, i.e.,

$$\min_{K} \sup_{P \in \Omega} \sup_{\|w\|_\infty \leq 1} \|y\|_\infty.$$

A detailed discussion of this problem is left to Chapter 7. An easier problem will be to satisfy both nominal stability and performance. Let $S_o = (I + P_o K)^{-1} W_1$ and $T_o = W_2 P_o K (I + P_o K)^{-1} W_3$. One set of design specifications can be: Find K such that

$$\|S_o\|_1 \leq \nu,$$
$$\|T_o\|_1 < 1,$$

which can be captured as a set of linear constraints. It is worthwhile commenting that the general robust performance problem may not yield a set of linear (or convex) constraints.

A multi-objective control problem. In almost all practical control problems, more than one objective must be met simultaneously. Perhaps one of the most attractive features of the approach presented in Chapter 7 is its ability to accommodate multiple objectives in a natural way. As an example of a multiple objective problem consider the system in Figure 3.13. In the figure the plant is subjected to multiplicative output perturbations. In addition, it has a saturation nonlinearity at its input of the type discussed earlier. A command input, w_2, is applied while a bounded disturbance, w_1, is acting at the plant output. The objectives in this problem are a combination of those objectives in the first three problems discussed earlier. Aside from stabilizing all admissible plants, the controller must also ensure that the plant input, u, never exceeds its maximum,

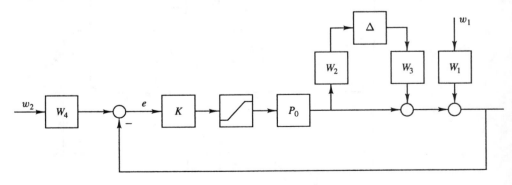

Figure 3.13 Multi-objective problem.

$U_{\max}$, despite the presence of the output disturbance, the command input, and the plant uncertainty. Furthermore, the tracking error in this unfriendly environment must be maintained at a minimum level for all time. These requirements on the controller are summarized as follows:

1. K stabilizes all plants in Ω.
2. K is chosen such that $\displaystyle\sup_{\|w_i\|_\infty \le 1}\ \sup_{P \in \Omega}\ \|u\|_\infty \le U_{\max}$.
3. K is chosen such that $\displaystyle\sup_{\|w_i\|_\infty \le 1}\ \sup_{P \in \Omega}\ \|e\|_\infty$ is minimized.

It is possible in this formulation to include time-varying weights with which one can emphasize certain periods of the time response. This may, however, complicate the solution.

3.8 DESIGN EXAMPLES

In future chapters we will develop a powerful and flexible theory for control system design in the presence of uncertainty. To demonstrate this theory, it is essential that a collection of meaningful and interesting problems be solved using the different concepts and techniques presented. The purpose of this is twofold: to illustrate the theory and to convey the general nature of such a design approach.

Two types of numerical examples will be presented: First, a collection of academic examples, consisting of simple numerical problems that are "constructed" so that the computations are (to some extent) tractable within the text. Many of these examples will be introduced throughout the book so as to illustrate specific aspects of the theory. Second, a number of design examples consisting of more realistic and complex problems that typically arise in control design. This section states the latter, postponing their solutions until the theory is developed. They are presented here for the purpose of motivating the reader.

3.8.1 Pitch Axis Control of the X29 Aircraft

The X29 aircraft poses an interesting control problem owing to its revolutionary forward-swept wing design. With such a configuration, the center of gravity lies behind the aerodynamic center of pressure, rendering the aircraft statically unstable. Consequently, a control system has to actively stabilize the aircraft during flight. These type of wings have some desirable aerodynamic characteristics such as better maneuverability and reduced drag when compared with the more classical wing design.

We are interested in designing a digital controller for a simple model of the pitch axis dynamics of the aircraft. The airplane has three types of control surfaces: canard wings, flaperons on the main wings, and strakes on the tail. To simplify the model, the action of these control surfaces are lumped into one equivalent actuator with first order dynamics. Similarly, the gyroscopes and accelerometers are modeled by an equivalent sensor with negligible dynamics. Thus, the system can be approximately represented by the following continuous-time SISO plant:

$$\hat{P}(s) = \underbrace{\frac{(s+3)}{(s+10)(s-6)}}_{\text{airframe}} \underbrace{\frac{20}{(s+20)}}_{\text{equiv. actuator}} \underbrace{\frac{(s-26)}{(s+26)}}_{\text{overhead}} \tag{3.18}$$

where s is the Laplace variable. The airframe factor corresponds to a simplified model of the pitch dynamics of the airplane (considered as a rigid body) flying at a low altitude and with an air speed of approximately 0.9 Mach. The overhead factor lumps the equivalent low frequency phase lag introduced by the dynamics that are neglected in deriving the reduced model (3.18). More specifically, this allpass factor is an approximate representation of the collected phase lag contributed by the gyroscopic sensor dynamics, the actuator servo dynamics, the airframe flexible modes, and the digital implementation of the controller (i.e., pre-filter, zero order hold, and computing delay) corresponding to a sampling period $T_s = 1/30$ seconds.

ℓ_1 performance objective.   Consider the following formal synthesis problem (see Figure 3.14) with ℓ_1 performance objectives: Find a stabilizing discrete-time controller such that the ℓ_1 norm of the discrete-time transfer function from the disturbance w, to the weighted control sequence z_1, and the weighted output z_2, is minimized. That is,

$$\inf_{K \text{ stab.}} \left\| \begin{matrix} W_1 K S \\ W_2 S \end{matrix} \right\|_1$$

where $S := (I - PK)^{-1}$ denotes the discrete-time sensitivity function. The above equation requires the discrete-time version of Equation (3.18) and two weighting transfer functions. The λ-domain model of the plant, $\hat{P}(\lambda)$, is obtained by discretizing Equation (3.18) assuming a zero order hold at the plant input and a synchronized sampling of the (pre-filtered) plant output. The weights are generally chosen to reflect the trade-offs between low frequency disturbance rejection and the control effort, as well as to emphasize those frequency regions corresponding to the spectral content of the exogenous disturbance. In this particular case we choose $\hat{W}_2$ to be the discrete-time form of the continuous time transfer function $(s + 1)/(s + 0.001)$, and $\hat{W}_1 = 0.01$.

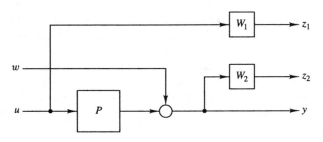

Figure 3.14 The X29 Problem in standard form.

The 2-input 2-output representation of this problem is given by

$$\begin{pmatrix} z_1 \\ z_2 \\ y \end{pmatrix} = \begin{pmatrix} 0 & W_1 \\ W_2 & W_2 P \\ I & P \end{pmatrix} \begin{pmatrix} w \\ u \end{pmatrix}.$$

Notice that a controller designed for the discrete-time model of a continuous-time plant completely ignores the inter-sampling behavior of the system. A controller designed to meet certain specifications on the sampled instances may violate these specifications in-between the samples. This suggests an alternate formulation in which a hybrid system is analyzed (i.e., a system with interconnected continuous-time and discrete-time components). A formal treatment of hybrid systems has been developed in the context of $\mathcal{H}_2$, $\mathcal{H}_\infty$ and ℓ_1 optimization. Although such theory is beyond the scope of this book, a few central ideas will be introduced in the next chapter.

ℓ_1 performance objective with frequency domain constraints. Consider the above problem augmented with $\mathcal{H}_\infty$-norm constraints on the transfer function $W_2 S$. That is,

$$\|\hat{W}_2 \hat{S}\|_\infty \leq \gamma_{\mathcal{H}_\infty}.$$

This situation may arise if the specifications include tracking performance for sinusoidal type inputs.

ℓ_1 performance objective with fixed input constraints: Trade-offs in design. Design specifications often include maximum deviation constraints on different signals of the closed-loop system. In a typical design scenario, the control signals must be uniformly bounded when the closed-loop system is disturbed with a specific signal such as a unit step. This specifications can be translated into a set of extra constraints and augmented into the standard ℓ_1 problem:

$$\inf_{K \text{ stab.}} \left\| \begin{matrix} W_1 K S \\ W_2 S \end{matrix} \right\|_1$$

subject to

$$\|K S w_{\text{step}}\|_\infty \leq U_{\max}$$

where w_{step} is a unit step input disturbance and $U_{\max}$ is the specified bound on the control signal.

It is sometimes interesting to analyze the trade-offs in such design. That is, how does the constraint on the control signal affect performance? It is clear that performance will deteriorate as $U_{\max}$ is reduced. However, we would like to answer more specific questions such as; How much can we gain in performance if we replace the actuator with a more expensive one with larger dynamic range?

3.8.2 Flexible beam

A flexible beam has one end pinned to the shaft of a high-torque DC motor and the
other end free. A position measurement of the free end of the beam is available (see
Figure 3.15). We would like to design a digital controller to regulate the position of the
tip of the beam. (Note: This setup corresponds to an actual experimental facility.)

A complete continuous-time model for this system is infinite dimensional due to the
distributed parameter nature of the flexible beam. Indeed, the continuous-time transfer
function from torque input to tip deflection has the generic form:

$$\sum_{i=0}^{\infty} \frac{c_i}{s^2 + 2\xi_i \omega_i s + \omega_i^2}$$

where c_i, ω_i, and ξ_i are constants. We will consider the following fourth order reduced
model for the purpose of design:

$$\hat{P}(s) = \frac{-6.475s^2 + 4.0302s + 175.77}{s(5s^3 + 3.5682s^2 + 139.5091s + 0.0929)}.$$

It represents the rigid body motion and the first flexible mode of the beam. The plant is
clearly unstable and nonminimum phase, with poles at

$$0, \ -0.0007, \ -0.3565 \pm j5.2700$$

and zeros at

$$-4.9081, \ 5.5308.$$

Notice that the rigid body motion is slightly damped by the back electromagnetic force
in the DC motor. The nonminimum phase character of the plant is a direct consequence
of the uncollocated sensor (i.e., the traveling-wave effect between the actuator and the
sensor).

ℓ_1 performance objective.   First we will consider a standard two-block nominal performance objective:

$$\inf_{K \text{ stab.}} \left\| \begin{matrix} W_1 S \\ W_2 K S \end{matrix} \right\|_1$$

where W_1 and W_2 are chosen to reflect the trade-offs between tracking performance and
control effort, and the expected spectral characteristics of the exogenous disturbance (or
commands).

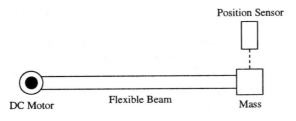

Position Sensor

DC Motor   Flexible Beam   Mass   **Figure 3.15**   Flexible beam.

ℓ_1 **performance objective with** ℓ_∞ **constraints.**   Flexible beams are typical building blocks of robotic manipulators and space structures. It is very common for these applications to have hard constraints on the amount of torque the actuator can exert. Therefore, the control of a simple beam is a good example that demonstrates the power of a design method in addressing time-domain constraints.

Let us say that the amount of torque is bounded by U_{max}, beyond which the motor saturates. Then, two different specifications may arise: First, when a certain level of tracking performance is required, expressed as $\|W_1 S\|_1 \leq \gamma$. Second, when the best achievable performance has to be determined without actuator saturation.

The first specifications are equivalent to the following optimization problem:

$$v^o = \inf_{K \text{ stab.}} \left\| \begin{matrix} W_1 S/\gamma \\ K S/U_{max} \end{matrix} \right\|_1 .$$

Clearly, if $v^o \leq 1$, then the design specifications can be met. On the other hand, if $v^o > 1$, then there is no linear controller that could achieve such design requirements.

In the second case the specifications translate into the following optimization problem:

$$\inf \|W_1 S\|_1$$

$$\text{subject to}$$

$$K \text{ stab.} \tag{3.19}$$

$$\|K S\|_1 \leq U_{max}$$

which is a slight variation of the standard ℓ_1 problem presented before and clearly falls into the same framework. Notice that the result of this optimization problem captures the fundamental limitations of a linear design (i.e., the best achievable performance within the region of linear operation).

ℓ_1 **performance objective with maximum deviation constraints.**   It is common practice in control design to specify the time-domain characteristics of a given system in terms of its response to a unit step input. As an example of this situation, consider the following specifications for the beam problem: obtain the best tracking performance in the ℓ_1 sense, subject to the magnitude of the control signal not exceeding a given value, say U_{max}, when the input disturbance is a unit step. In other words,

$$\inf \|W_1 S\|_1$$

$$\text{subject to}$$

$$K \text{ stab.}$$

$$\|K S w_{step}\|_\infty \leq U_{max}$$

where w_{step} is a unit step. Notice that this problem is less stringent than Problem (3.19) since the design avoids saturation for a single input (i.e., a step input) instead of guarding against a whole class of inputs (i.e., all magnitude bounded signals).

A robust performance problem.   Given a physical system, it is generally hard to obtain an accurate mathematical model of its high-frequency behavior. This is particularly true with flexible structures, where flexible modes of higher order are typically hard to identify (particularly their damping ratios).

Assume, for the sake of illustration, that only the rigid body dynamics of the beam have been reliably identified, and that the unmodeled dynamics are accounted for by adding a bounded perturbation to the nominal plant. That is,

$$P = P_0 + W_1 \Delta_1, \quad \|\Delta_1\|_{\ell_\infty - \text{ind}} \leq 1$$

where

$$P_0(s) = \frac{-0.0032s + 1.26}{s(s + 0.0007)}$$

is the nominal plant, and

$$W_1(s) = \frac{s + 0.5}{s + 5}$$

reflects the validity of the model over frequency. Then, the robust performance problem can be stated as follows: Find a controller that stabilizes the class of plants $P = P_0 + W_1 \Delta_1$ and guarantees a given level of performance, say $\|W_2 S\|_1 < \gamma$, for all members of such class. Indeed, we seek performance guarantees for a whole class of plants.

With this we complete the presentation of design examples. They encompass a fairly broad class of relevant problems in control design. After the theory is developed in the coming chapters, the design examples will be solved in detail, so that the reader can have the general flavor of the theory at work.

3.9 SUMMARY

We have proposed a general framework for posing performance specifications. We put special emphasis on minimax performance objectives, particularly ℓ_1 and $\mathcal{H}_\infty$. We have shown that other performance objectives can be posed as sets of linear constraints, or approximated by linear constraints. Robust stability problems were shown to be equivalent to performance problems measured by either the ℓ_1 norm or the $\mathcal{H}_\infty$ norm. Prototype robust performance problems were discussed. We presented design examples to motivate the general framework and, later on, to demonstrate the design methodology developed.

EXERCISES

3.1. Let $\hat{G}$ be a strictly causal linear time-invariant system. Show that $(I - \hat{\Delta}\hat{G})^{-1}$ is stable for all LTI $\hat{\Delta}$ with $\|\hat{\Delta}\|_\infty < 1$ if $\|\hat{G}\|_\infty \leq 1$.

Show that if $\|\hat{G}\|_\infty > 1$, then there exists a $\hat{\Delta}$, with $\|\hat{\Delta}\|_\infty < 1$ such that the operator $(I - \hat{\Delta}\hat{G})^{-1}$ is not stable.

3.2. Assume in the previous problem that Δ, LTI, satisfies $\|\Delta\|_1 < 1$. Show that the $\|G\|_1 \leq 1$ is sufficient to guarantee that $(I - \Delta G)^{-1}$ is ℓ_∞ stable.

If $\|G\|_1 > 1$, does there exist a Δ, LTI, with $\|\Delta\|_1 < 1$ such that the operator $(I - \Delta G)^{-1}$ is not ℓ_∞ stable? If not, show a counter example.

3.3. This problem is concerned with model invalidation.

(a) A class of uncertain systems is modeled as

$$P = P_0 + \Delta W$$

where P_o is a time invariant nominal plant, W is a stable weight, and Δ is an unknown time-varying perturbation with $\|\Delta\|_{\ell_\infty - \text{ind}} < 1$. A designer wants to validate this model. He/She conducts an experiment of length N and records the set of data:

$$[u(k), \ y(k)|k = 1, \ 2, \ldots N - 1].$$

Show how the designer can validate, or invalidate the model. (To invalidate means to show that the measured data is inconsistent with the uncertainty description. Clearly, the designer can only invalidate a model.)

(b) Suppose that the output is corrupted with bounded noise, $w \in \ell_\infty$, with $\|w\|_\infty \leq \epsilon$. Show that the validation problem is equivalent to finding an element in a convex set.

3.4. Let a model be given by the stable plant:

$$\hat{P}_0(\lambda) = \frac{1}{\lambda - (1 + a_0)}, \quad 1 >> a_0 > 0.$$

Consider the class of plants given by

$$\Omega = \left\{ \hat{P}(\lambda) = \frac{1}{\lambda - (1 + b)} | \ -2a_0 \leq b \leq 2a_0 \right\}.$$

(a) Can the set Ω be embedded in a set of additive or multiplicative norm-bounded perturbations, with nominal plant $\hat{P}_0$? Show how or explain your answer.

(b) If your answer to the previous part is NO, show that the class Ω can be embedded in some other larger set characterized by norm-bounded perturbations. Give a sufficient condition for stability using the small gain theorem.

(c) Improve your earlier condition so that it captures the fact that the unknown is a real parameter. (The condition does not have to be necessary, but should still take into consideration the phase information!)

3.5. An engineer wanted to estimate the peak-to-peak gain of a closed-loop system h (the input-output map). The controller was designed so that the system tracks a step input in the steady state. The designer simulated the step response of the system and computed the amount of overshoot (e_1) and undershoot (e_2) of the response and immediately concluded that

$$\|h\|_1 \geq 1 + 2e_1 + 2e_2.$$

Is this a correct conclusion? Verify.

3.6. The design of a controller should take into consideration quantization effects. Let us assume that the only variable in the closed loop that is subject to quantization is the output of the plant. Two very simple schemes are proposed:

(a) Assume that the output is passed through a quantization operator Q defined as

$$Q(x) = a \lfloor \frac{|x|}{.5 + a} \rfloor sgn(x), \quad a > 0,$$

where $\lfloor r \rfloor$ denotes the largest integer smaller than r. The output of this operator feeds into the controller, as in Figure 3.16. Derive a sufficient condition that guarantees stability in the presence of Q.

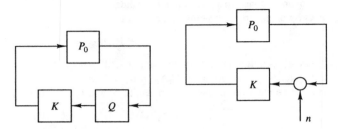

Figure 3.16 Quantization in the closed loop.

Figure 3.17 Quantization modeled as bounded noise.

(b) Assume that the input of the controller is corrupted with an unknown but bounded signal, with a small bound as in Figure 3.17. Argue that the controller should be designed so that it does not amplify this disturbance at its input.

Compare the two schemes, i.e., do they yield the same result? Is there a difference?

NOTES AND REFERENCES

The formulation of control problems using linear fractional transformations can be found in [BB91, DFT92, Fra87, MZ89, Mac89]. Background and more elaborate properties on LFT's can be found in [Red60]. The $\mathcal{H}_\infty$ problem was originally formulated in [Zam81] and developed later on in [Doy83, Doy84, DGKF89, FHZ84, Fra87, Glo84, GD88, Kwa85, SV83, Vid85]. The ℓ_1 problem was formulated in [Vid86] and developed later on in [DP87a, DP87b, DP88a, DBD92, DBD93, MP91, Men89, Sta90, Sta91, Vid91]. The prototypes on the ℓ_1 problem are taken from [DK93]. Minimizing the ℓ_∞ norm of the response to a fixed signal is studied in [DP88b].

For more details on stability robustness in the presence of additive, multiplicative, coprime-factor perturbations see [DS81, GM89, Zam66] for the ℓ_2 case, and [DO88, Dah92] for the ℓ_∞ case. The beam example is taken from [DFT92]. For more details on model validation, as in Problem (3.3), see [PKT$^+$92].

CHAPTER 4 ———————————

Finite Dimensional LTI Systems

In the previous chapters we discussed general systems as operators on classes of inputs without giving much attention to their internal structures. In particular, LTI systems were represented as convolution operators. In many situations, these systems are described as a set of ordinary linear difference equations, or linear partial difference equations. A particular class of such descriptions is the set of LTI systems with finitely many states, known as finite-dimensional LTI systems or (lumped) systems. Not all LTI systems are finite dimensional. However, many systems are modeled as such mainly for simplicity and for the fact that they are much easier to use for design.

This chapter is devoted to the study of finite-dimensional linear time-invariant systems (FDLTI). We first review some of the basic algebraic properties of such systems, which include poles and zeros, observability and reachability, and matrix fraction description of rational functions. This is followed by methods for computing norms of FDLTI systems—in particular, the $\mathcal{H}_2$, $\mathcal{H}_\infty$ and ℓ_1 norms.

Then, the concept of the Hankel operator associated with a system is introduced. We will see that the Hankel operator plays an important role in providing upper and lower bounds for various gains of systems. Furthermore, it highlights the fundamental limitations of model reduction problems. Along the way we will lay out the theoretical foundation for the solution of the SISO $\mathcal{H}_\infty$ optimal control problem from an operator viewpoint.

Finally, the last part of this chapter contains a brief but relevant discussion on sampled-data systems.

4.1 REVIEW OF FDLTI SYSTEMS

Finite-dimensional linear time-invariant (FDLTI) causal systems are elements in $\mathcal{L}$ that have state-space realizations of the form

$$\begin{aligned} x(t+1) &= Ax(t) + Bu(t), \\ y(t) &= Cx(t) + Du(t), \end{aligned} \tag{4.1}$$

where $(A, B, C, D) \in \mathbb{R}^{n \times n} \times \mathbb{R}^{n \times m} \times \mathbb{R}^{p \times n} \times \mathbb{R}^{p \times m}$. This description defines a linear operator on ℓ^m, with inputs $u(t) \in \mathbb{R}^m$ and outputs $y(t) \in \mathbb{R}^p$ and states $x(t) \in \mathbb{R}^n$. To find the output for all time, both the inputs and the initial condition have to be specified. If only the input-output map is of interest, then the initial condition is taken as zero.

Systems described by Equation (4.1) are completely determined from the matrices (A, B, C, D) and are given the notation

$$\left[\begin{array}{c|c} A & B \\ \hline C & D \end{array} \right]. \tag{4.2}$$

The pulse response of the system is given by

$$G(k) = \begin{cases} D & \text{for } k = 0 \\ CA^{k-1}B & \text{for } k > 0 \end{cases} \tag{4.3}$$

and the λ-transform is given by

$$\hat{G}(\lambda) = \left(C \frac{1}{\lambda} I - A \right)^{-1} B + D. \tag{4.4}$$

The system $\hat{G}(\lambda)$ is a real rational matrix function. Its poles are a subset of the reciprocal of the eigenvalues of the matrix A.

4.1.1 Reachability and Observability

A system is reachable if there always exists an input that takes the states from any initial condition to any prespecified final state in finite time. It is observable if an initial condition can be exactly determined from observing the output for some finite time. Equivalent mathematical definitions are given below.

Definition 4.1.1.   A FDLTI system is reachable if and only if the matrix $\mathcal{C}_n$ defined as

$$\mathcal{C}_n := \left(B \; AB \; A^2B \; \ldots \; A^{n-1}B \right)$$

has rank equal to n. The system is observable if and only if the matrix $\mathcal{O}_n$ defined as

$$\mathcal{O}_n := \begin{pmatrix} C \\ CA \\ CA^2 \\ \vdots \\ CA^{n-1} \end{pmatrix}$$

has rank equal to n.   ∎

We will also refer to the degree (i.e., order) of a reachable and observable realization (i.e., a minimal realization) as the McMillan degree of the system.

4.1.2 Stability

For FDLTI systems, with reachable and observable realizations, stability is related to the eigenvalues of the matrix A:

$$\ell_p \text{ stability} \iff \rho(A) < 1$$

where $\rho(A)$ denotes the spectral radius (i.e., absolute value of the maximum eigenvalue of A). Since the stability of a real rational function is independent of the space it operates on, we use the following definition to denote stable real rational functions.

Definition 4.1.2.   The space of all stable finite-dimensional systems will be denoted by $\mathcal{RH}_\infty$ (with appropriate dimension).   ∎

It follows that the space $\mathcal{RH}_\infty$ lies in the intersection of ℓ_1 and $\mathcal{H}_\infty$. We will think of $\mathcal{RH}_\infty$ either in time domain or in transform domain (i.e., we will write $Q \in \mathcal{RH}_\infty$ or $\hat{Q} \in \mathcal{RH}_\infty$).

Stability vs. causality.   Denote the space of real rational functions that have no poles on the unit circle by $\mathcal{RL}_\infty$ ($\mathcal{RL}_\infty \subset \mathcal{L}_\infty$) and let $\hat{G}(\lambda) \in \mathcal{RL}_\infty$. We observe the following:

1. If all the poles of $\hat{G}$ are outside the disc, then G can be viewed as a stable, causal operator on ℓ_p. This interpretation is obtained if the region of convergence of $\hat{G}$ includes the unit disc.

2. If all the poles of $\hat{G}$ are inside the unit circle, then G can be viewed as a causal operator on ℓ_p, which is not stable. This interpretation is obtained if the region of convergence of $\hat{G}$ is inside the unit disc not including the unit circle. On the other hand, G can be viewed as a stable anti-causal operator, i.e., a bounded operator on the space $\ell_p(\mathbf{Z}_-)$. This interpretation is obtained if the region of convergence of $\hat{G}$ includes the complement of the open unit disc. In this case, the pulse response is supported on the negative integers.

Thus for a given rational function in $\mathcal{RL}_\infty$ it is sometimes convenient to split the function as the sum of two stable systems, a causal one and an anti-causal one. The reason behind this will be evident later on.

Given a real rational function $\hat{G}(\lambda)$, we will define the adjoint of that function as

$$\hat{G}^\sim(\lambda) = \hat{G}^T\left(\frac{1}{\lambda}\right). \tag{4.5}$$

Indeed, $\hat{G}^{\sim}$ is the Hilbert space adjoint of the operator $\hat{G}$ as it operates on $\mathcal{L}_2[0, 2\pi)$, i.e.,

$$\langle \hat{G}\hat{f}, \hat{h} \rangle = \langle \hat{f}, \hat{G}^{\sim}\hat{h} \rangle \quad \text{for all } \hat{f}, \hat{h} \in \mathcal{L}_2[0, 2\pi).$$

If A^{-1} exists, then a state-space realization for $\hat{G}^{\sim}$ is given by

$$G^{\sim} = \left[\begin{array}{c|c} A^{-T} & A^{-T}C^T \\ \hline -B^T A^{-T} & D^T - B^T A^{-T} C^T \end{array} \right].$$

If $\hat{G}$ is stable and causal, then $\hat{G}^{\sim}$ is stable and anti-causal.

4.1.3 Rational Matrices

This section reviews some well-known concepts and results from the theory of rational matrices. Here, LTI finite dimensional systems are viewed as algebraic objects, represented by matrices of rational functions of λ (i.e., λ-transform of the system pulse response).

The theory is needed for several reasons. The first is to quantify precisely the notion of zeros for MIMO systems. While zeros of SISO systems are simply complex numbers that annihilate the transfer function, MIMO zeros also have directions associated with them. In Chapter 6 these concepts will be utilized to characterize all feasible closed-loop maps for a given plant in terms of linear equations.

The second reason is to introduce the concept of factorization of a rational function over polynomials. It is evident that a SISO rational system can be written as the ratio of two polynomials in λ which are coprime, i.e., have no common zeros. Such a factorization is also possible for matrix transfer functions. This kind of factorization turns out to be a special case of a more general factorization known as *factorization over stable functions*, developed in Chapter 5.

Finally, we should emphasize that polynomial matrices are stable functions, with all the poles located at infinity (eigenvalues of A are at the origin). These are precisely the finite pulse response systems.

Definition 4.1.3. A square polynomial matrix $\hat{P}(\lambda) = P(0) + P(1)\lambda + \cdots + P(k)\lambda^k$ is said to be unimodular if its determinant is a nonzero constant independent of λ. ∎

Therefore, unimodular matrices have polynomial inverses, and have full rank at all points of the complex plane. Also, the product of two unimodular matrices is a unimodular matrix.

Theorem 4.1.1. Let $\hat{G}(\lambda)$ be an $m \times n$ rational matrix of normal rank r (i.e., of rank r for almost all λ). Then $\hat{G}(\lambda)$ can always be factored in the following way:

$$\hat{G}(\lambda) = \hat{L}(\lambda)\hat{M}(\lambda)\hat{R}(\lambda) \tag{4.6}$$

where $\hat{L}(\lambda)$ and $\hat{R}(\lambda)$ are unimodular matrices of the appropriate dimensions, and

$$\hat{M}(\lambda) = \begin{pmatrix} \frac{\hat{\epsilon}_1(\lambda)}{\hat{\psi}_1(\lambda)} & & & 0 & \cdots & 0 \\ & \ddots & & \vdots & \ddots & \vdots \\ & & \frac{\hat{\epsilon}_r(\lambda)}{\hat{\psi}_r(\lambda)} & 0 & \cdots & 0 \\ 0 & \cdots & 0 & 0 & \cdots & 0 \\ \vdots & \ddots & \vdots & \vdots & \ddots & \vdots \\ 0 & \cdots & 0 & 0 & \cdots & 0 \end{pmatrix} \tag{4.7}$$

is $m \times n$, in which the monic polynomials $\{\hat{\epsilon}_i(\lambda), \hat{\psi}_i(\lambda)\}$ are coprime for all $i = 1, 2, \ldots, r$ (no common zeros) and have the following divisibility property: $\hat{\epsilon}_i(\lambda)$ divides $\hat{\epsilon}_{i+1}(\lambda)$ without remainder and $\hat{\psi}_{i+1}(\lambda)$ divides $\hat{\psi}_i(\lambda)$ without remainder for $i = 1, 2, \ldots, r - 1$. ∎

$\hat{M}(\lambda)$ is a general canonical form for rational matrices known as the Smith-McMillan form. It is unique, while $\hat{L}(\lambda)$ and $\hat{R}(\lambda)$ are not.

Definition 4.1.4.   Let $\hat{G}(\lambda)$ be a rational transfer-function matrix with Smith-McMillan form $\hat{M}(\lambda)$, then the roots of $\prod_{i=1}^{r} \hat{\epsilon}_i(\lambda)$ are the zeros of $\hat{G}(\lambda)$, and the roots of $\prod_{i=1}^{r} \hat{\psi}_i(\lambda)$ are the poles of $\hat{G}(\lambda)$. ∎

Definition 4.1.5.   Let λ_0 be a zero of $\hat{G}(\lambda)$. Let $\sigma_i(\lambda_0)$ denote the multiplicity of λ_0 as a root of $\hat{\epsilon}_i(\lambda)$, then $\{\sigma_i(\lambda_0)\}_{i=1}^{r}$ defines a nondecreasing sequence of nonnegative integers. For a given $i \in \{1, 2, \ldots, r\}$, $\sigma_i(\lambda_0)$ is known as the algebraic multiplicity of λ_0. The total number of indices i for which $\sigma_i(\lambda_0)$ is strictly positive is known as the geometric multiplicity of λ_0. Finally, $\sum_{i=1}^{r} \sigma_i(\lambda_0)$ defines the total multiplicity of λ_0. ∎

Analogous definitions can be applied to the poles of $\hat{G}(\lambda)$ by allowing the σ_i's to take nonpositive integer values. The $\sigma_i(\lambda_0)$'s are known as the structural indices of $\hat{G}(\lambda)$ at λ_0.

Next we give a definition of coprimeness of polynomial matrices.

Definition 4.1.6.   Two polynomial matrices, $\hat{N}(\lambda)$ and $\hat{D}(\lambda)$, of full normal row rank are left coprime if the composite matrix

$$\left(\hat{N}(\lambda) \quad \hat{D}(\lambda) \right)$$

has no zeros (or equivalently has full row rank for all λ). They are right coprime if $\hat{N}^T(\lambda)$ and $\hat{D}^T(\lambda)$ are left coprime. ∎

From the above definition, two polynomial matrices can have a zero at the same location, λ_0, and still be coprime. This, of course, does not occur for SISO polynomials.

Example 4.1.1

Let

$$\hat{N} = \begin{pmatrix} \lambda - 2 & 0 \\ \lambda - .5 & 1 \end{pmatrix}, \quad \hat{D} = \begin{pmatrix} 1 & 1 \\ 0 & \lambda - 2 \end{pmatrix}.$$

The matrix $\hat{N}$ has a zero at $\lambda = 2$ with left direction $(1, 0)$, i.e., $(1 \quad 0)\hat{N}(2) = 0$. The matrix $\hat{D}$ has a zero at $\lambda = 2$ with left direction $(0, 1)$. The two matrices have no other zeros. However, the composite matrix

$$\begin{pmatrix} \lambda - 2 & 0 & 1 & 1 \\ \lambda - .5 & 1 & 0 & \lambda - 2 \end{pmatrix}$$

has full row rank for all λ since the left directions are different.

The following theorem gives an equivalent condition for coprimeness.

Theorem 4.1.2.   Two polynomial matrices $\hat{N}(\lambda)$ and $\hat{D}(\lambda)$ are left coprime if and only if there exist two polynomial matrices $\hat{X}$ and $\hat{Y}$ such that

$$\hat{N}(\lambda)\hat{X}(\lambda) + \hat{D}(\lambda)\hat{Y}(\lambda) = I.$$

The last equation is known as the Bezout identity.

Proof. The "if" part is immediate since the lack of left coprimeness implies that the matrix $(\hat{N}(\lambda) \quad \hat{M}(\lambda))$ loses rank at some λ_0 contradicting the Bezout identity. To show the "only if" part we perform a Smith-McMillan decomposition of the matrix $(\hat{N} \quad \hat{D})$. Since it has no zeros, the decomposition takes the form (assuming there are more columns than rows)

$$(\hat{N} \quad \hat{D}) = \hat{L}(I \quad 0 \quad \ldots \quad 0)\hat{R}.$$

Define

$$\hat{B}(\lambda) = 1^{\text{st}} \text{ set of columns of } \hat{R}^{-1}(\lambda) \text{ compatible with } I.$$

Then, it follows directly that

$$(\hat{N} \quad \hat{D})\hat{B}\hat{L}^{-1} = I,$$

which proves the claim.   ∎

Using the coprimeness definitions, we can factor a rational matrix in the following way:

$$\hat{G} = \hat{D}^{-1}\hat{N}$$

where both $\hat{N}$ and $\hat{D}$ are left coprime polynomials. One way of doing this is shown in the next corollary.

Corollary 4.1.1.   Let $\hat{G}(\lambda)$ be as in Theorem 4.1.1, then $\hat{G}(\lambda)$ has the following left and right coprime polynomial factorizations:

$$\hat{G}(\lambda) = \hat{D}_L^{-1}(\lambda)\hat{N}_L(\lambda) = \hat{N}_R(\lambda)\hat{D}_R^{-1}(\lambda)$$

where

$$\hat{N}_L(\lambda) = \hat{\mathcal{E}}(\lambda)\hat{R}(\lambda),$$

$$\hat{D}_L(\lambda) = \hat{\Psi}_L(\lambda)\hat{L}^{-1}(\lambda),$$

$$\hat{N}_R(\lambda) = \hat{L}(\lambda)\hat{\mathcal{E}}(\lambda),$$

$$\hat{D}_R(\lambda) = \hat{R}^{-1}(\lambda)\hat{\Psi}_R(\lambda),$$

and

$$\hat{\mathcal{E}}(\lambda) = \begin{pmatrix} \hat{\epsilon}_1(\lambda) & & & 0 & \cdots & 0 \\ & \ddots & & \vdots & \ddots & \vdots \\ & & \hat{\epsilon}_r(\lambda) & 0 & \cdots & 0 \\ 0 & \cdots & 0 & 0 & \cdots & 0 \\ \vdots & \ddots & \vdots & \vdots & \ddots & \vdots \\ 0 & \cdots & 0 & 0 & \cdots & 0 \end{pmatrix} \quad \text{is } (m \times n),$$

$$\hat{\Psi}_R(\lambda) = \begin{pmatrix} \hat{\psi}_1(\lambda) & & & 0 & \cdots & 0 \\ & \ddots & & \vdots & \ddots & \vdots \\ & & \hat{\psi}_r(\lambda) & 0 & \cdots & 0 \\ 0 & \cdots & 0 & 1 & & 0 \\ \vdots & \ddots & \vdots & & \ddots & \\ 0 & \cdots & 0 & 0 & & 1 \end{pmatrix} \quad \text{is } (n \times n),$$

$$\hat{\Psi}_L(\lambda) = \begin{pmatrix} \hat{\psi}_1(\lambda) & & & 0 & \cdots & 0 \\ & \ddots & & \vdots & \ddots & \vdots \\ & & \hat{\psi}_r(\lambda) & 0 & \cdots & 0 \\ 0 & \cdots & 0 & 1 & & 0 \\ \vdots & \ddots & \vdots & & \ddots & \\ 0 & \cdots & 0 & 0 & & 1 \end{pmatrix} \quad \text{is } (m \times m).$$

Proof. For the left coprime factorization, notice that

$$\begin{pmatrix} \hat{\mathcal{E}}\hat{R} & \hat{\Psi}_L\hat{L}^{-1} \end{pmatrix} = \begin{pmatrix} \hat{\mathcal{E}} & \hat{\Psi}_L \end{pmatrix} \begin{pmatrix} \hat{R} & 0 \\ 0 & \hat{L}^{-1} \end{pmatrix}.$$

Since $\hat{\epsilon}_i$, $\hat{\psi}_i$ are coprime, the result follows immediately. A similar argument can be used with the right coprime factors. ∎

Note that the complete zero structure of $\hat{G}(\lambda)$ (i.e., structural indices and directional information) is captured by $\hat{N}_R(\lambda)$ and $\hat{N}_L(\lambda)$. Moreover, any right and left polynomial coprime factorization of $\hat{G}(\lambda)$ will have this property.

4.2 THE HANKEL OPERATOR AND ITS SINGULAR VALUES

In this section we study in detail the Hankel operator and its relationship to the Hankel singular values. We will utilize this study in different ways: First, we show how this operator contributes to finding upper and lower bounds on induced norms such as the ℓ_1 and the $\mathcal{H}_\infty$ norms. Using these bounds, it is possible to provide an efficient algorithm for computing the ℓ_1 norm to any desired degree of accuracy. Second, we show how the Hankel operator is related to the model reduction problem. More precisely, we will show that the Hankel singular values give lower bounds on the best approximation of a system by another system of lower McMillan degree. Interestingly, the theory also provides a method for computing good approximants. Finally, in Chapter 10, the theory of Hankel operators is shown to be instrumental in solving the $\mathcal{H}_\infty$ model matching problem.

Definition 4.2.1.   The Reachability and Observability Gramians **P** and **Q**, respectively, are defined for a stable FDLTI system as

$$\mathbf{P} = \sum_{k=0}^{\infty} A^k B B^T (A^T)^k, \quad \mathbf{Q} = \sum_{k=0}^{\infty} (A^T)^k C^T C A^k. \tag{4.8}$$

It follows that **P** and **Q** are the positive semi-definite solutions of the Lyapunov equations:

$$A\mathbf{P}A^T - \mathbf{P} = -BB^T, \tag{4.9}$$

$$A^T\mathbf{Q}A - \mathbf{Q} = -C^T C. \tag{4.10}$$

∎

A system is reachable if and only if **P** is nonsingular, and observable if and only if **Q** is nonsingular.

Definition 4.2.2.   The Hankel singular values of a stable FDLTI system are the square roots of the eigenvalues of **PQ**, i.e.,

$$i\text{th Hankel Singular Value} = \sigma_i := \sqrt{\lambda_i(\mathbf{PQ})}. \tag{4.11}$$

It will be assumed that the σ_i's are ordered such that $\sigma_1 \geq \sigma_2 \ldots \geq \sigma_n$. ∎

It can be easily verified (see Exercise 4.1) that the eigenvalues of **PQ** are nonnegative, and thus the Hankel singular values are well defined. Also, the Hankel singular values are independent of the particular realization of the system. In fact, they are invariant under orthonormal transformations of the inputs and the outputs.

4.2.1 The Hankel Operator

The Hankel singular values are related to the following operator known as the Hankel operator. If G is a finite-dimensional LTI operator with $\hat{G} \in \mathcal{L}_\infty$, then G maps sequences supported on the negative integers (past inputs) into two-sided sequences. The Hankel operator is simply the portion of the operator G that maps past inputs to future outputs (sequences supported on the positive integers). For simplicity, we drop the dimensions of the spaces as they are readily understood from the context.

Definition 4.2.3.   The Hankel operator associated with a finite-dimensional LTI system G with $\hat{G} \in \mathcal{L}_\infty$, denoted by Γ_G, is defined as

$$
\begin{aligned}
\Gamma_G &:= \ell_2(\mathbf{Z}_-) \to \ell_2(\mathbf{Z}_+), \\
f &\to \Pi_{\ell_2(\mathbf{Z}_+)} G f
\end{aligned}
\tag{4.12}
$$

where $\Pi_{\ell_2(\mathbf{Z}_+)}$ denotes the projection on the space $\ell_2(\mathbf{Z}_+)$. We will refer to Γ_G as the Hankel operator with symbol G.   ∎

Linearity in the symbol G.   If $G = G_1 + G_2$, then

$$
\Gamma_G = \Gamma_{G_1} + \Gamma_{G_2}
$$

as is easily seen from the definition.

Assume that for a given G, $\hat{G} \in \mathcal{L}_\infty$. Then G can be written as the sum of a stable causal and anti-causal transfer functions:

$$
G = G_- + G_+ \text{ where } G_- \text{ is anti-causal, } G_+ \text{ is strictly causal.}
$$

G_- can be interpreted as a bounded anti-causal operator, with impulse response equal to zero on the nonnegative integers. With this interpretation, the value:

$$
\Gamma_{G_-} f = 0 \quad \text{for all } f \in \ell_2(\mathbf{Z}_-).
$$

In other words, the Hankel operator depends only on the causal part of the system, i.e.,

$$
\Gamma_G = \Gamma_{G_+}
$$

Relation to the $\mathcal{L}_\infty$ norm.   The induced norm of the Hankel operator is defined in the standard way:

$$
\|\Gamma_G\| = \sup_{0 \neq f \in \ell_2(\mathbf{Z}_-)} \frac{\|\Gamma_G f\|_2}{\|f\|_2}.
$$

Recall that the $\mathcal{L}_\infty$ norm is the induced operator norm on the space $\ell_2(\mathbf{Z})$. Then, it follows directly from the definition of the Hankel operator that

$$
\|\hat{G}\|_\infty \geq \|\Gamma_G\|,
\tag{4.13}
$$

i.e., the norm of the Hankel operator is a lower bound on the $\mathcal{L}_\infty$ norm.

4.2.2 Decomposition of the Hankel Operator

Let G be stable with a minimal state-space description:

$$
\left[\begin{array}{c|c} A & B \\ \hline C & 0 \end{array} \right].
$$

Let $h = \Gamma_G f$. Then

$$
\begin{aligned}
h(t) &= \sum_{k=-1}^{-\infty} CA^{t-k-1} Bf(k) \qquad \text{for all } t \geq 0 \\
&= CA^t \sum_{k=-1}^{-\infty} A^{-k-1} Bf(k) \qquad \text{for all } t \geq 0
\end{aligned}
$$
$$
h =: \mathcal{OC}f
$$

where $\mathcal{O}$ and $\mathcal{C}$ are the observability and reachability operators:

$$
\begin{aligned}
\mathcal{O}: \quad &\mathbf{C}^n \quad \to \ell_2(\mathbf{Z}_+), \\
&x \quad \to CA^t x,
\end{aligned}
$$

$$
\begin{aligned}
\mathcal{C}: \ &\ell_2(\mathbf{Z}_-) \to \mathbf{C}^n, \\
&f \quad \to \sum_{k=-1}^{-\infty} A^{-k-1} Bf(k).
\end{aligned}
$$

To put this in a more familiar form, let $f = \{\ldots,\ f(-3),\ f(-2),\ f(-1)\}$, then

$$
\Gamma_G f = \begin{pmatrix} C \\ CA \\ CA^2 \\ \vdots \end{pmatrix} \left(B\ AB\ A^2B\ \ldots \right) \begin{pmatrix} f(-1) \\ f(-2) \\ f(-3) \\ \vdots \end{pmatrix}
$$

$$
= \begin{pmatrix} CB & CAB & CA^2B & \ldots \\ CAB & CA^2B & CA^3B & \ldots \\ CA^2B & CA^3B & CA^4B & \ldots \\ \vdots & \vdots & \vdots & \ddots \end{pmatrix} \begin{pmatrix} f(-1) \\ f(-2) \\ f(-3) \\ \vdots \end{pmatrix}. \tag{4.14}
$$

The above decomposition shows that even though the Hankel operator is a map between two infinite-dimensional spaces, it is essentially a finite-dimensional one since the dimension of its range cannot exceed n, (see Figure 4.1). In fact, it is straightforward to verify that the dimension of the range is equal to n, if the realization is both reachable and observable. Hence, by using a different basis, we can rewrite the Hankel operator as a finite-dimensional operator. This will be done later on.

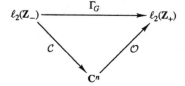

Figure 4.1 Decomposition of the Hankel operator.

4.2.3 The Norm of the Hankel Operator

The singular values of Γ_G are the same as the square-root of the eigenvalues of $\Gamma_G{}^*\Gamma_G$. The following result relates these singular values to the previously defined Hankel singular values.

Theorem 4.2.1.   The Hankel singular values in Equation (4.11) are precisely the square roots of the nonzero eigenvalues of $\Gamma_G{}^*\Gamma_G$.

Proof. From the decomposition of Γ_G we have

$$\Gamma_G{}^*\Gamma_G = \mathcal{C}^*\mathcal{O}^*\mathcal{O}\mathcal{C}.$$

Directly from the definition, these adjoint operators are given by

$$\begin{aligned}
\mathcal{C}^* \; : \quad &\mathbf{C}^n \quad \rightarrow \ell_2(\mathbf{Z}_-), \\
&x \quad \rightarrow B^T (A^{-T})^{(t+1)} x, \quad t \leq -1,
\end{aligned}$$

$$\begin{aligned}
\mathcal{O}^* : \; &\ell_2(\mathbf{Z}_+) \rightarrow \mathbf{C}^n, \\
&f \quad \rightarrow \sum_{k=0}^{\infty} (A^T)^k C^T f(k).
\end{aligned}$$

The matrix representations of these operators are precisely the transposition of the matrices in Equation (4.14). Also, note that

$$\mathbf{Q} = \mathcal{O}^*\mathcal{O}, \qquad \mathbf{P} = \mathcal{C}\mathcal{C}^*.$$

The main idea of the proof is to show that the nonzero eigenvalues of $\mathcal{C}^*\mathcal{O}^*\mathcal{O}\mathcal{C}$ are the same as the nonzero eigenvalues of $\mathcal{C}\mathcal{C}^*\mathcal{O}^*\mathcal{O} = \mathbf{PQ}$. For finite matrices, this is a standard result from linear algebra. In the sequel, we will prove this result and provide an explicit construction for the right and left singular vectors of the Hankel operator. We assume that the system is both reachable and observable.

Let x_i be the right eigenvector of $\mathbf{PQ}$ associated with σ_i, i.e.,

$$\mathbf{PQ}x_i = \sigma_i^2 x_i.$$

Define the vector y_i as

$$y_i = \frac{1}{\sigma_i}\mathbf{Q}x_i.$$

Then x_i and y_i satisfy

$$\mathbf{P}y_i = \sigma_i x_i,$$

$$\mathbf{Q}x_i = \sigma_i y_i.$$

It follows that $y_j^T x_i = 0$ whenever $\sigma_j \neq \sigma_i$. Also, the x_i's can be chosen such that (see Exercise 4.1)

$$y_j^T x_i = \begin{cases} 0 & i \neq j \\ \frac{1}{\sigma_i} & i = j \end{cases}. \tag{4.15}$$

Define the functions:

$$w_i := \mathcal{O}x_i = CA^t x_i, \quad t \geq 0,$$

$$v_i := \mathcal{C}^* y_i = B^T (A^{-T})^{(t+1)} y_i, \quad t \leq -1. \tag{4.16}$$

From the definition, the collection $\{w_i\}$ form an orthonormal set, similarly the collection $\{v_i\}$. Indeed,

$$\langle w_i, w_j \rangle = \langle \mathcal{O}x_i, \mathcal{O}x_j \rangle = \langle Qx_i, x_j \rangle = \delta_{ij}$$

where δ_{ij} is the Kronecker delta. The functions w_i, v_i are called the *Schmidt pair* associated with σ_i. Also, note that $w_i \in \ell_2(\mathbf{Z}_+)$ and $v_i \in \ell_2(\mathbf{Z}_-)$.

It follows by direct substitution that

$$\Gamma_G v_i = \mathcal{O}\mathcal{C}\mathcal{C}^* y_i = \mathcal{O}\mathbf{P}y_i = \sigma_i w_i,$$

$$\Gamma_G^* w_i = \mathcal{C}^* \mathcal{O}^* \mathcal{O}x_i = \mathcal{C}^* \mathbf{Q}x_i = \sigma_i v_i,$$

which implies that

$$\Gamma_G^* \Gamma_G v_i = \sigma_i^2 v_i,$$

$$\Gamma_G \Gamma_G^* w_i = \sigma_i^2 w_i.$$

This shows that every eigenvalue of $\mathbf{PQ}$ is also an eigenvalue of $\Gamma_G^* \Gamma_G$. The converse is true in the same way. ∎

As a consequence of this result, we obtain a singular value decomposition of Γ_G, and compute its norm. This is given in the corollaries that follow.

Corollary 4.2.1. The Hankel operator Γ_G can be written as

$$\Gamma_G f = \sum_{i=1}^{n} \sigma_i \langle v_i, f \rangle w_i \tag{4.17}$$

where v_i and w_i are as in Equation (4.16) and σ_i are the Hankel singular values of G.

Proof. Let $\{v_{i+n}, i \geq 1\}$ be a set of signals in $\ell_2(\mathbf{Z}_-)$ such that $\{v_i, i \geq 1\}$ forms an orthonormal basis for $\ell_2(\mathbf{Z}_-)$. Notice that the dimension of $\mathcal{R}(\Gamma_G^* \Gamma_G)$ is equal to n. From this, it follows that

$$\mathcal{R}(\Gamma_G^* \Gamma_G) = \mathrm{Span}\{v_i, i = 1, \ldots, n\}.$$

Next we claim that $\Gamma_G v_{i+n} = 0$ for all $i \geq 1$. This follows from the fact that for any $f \in \ell_2(\mathbf{Z}_-)$, we have

$$\langle f, \Gamma_G^* \Gamma_G v_{i+n} \rangle = \langle \Gamma_G^* \Gamma_G f, v_{i+n} \rangle = 0 \text{ for all } i \geq 1.$$

The last equality holds since the set of vectors $\{v_{i+n}, i \geq 1\}$ is orthogonal to $\mathcal{R}(\Gamma_G^* \Gamma_G)$. Then

$$\|\Gamma_G v_{n+i}\|_2^2 = \langle \Gamma_G v_{n+i}, \Gamma_G v_{n+i} \rangle = \langle \Gamma_G^* \Gamma_G v_{i+n}, v_{i+n} \rangle = 0 \text{ for all } i \geq 1.$$

Given any $f \in \ell_2(\mathbf{Z}_-)$, f can be written as

$$f = \sum_{i=1}^{\infty} \langle v_i, f \rangle v_i.$$

Then, it follows that

$$\Gamma_G f = \Gamma_G \left(\sum_{i=1}^{\infty} \langle v_i, f \rangle v_i \right) = \sum_{i=1}^{n} \langle v_i, f \rangle \Gamma_G v_i = \sum_{i=1}^{n} \langle v_i f \rangle \sigma_i w_i. \qquad \blacksquare$$

Corollary 4.2.2

$$\|\Gamma_G\| = \sigma_1.$$

It is sometimes convenient to refer to this quantity as the Hankel norm of G. We will denote that by $\|G\|_H$. $\qquad \blacksquare$

4.3 BOUNDS ON INDUCED NORMS

Earlier, we have shown that the ℓ_1 norm gives an upper bound on the $\mathcal{H}_\infty$ norm. For stable finite-dimensional systems both norms are finite, and for systems with fixed McMillan degrees they are equivalent. Next we will verify this fact by deriving an upper bound on the ℓ_1 norm using the $\mathcal{H}_\infty$ norm. As a by-product, we will exhibit easily computable upper and lower bounds on both norms using the Hankel singular values.

Theorem 4.3.1. Let $\hat{R}$ be a $p \times m$ finite dimensional LTI stable system (i.e., $\hat{R} \in \mathcal{RH}_\infty^{p \times m}$), with McMillan degree equal to n. Then

1. $\|\hat{R}\|_\infty \geq \sigma_1$.
2. $\|R\|_1 \leq |D|_1 + 2\sqrt{m} \sum_{i=1}^{n} \sigma_i$.
3. $\|R\|_1 \leq (2n+1)\sqrt{m}\|\hat{R}\|_\infty$.

Proof. We will prove this theorem for the SISO case. Part 1 follows from standard properties of Hankel operators. The ℓ_1 norm of R is given by

$$\|R\|_1 = |d| + \sum_{k=0}^{\infty} |cA^k b|$$
$$= |d| + \sum_{k=0}^{\infty} |cA^{2k}b| + \sum_{k=0}^{\infty} |cA^{2k+1}b|.$$

Consider the second term in the above equation. Using Cauchy-Schwartz inequality we get

$$|cA^{2k}b| \leq |(A^T)^k c^T|_2 |A^k b|_2.$$

Summing both sides we get

$$\sum_{k=0}^{\infty} |cA^{2k}b| \le \sum_{k=0}^{\infty} |(A^T)^k c^T|_2 |A^k b|_2$$

$$\le \left(\sum_{k=0}^{\infty} |(A^T)^k c^T|_2^2 \right)^{\frac{1}{2}} \left(\sum_{k=0}^{\infty} |A^k b|_2^2 \right)^{\frac{1}{2}}$$

$$= (\text{Trace}\mathbf{Q})^{\frac{1}{2}} (\text{Trace}\mathbf{P})^{\frac{1}{2}}.$$

The second term can be bound in the same way, giving

$$\|R\|_1 \le |d| + 2(\text{Trace}\mathbf{Q})^{\frac{1}{2}} (\text{Trace}\mathbf{P})^{\frac{1}{2}}.$$

To get the explicit bound, let the realization (A, b, c, d) be such that

$$\mathbf{P} = \mathbf{Q} = \text{diag}(\sigma_1, \ldots \sigma_n).$$

This is known as a balanced realization and can always be found (see Exercise 4.2). The result follows directly by substitution. The multivariable result is left as an exercise (see Exercise 4.4). Finally, parts 1 and 2 directly imply part 3. ∎

The above theorem shows that the Hankel singular values provide lower and upper bounds on both the $\mathcal{H}_\infty$ and ℓ_1 norms of any stable linear time-invariant system. As a by-product, it shows that the ℓ_1 norm is bounded by a constant multiplying the $\mathcal{H}_\infty$ norm, where the constant is a linear function of the degree of the system. So, for low-order systems, it is possible to get a good bound on the ℓ_∞ gain of the system from observing the ℓ_2 gain. For higher-order systems the $\mathcal{H}_\infty$ norm and the ℓ_1 norm can be quite far apart. Notice that the bound derived in the above theorem is quite tight. This is shown in the following example.

Example 4.3.1

Let the transfer function of R be given by the following allpass function:

$$\hat{R}(\lambda) = \prod_{i=1}^{n} \frac{\lambda - (1 - \alpha^i)}{(1 - \alpha^i)\lambda - 1}, \qquad 0 < \alpha < 1.$$

The $\mathcal{H}_\infty$ norm of $\hat{R}$ is equal to 1. As $\alpha \to 0$, $\|R\|_1 \to 2n + 1$. This shows that the bound in Theorem 4.3.1 is tight.

4.4 COMPUTING NORMS

In here, we discuss the computation of the $\mathcal{H}_2$, $\mathcal{H}_\infty$, and ℓ_1 norms for stable finite-dimensional systems.

4.4.1 Computing the $\mathcal{H}_2$ Norm

Let G be a stable transfer function with state-space realization:

$$\left[\begin{array}{c|c} A & B \\ \hline C & D \end{array}\right].$$

We can compute the $\mathcal{H}_2$ norm by substituting Equation (4.3) into the definition of the $\mathcal{H}_2$ norm given in Equation (2.45). The result is given by

$$\|\hat{G}\|_2 = \sqrt{\text{Trace}(D^T D + C P C^T)} = \sqrt{\text{Trace}(D^T D + B^T \mathbf{Q} B)}. \tag{4.18}$$

In other words, the $\mathcal{H}_2$ norm can be computed exactly for a finite-dimensional system.

4.4.2 Computing the $\mathcal{H}_\infty$ Norm

The computation of the $\mathcal{H}_\infty$ norm involves computing the maximum singular value of the matrix transfer function evaluated at each frequency point and then finding its maximum value over all frequencies. This computation cannot be done exactly and in general it is done using one of the following methods:

1. A direct evaluation on samples of the unit circle.
2. A bisection method utilizing the state-space realization.

The first method is clear: N samples from the unit circle are selected, namely the points corresponding to $\theta_k = 2\pi k/N$. The $\mathcal{H}_\infty$ norm is then given by

$$\|\hat{G}\|_\infty = \lim_{N\to\infty} \max_{k \le N} \sigma_{\max}[\hat{G}(e^{i\theta_k})]. \tag{4.19}$$

If a large N is selected, the $\mathcal{H}_\infty$ norm can be approximated arbitrarily closely.

For the second method, we will derive an eigenvalue test that can be performed immediately as a function of the state-space data. First, we will make the following definitions:

$$R := (I - \tfrac{1}{\gamma^2} D^T D)^{-1}, \tag{4.20}$$

$$S := (I - \tfrac{1}{\gamma^2} D D^T)^{-1}, \tag{4.21}$$

$$E := A + \tfrac{1}{\gamma^2} B R D^T C, \tag{4.22}$$

where γ is a positive real number and is not a singular value of D (this guarantees the existence of the above inverses). Also define the following symplectic pair of matrices (see Appendix B):

$$P_1 = \begin{pmatrix} E & 0 \\ \tfrac{1}{\gamma^2} C^T S C & I \end{pmatrix}, \qquad P_2 = \begin{pmatrix} I & B R B^T \\ 0 & E^T \end{pmatrix}. \tag{4.23}$$

The algorithm for computing the $\mathcal{H}_\infty$ norm is given in the following theorem.

Theorem 4.4.1. Assume that $\rho(A) < 1$. Given a positive number γ such that

1. There exists $\omega_0 \in [0, 2\pi)$ with $\sigma_{\max}[\hat{G}(e^{i\omega_0})] < \gamma$,
2. γ is not a singular value of D.

Then

$$\|\hat{G}\|_\infty < \gamma \iff \det(P_1 - e^{-i\omega} P_2) \neq 0 \quad \text{for all } \omega \in [0, 2\pi).$$

Proof. First we show that γ is a singular value of $\hat{G}(e^{i\omega_0})$ if and only if $\det(P_1 - e^{-i\omega_0} P_2) = 0$.

Let γ be a singular value of $\hat{G}(e^{i\omega_0})$. Then, there exist nonzero vectors u and v such that

$$\hat{G}(e^{i\omega_0})u = \gamma v, \qquad \hat{G}^*(e^{i\omega_0})v = \gamma u,$$

that is

$$\begin{aligned}
\{C(e^{-i\omega_0} I - A)^{-1} B + D\}u &= \gamma v, \\
\{B^T (e^{i\omega_0} I - A^T)^{-1} C^T + D^T\}v &= \gamma u.
\end{aligned} \tag{4.24}$$

Define

$$\begin{aligned}
r &= (A - e^{-i\omega_0} I)^{-1} Bu, \\
s &= \tfrac{1}{\gamma}(I - e^{-i\omega_0} A^T)^{-1} C^T v.
\end{aligned} \tag{4.25}$$

It follows that

$$\begin{aligned}
-\tfrac{1}{\gamma} Cr + \tfrac{1}{\gamma} Du &= v, \\
e^{-i\omega_0} B^T s + \tfrac{1}{\gamma} D^T v &= u.
\end{aligned} \tag{4.26}$$

Now solving for u and v in terms of r and s

$$\begin{pmatrix} u \\ v \end{pmatrix} = \begin{pmatrix} -\tfrac{1}{\gamma} D & I \\ I & -\tfrac{1}{\gamma} D^T \end{pmatrix}^{-1} \begin{pmatrix} -\tfrac{1}{\gamma} C & 0 \\ 0 & e^{-i\omega_0} B^T \end{pmatrix} \begin{pmatrix} r \\ s \end{pmatrix}. \tag{4.27}$$

Note that the assumption on γ guarantees the existence of the inverse and that $\begin{pmatrix} r \\ s \end{pmatrix} \neq 0$ follows from above. From (4.25)

$$\begin{pmatrix} Bu \\ \tfrac{1}{\gamma} C^T v \end{pmatrix} = \begin{pmatrix} (A - e^{-i\omega_0} I)r \\ (I - e^{-i\omega_0} A^T)s \end{pmatrix} \tag{4.28}$$

and from (4.27) and (4.28) we get

$$\begin{pmatrix} B & 0 \\ 0 & \tfrac{1}{\gamma} C^T \end{pmatrix} \begin{pmatrix} -\tfrac{1}{\gamma} D & I \\ I & -\tfrac{1}{\gamma} D^T \end{pmatrix}^{-1} \begin{pmatrix} -\tfrac{1}{\gamma} C & 0 \\ 0 & e^{-i\omega_0} B^T \end{pmatrix} \begin{pmatrix} r \\ s \end{pmatrix} = \begin{pmatrix} (A - e^{-i\omega_0} I)r \\ (I - e^{-i\omega_0} A^T)s \end{pmatrix}.$$

It is straightforward to show

$$\begin{pmatrix} -\tfrac{1}{\gamma} D & I \\ I & -\tfrac{1}{\gamma} D^T \end{pmatrix}^{-1} = \begin{pmatrix} \tfrac{1}{\gamma} R D^T & R \\ S & \tfrac{1}{\gamma} DR \end{pmatrix}.$$

Substitute this to get

$$\left\{\begin{pmatrix} A - e^{-i\omega_0}I & 0 \\ 0 & I - e^{-i\omega_0}A^T \end{pmatrix}\right.$$

$$\left.- \begin{pmatrix} B & 0 \\ 0 & \frac{1}{\gamma}C^T \end{pmatrix}\begin{pmatrix} \frac{1}{\gamma}RD^T & R \\ S & \frac{1}{\gamma}DR \end{pmatrix}\begin{pmatrix} -\frac{1}{\gamma}C & 0 \\ 0 & e^{-i\omega_0}B^T \end{pmatrix}\right\}\begin{pmatrix} r \\ s \end{pmatrix} = 0.$$

Then,

$$\left\{\begin{pmatrix} A - e^{-i\omega_0}I & 0 \\ 0 & I - e^{-i\omega_0}A^T \end{pmatrix} + \begin{pmatrix} \frac{1}{\gamma^2}BRD^T C & -e^{-i\omega_0}BRB^T \\ \frac{1}{\gamma^2}C^T SC & -e^{-i\omega_0}\frac{1}{\gamma^2}C^T DRB^T \end{pmatrix}\right\}\begin{pmatrix} r \\ s \end{pmatrix} = 0.$$

Equivalently,

$$\left\{\begin{pmatrix} A + \frac{1}{\gamma^2}BRD^T C & 0 \\ \frac{1}{\gamma^2}C^T SC & I \end{pmatrix} - e^{-i\omega_0}\begin{pmatrix} I & BRB^T \\ 0 & A^T + \frac{1}{\gamma^2}C^T DRB^T \end{pmatrix}\right\}\begin{pmatrix} r \\ s \end{pmatrix} = 0. \quad (4.29)$$

The converse follows exactly by reversing the argument above. So far, only the assumption that γ is not a singular value of D has been used.

To prove the theorem, notice that if $\|\hat{G}\|_\infty$ is greater than γ, then γ must be a singular value of $\hat{G}$ at some frequency. This follows from the first assumption and the continuity of $\hat{G}$. ∎

In words, given that at some frequency ω_0, $\sigma_{max}[\hat{G}(e^{i\omega_0})] < \gamma$, then $\|\hat{G}\|_\infty$ is guaranteed to be less than γ if and only if the pair of matrices (P_1, P_2) have no generalized eigenvalues on the unit circle.

If E is nonsingular, the generalized eigenvalue test becomes a standard eigenvalue test. Let P denote the matrix $P_2^{-1}P_1$ (note that P is a function of the parameter γ):

$$P(\gamma) = \begin{pmatrix} E - \frac{1}{\gamma^2}BRB^T E^{-T}C^T SC & BRB^T E^{-T} \\ -\frac{1}{\gamma^2}E^{-T}C^T SC & E^{-T} \end{pmatrix}. \quad (4.30)$$

Then, the test simplifies to

$$\|\hat{G}\|_\infty < \gamma \iff \begin{cases} P \text{ has no eigenvalues on the unit circle, and} \\ \sigma_{max}[C(I - A)^{-1}B + D] < \gamma \end{cases} \quad (4.31)$$

where ω_0 in Theorem 4.4.1 is taken as $\omega_0 = 0$. The computation of the $\mathcal{H}_\infty$ norm is done following a bisection algorithm:

1. Pick a lower bound on $\|\hat{G}\|_\infty$, say γ_ℓ. This can be done by computing the maximum singular value of $\hat{G}$ evaluated at an arbitrary frequency point.

2. Guess an upper bound on $\|\hat{G}\|_\infty$, say γ_u. Check that γ_u is in fact an upper bound by verifying that $P(\gamma_u)$ has no eigenvalues on the unit circle.

3. If $\gamma_u - \gamma_\ell$ is within the desired accuracy, stop.

4. Let $\gamma = \frac{1}{2}(\gamma_\ell + \gamma_u)$ and compute the eigenvalues of $P(\gamma)$. If there are eigenvalues on the unit circle, then let $\gamma_\ell = \gamma$ (i.e., γ is a tighter lower bound), otherwise let $\gamma_u = \gamma$ (i.e., γ is a tighter upper bound).

5. Go to step 3.

The above algorithm has to be slightly adjusted when, at some iteration, γ turns out to be equal to a singular value of D.

The following corollary can also be shown from Theorem 4.4.1 and Appendix B. We will omit the proof.

Corollary 4.4.1. Given the assumption of Theorem 4.4.1, it follows that $\|\hat{G}\|_\infty < \gamma$ if and only if the pair $P = (P_1, P_2) \in \mathrm{dom(Ric)}$ and if $X = \mathrm{Ric}(P)$, then $X \leq 0$ and $R^{-1} - \frac{1}{\gamma^2} B^T X B > 0$. ∎

4.4.3 Computing the ℓ_1 Norm

The simplest method for computing the ℓ_1 norm is by expansion. The pulse response of the system G is given by

$$G(k) = \begin{cases} D & \text{for } k = 0 \\ CA^{k-1}B & \text{for } k \geq 1 \end{cases}. \tag{4.32}$$

Since all the eigenvalues of A are inside the unit disc, there exists an N such that $\sum_{k>N} |A|_1^k$ is negligible. The ℓ_1 norm of G is then approximated by the norm of the FIR system $G(k)$ for $0 \leq k \leq N$. To find the degree of accuracy of this estimate, the norm of the tail of the impulse response can be estimated using the maximum eigenvalue of A. This can be used to find N for a given degree of accuracy ϵ. Such an estimate of N may very well be conservative, resulting in unnecessary computations. Next we present an alternate method for bounding the error. For that purpose, it is sufficient to consider multi-input single-output systems. The ℓ_1-norm of a MIMO system is given by the maximum of the norms of such systems.

For any MISO system G, the ℓ_1 norm can be written as

$$\|G\|_1 = \|P_N G\|_1 + \|(I - P_N)G\|_1 = \|P_N G\|_1 + \|S_{-N+1}(I - P_N)G\|_1.$$

In particular, let $\hat{G} \in \mathcal{RH}_\infty^{1 \times m}$, with nth order state-space realization (A, B, C, D). Then, the "tail" of the pulse response is given by

$$\tilde{G} = S_{-N+1}(I - P_N)G = \{0, CA^N B, CA^{N+1}B, \ldots\}$$

and has the state-space realization:

$$\left[\begin{array}{c|c} A & A^N B \\ \hline C & 0 \end{array} \right].$$

To get a good estimate of $\|\tilde{G}\|_1$, we use the result of Theorem 4.3.1. Namely, that

$$\tilde{\sigma}_1 \leq \|\tilde{G}\|_1 \leq 2\sqrt{m} \sum_{i=1}^n \tilde{\sigma}_i$$

where the $\tilde{\sigma}_i$'s are the Hankel singular values associated with $\tilde{G}$. Let $\mathbf{P}$ and $\mathbf{Q}$ be the reachability and observability gramians of G; then the corresponding gramians for $\tilde{G}$ are

$$\tilde{\mathbf{P}} = A^N \mathbf{P}(A^T)^N, \qquad \tilde{\mathbf{Q}} = \mathbf{Q}.$$

The Hankel singular values of $\tilde{G}$ are given by

$$\tilde{\sigma}_i = \sqrt{\lambda_i(\mathbf{Q}A^N\mathbf{P}(A^T)^N)}.$$

Clearly, these singular values are approaching zero as N approaches infinity. Also, the ℓ_1 norm of G satisfies

$$\|P_N G\|_1 + \tilde{\sigma}_1 \leq \|G\|_1 \leq \|P_N G\|_1 + 2\sqrt{m} \sum_{i=1}^{n} \tilde{\sigma}_i. \qquad (4.33)$$

This bound provides us with an algorithm for computing the ℓ_1 norm of G to a degree of accuracy better than ϵ. Let $N_{\min}$ denote the smallest integer such that the difference between the upper and lower bound is less than ϵ, i.e.,

$$N_{\min} = \min\left\{N \,\middle|\, 2\sqrt{m}\sum_{i=1}^{n}\tilde{\sigma}_i - \tilde{\sigma}_1 < \epsilon\right\}. \qquad (4.34)$$

Then Equation (4.33), with $N = N_{\min}$ gives the value of $\|G\|_1$ within an accuracy equal to ϵ. There are three advantages in using this method to estimate the ℓ_1 norm:

1. The ratio between the upper and lower bound is bounded from above by $2n\sqrt{m}$, which is independent of N. This contrasts other methods of approximation using dominant poles in which the bounds can be quite conservative.
2. Each of the Hankel singular values is a nonincreasing function of N.
3. The bounds are not affected by nonminimal state-space realizations, with possibly slowly decaying unobservable or unreachable modes.

Example 4.4.1

Let us illustrate the results of this section with a simple academic example. Consider the following 2×2 MIMO transfer function given in state-space form:

$$G = \left[\begin{array}{ccc|cc} 1 & -0.6 & 0.8 & 0 & 0 \\ 0.6 & 0.4 & -0.5 & 1 & 0 \\ -0.3 & 0.1 & -0.9 & 0 & 1 \\ \hline 1 & 0 & 0 & 0.5 & 1 \\ 0 & 1 & 0 & -0.5 & 0 \end{array}\right]. \qquad (4.35)$$

It is easy to check that G is stable but quite oscillatory. We want to compute different system norms on G.

$\mathcal{H}_2$ **Norm:** This is the most straightforward computation. After solving the two Lyapunov equations corresponding to the observability and reachability gramians (see Definition 4.2.1), we apply Equation (4.18) to get an exact answer:

$$\|G\|_2 = 2.909.$$

$\mathcal{H}_\infty$ **Norm:** First, we proceed by taking 400 samples of the unit circle and computing the maximum singular value of $\hat{G}(e^{i\omega_k})$ for each sample ω_k. In fact, only 200 samples between 0 and π are required, since $\hat{G}$ is real. The results are displayed in Figure 4.2. The maximum within this set is given by

$$\|\hat{G}\|_\infty \approx 6.5876.$$

Note that the above estimate does not come with error bounds.

Next, we apply the γ bisection method described before. We initialize the algorithm with the value of $\hat{G}$ at $\omega = 0$ as a lower bound and ten times that as an upper bound. After 18 iterations the following result is obtained:

$$\|\hat{G}\|_\infty = 6.5902 \pm 10^{-4}.$$

ℓ_1 **Norm:** To compute this norm we take a finite truncation of the system's pulse response sequence, i.e., $\{G(k)\}_{k=0}^{N_{\min}}$. The value of $N_{\min}$ is determined from Equation (4.34) to satisfy a given error bound, ϵ. In this particular case we choose $\epsilon = 10^{-4}$, resulting in $N_{\min} = 90$ and

$$\|G\|_1 = 9.7441 \pm 10^{-4}.$$

In fact, examining the ℓ_1 norm of the individual entries of G shows that $\|G\|_1$ is achieved by the second row but not by the first one:

$$\begin{pmatrix} \|g_{11}\|_1 & \|g_{12}\|_1 \\ \|g_{21}\|_1 & \|g_{22}\|_1 \end{pmatrix} = \begin{pmatrix} 5.0270 & 4.6299 \\ 5.8409 & 3.9032 \end{pmatrix}.$$

Let us construct an input-output pair of sequences so that the ℓ_∞ gain of G is exhibited. For this, we need an input vector sequence $(u_1 \ u_2)^T$ that captures the ℓ_∞ gain of g_{21} and g_{22}, respectively. First, consider the case of g_{21} acting on u_1, that is,

$$(g_{21}u_1)(t) = \sum_{k=0}^{t} g_{21}(t-k)u_1(k).$$

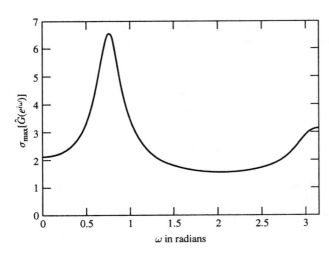

Figure 4.2 Maximum singular value plot of $\hat{G}$.

Clearly, the maximum attainable output at any given time t is obtained by applying the following input:

$$u_1(k) = \operatorname{sgn}\left(g_{21}(t-k)\right) \quad \text{for } 0 \le k \le t$$

and $|(g_{21}u_1)(t)|$ approaches $\|g_{21}\|_1$ as t approaches infinity. Note that such a maximizing input signal has an absolute value of one at all times, and switches signs as many times as the impulse response of g_{12}. Hence, an oscillatory system such as this (see Figure 4.3) shows its maximum ℓ_∞ gain with square-wave type of inputs. On the other hand, overdamped systems exhibit their ℓ_∞ gain with step-like inputs with little or no switching.

Figure 4.4 shows a simulation of of the input-output pair that approximately achieves $\|g_{21}\|_1$ (maximizing the output at t=59). The square-wave form of u_1 dominates up to the end, where the pattern is broken in order to exploit the initial transients of the system.

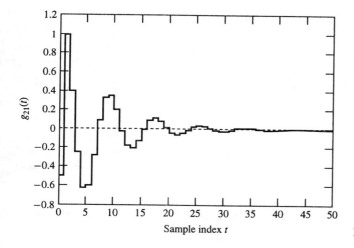

Figure 4.3   Impulse response of g_{21}.

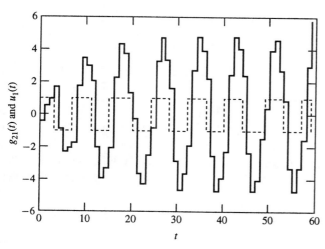

Figure 4.4   Input (dash) and output (solid) signals exhibiting the ℓ_∞ gain of g_{21}.

4.5 ELEMENTS OF MODEL REDUCTION

The model reduction problem arises in many control applications where it is desirable to reduce the order of the plant and/or the controller. The theory of model reduction is quite elaborate; however, in this section we present some of the highlights of the theory related to Hankel operators.

The model reduction problem for stable causal systems can be stated as follows: Given a stable causal system G, find a stable causal system $\tilde{G}_k$ with McMillan degree equal to k such that $\|G - \tilde{G}_k\|$ is minimized. The last norm may be any induced operator norm, e.g., $\mathcal{H}_\infty$ norm or ℓ_1 norm. This is a hard problem to which no analytical solutions are known. Our objective in this section is to develop a theory in which we can furnish:

1. Lower bounds on the minimal approximant.
2. A reduction method that gives "good" approximants.

The Hankel operator is intimately related to issues of model reduction. We have already seen that the rank of the Hankel operator is equal to the degree of the minimal realization of a function G. In here, we show that the Hankel singular values give lower bounds on the norm of the difference between the plant G and any lower degree approximant of G.

Theorem 4.5.1. Let G be in $\mathcal{R}\mathcal{H}_\infty$ with McMillan degree equal to n and with Hankel singular values σ_i. Let $\tilde{G}_k$ be any function in $\mathcal{R}\mathcal{H}_\infty$ with McMillan degree equal to $k < n$. Then:

1.
$$\|G - \tilde{G}_k\|_H \geq \sigma_{k+1}. \tag{4.36}$$

2. If $\|G - \tilde{G}_k\|_H = \sigma_{k+1}$ and $\sigma_k > \sigma_{k+1}$, then

$$(\Gamma_G - \Gamma_{\tilde{G}_k})v_{k+1} = \sigma_{k+1}w_{k+1} \tag{4.37}$$

where v_{k+1}, w_{k+1} are the Schmidt pair associated with σ_{k+1}.

Proof. Since the McMillan degree of $\tilde{G}_k$ is equal to k, then the rank of $\Gamma_{\tilde{G}_k}$ is equal to k. Consider the subspace

$$M = \left\{ f \mid f = \sum_{i=1}^{k+1} a_i v_i \text{ for } a_i \in \mathbf{C} \right\}$$

where the v_i's are as in Equation (4.16). The dimension of this subspace is $k + 1$. This implies that there exists an element $f \in M$ such that $\Gamma_{\tilde{G}_k} f = 0$. Normalize f such that

$$f = \sum_{i=1}^{k+1} a_i v_i \text{ with } \sum_{i=1}^{k+1} |a_i|^2 = 1.$$

It follows that

$$\|G - \tilde{G}_k\|_H \geq \|(\Gamma_G - \Gamma_{\tilde{G}_k})f\|_2$$

$$\geq \|\Gamma_G f\|_2$$

$$\geq \left\| \sum_{i=1}^{k+1} a_i \sigma_i w_i \right\|_2$$

$$= \left(\sum_{i=1}^{k+1} |a_i|^2 \sigma_i^2 \right)^{\frac{1}{2}}$$

$$\geq \sigma_{k+1}$$

which shows the result in part 1. To show the second part of the theorem, notice that for equality in the last equation to hold, we have to have $a_1, \ldots, a_k = 0$, which gives the result. ∎

One approach to model reduction is to minimize the Hankel norm, rather than the $\mathcal{H}_\infty$ or ℓ_1 norms. In the sequel, we show that the lower bound in the above theorem can be achieved, and the function $\tilde{G}_k$ can be constructed directly from the Schmidt pair.

It should be noted at this point that minimizing the Hankel norm results in minimizing a lower bound on both $\mathcal{H}_\infty$ and ℓ_1 norms. However, it is possible to derive an upper bound on these norms using the optimal Hankel norm approximant.

Optimal Hankel norm approximation. To develop the results on optimal Hankel norm approximation, we will present an important theorem known as *Nehari's theorem*. This theorem establishes the existence of an optimal solution of an $\mathcal{L}_\infty$ minimization problem. Once the existence is established, the construction of the solution is immediate. The results shown in this section are the basics behind the solution of the $\mathcal{H}_\infty$ model matching problem (see Chapter 10).

Theorem 4.5.2. Let $\hat{G} \in \mathcal{RH}_\infty$ with Hankel singular values σ_i, $i = 1, \ldots n$. Consider the following minimization problem:

$$\inf_{\hat{X}} \|\hat{G} + \hat{X}\|_\infty$$

subject to

$$X = \tilde{X}_k + X_-$$
$\tilde{X}_k$ causal with McMillan degree k
X_- anti-causal

Then it follows that a minimizing solution $\hat{X}$ always exists and the minimum value of the optimization problem is equal to σ_{k+1}.

Proof. We prove this theorem for the special case where $k = 0$, i.e., $\hat{X}$ is anti-causal. Let both $\hat{G}$ and $\hat{X}$ be written as

$$\hat{G}(\lambda) = \sum_{t=1}^{\infty} G(t)\lambda^t, \quad \hat{X}(\lambda) = \sum_{t=-\infty}^{0} X(t)\lambda^t.$$

First, notice that

$$\|\hat{G} + \hat{X}\|_{\infty} = \sup_{\hat{f} \in \mathcal{H}_2^{\perp}} \frac{\|(\hat{G} + \hat{X})\hat{f}\|_2}{\|\hat{f}\|_2}.$$

In other words, the $\mathcal{L}_{\infty}$ norm can be computed as the supremum over functions in $\mathcal{H}_2^{\perp}$ (or $\mathcal{H}_2$). This follows from the fact that these operators are time invariant; thus, if $f \in \ell_2(\mathbf{Z})$ achieves the norm arbitrarily closely, so does $S_{-N}f$ (S_{-N} denotes the shift operator). If we take N large enough and project the function on $\ell_2(\mathbf{Z}_-)$, the resulting function will also achieve the norm arbitrarily closely.

Consider the matrix representation of the operator $\hat{G} + \hat{X}$:

$$\begin{pmatrix} \vdots \\ y(-3) \\ y(-2) \\ y(-1) \\ y(0) \\ y(1) \\ \vdots \end{pmatrix} = \begin{pmatrix} \cdot & \cdot & \cdot & \cdot & & \\ X(-2) & X(-1) & X(0) & G(1) & \cdot & \cdot \\ X(-1) & X(0) & G(1) & G(2) & \cdot & \cdot \\ X(0) & G(1) & G(2) & G(3) & \cdot & \cdot \\ G(1) & G(2) & G(3) & G(4) & \cdot & \cdot \\ G(2) & G(3) & G(4) & G(5) & \cdot & \cdot \\ \cdot & \cdot & \cdot & \cdot & & \end{pmatrix} \begin{pmatrix} u(-1) \\ u(-2) \\ u(-3) \\ u(-4) \\ \vdots \end{pmatrix}.$$

Notice that

$$\sigma_1 = \|\Gamma_{\hat{G}}\| = \left\| \begin{pmatrix} G(1) & G(2) & G(3) & G(4) & \cdot & \cdot \\ G(2) & G(3) & G(4) & G(5) & \cdot & \cdot \\ \cdot & \cdot & \cdot & \cdot & \cdot & \end{pmatrix} \right\|.$$

To prove the result we apply Theorem A.9.1 iteratively. Consider the minimization problem:

$$\min_{X(0)} \left\| \begin{pmatrix} X(0) & G(1) & G(2) & G(3) & \cdot & \cdot \\ G(1) & G(2) & G(3) & G(4) & \cdot & \cdot \\ G(2) & G(3) & G(4) & G(5) & \cdot & \cdot \\ \cdot & \cdot & \cdot & \cdot & \cdot & \end{pmatrix} \right\|.$$

From Theorem A.9.1 it follows that the value of this minimum is equal to σ_1, and a constant $\tilde{X}(0)$ can be constructed to achieve this value. Similarly, it follows that

$$\min_{X(-1)} \left\| \begin{pmatrix} X(-1) & \tilde{X}(0) & G(1) & G(2) & \cdot & \cdot \\ \tilde{X}(0) & G(1) & G(2) & G(3) & \cdot & \cdot \\ G(1) & G(2) & G(3) & G(4) & \cdot & \cdot \\ G(2) & G(3) & G(4) & G(5) & \cdot & \cdot \\ \cdot & \cdot & \cdot & \cdot & \cdot & \end{pmatrix} \right\| =$$

$$\left\| \begin{pmatrix} \tilde{X}(0) & G(1) & G(2) & G(3) & \cdot & \cdot \\ G(1) & G(2) & G(3) & G(4) & \cdot & \cdot \\ G(2) & G(3) & G(4) & G(5) & \cdot & \cdot \\ \cdot & \cdot & \cdot & \cdot & \cdot & \end{pmatrix} \right\|$$

where the right-hand side has value σ_1. A constant $\tilde{X}(-1)$ can be constructed using Theorem A.9.1. Following this procedure iteratively, a minimizing $\tilde{X}$ is constructed as desired.

∎

Next, we present the general result on model reduction.

Theorem 4.5.3.   Let $G \in \mathcal{RH}_\infty$ with Hankel singular values $\sigma_1 \geq \sigma_2 \geq \ldots \sigma_k > \sigma_{k+1} \geq \ldots \sigma_n$.

1. The optimal $\hat{X}$ that solves the problem in Theorem 4.5.2 satisfies

$$(\hat{G} + \hat{X})(\lambda)\hat{v}_{k+1}(\lambda) = \sigma_{k+1}\hat{w}_{k+1}(\lambda).$$

2. The optimal Hankel norm approximant, $\hat{\tilde{G}}_k$, is given by the causal part of $\hat{X}$.

3. If σ_i^e are the Hankel Singular Values of $G + \tilde{G}_k$. Then

$$\sum_i \sigma_i^e \leq 2k\sigma_{k+1} + \sum_{i=k+1}^{n} \sigma_i.$$

Proof. We will only present proofs for parts 1 and 2. The existence of $\hat{X}$ is guaranteed from Theorem 4.5.2. To prove part 1, notice that from Theorem 4.5.1, we have

$$\Gamma_{G+X}v_{k+1} = \sigma_{k+1}w_{k+1}.$$

Define:

$$\hat{h} := (\hat{G} + \hat{X})(\lambda)\hat{v}_{k+1}(\lambda) - \sigma_{k+1}\hat{w}_{k+1}(\lambda),$$

then it follows that

$$\|\hat{h}\|_2^2 = \langle \hat{h}, \hat{h} \rangle$$

$$= \langle (\hat{G} + \hat{X})\hat{v}_{k+1}, (\hat{G} + \hat{X})\hat{v}_{k+1} \rangle - \langle (\hat{G} + \hat{X})\hat{v}_{k+1}, \sigma_{k+1}\hat{w}_{k+1} \rangle -$$

$$\langle \sigma_{k+1}\hat{w}_{k+1}, (\hat{G} + \hat{X})\hat{v}_{k+1} \rangle + \langle \sigma_{k+1}\hat{w}_{k+1}, \sigma_{k+1}\hat{w}_{k+1} \rangle.$$

Notice that:

$$\langle (\hat{G} + \hat{X})\hat{v}_{k+1}, \sigma_{k+1}\hat{w}_{k+1} \rangle = \langle \Pi_{\mathcal{H}_2}(\hat{G} + \hat{X})\hat{v}_{k+1}, \sigma_{k+1}\hat{w}_{k+1} \rangle$$

$$= \langle \Gamma_{G+X}v_{k+1}, \sigma_{k+1}w_{k+1} \rangle$$

$$= \sigma_{k+1}^2.$$

Thus, we have:

$$\|\hat{h}\|_2^2 = \langle (\hat{G} + \hat{X})\hat{v}_{k+1}, (\hat{G} + \hat{X})\hat{v}_{k+1} \rangle - \sigma_{k+1}^2 - \sigma_{k+1}^2 + \sigma_{k+1}^2$$

$$\leq \|\hat{G} + \hat{X}\|_\infty^2 - \sigma_{k+1}^2$$

$$= \sigma_{k+1}^2 - \sigma_{k+1}^2 \quad \text{since } X \text{ is optimal}$$

$$= 0.$$

Hence $\hat{h} = 0$ and the assertion follows. Part 2 is immediate since the Hankel operator does not depend on the anti-causal part of the system. The proof of part 3 is omitted. ∎

Using this theorem, we can provide a constructive procedure for computing the optimal solution, for SISO systems.

Corollary 4.5.1.   The optimal X for SISO systems is given uniquely by

$$(\hat{G} + \hat{X})(\lambda) = \sigma_{k+1} \frac{\hat{w}_{k+1}(\lambda)}{\hat{v}_{k+1}(\lambda)}. \tag{4.38}$$

∎

To compute the optimal Hankel norm approximant, the following computations are needed:

1. Compute the gramians **P**, **Q**.
2. Compute the Schmidt pair w_{k+1}, v_{k+1} associated with σ_{k+1}.
3. Compute $\hat{X}$ from Equation (4.38).
4. Define $\hat{\tilde{G}}_k$ to be the causal part of $\hat{X}$.

Part 3 of Theorem 4.5.3 provides an upper bound on the sum of the Hankel singular values of the error, which in turn gives a bound on the $\mathcal{H}_\infty$ and ℓ_1 norms.

Example 4.5.1

Let G be given by

$$G \; : \; \left[\begin{array}{cc|c} .5 & 0 & 1 \\ 0 & .2 & 1 \\ \hline 1 & 1 & 0 \end{array} \right].$$

We would like to find the optimal Hankel norm approximant of G, with McMillan degree equal to 1. The observability and reachability gramians are given by

$$\mathbf{P} = \mathbf{Q} = \begin{pmatrix} 1.333 & 1.111 \\ 1.111 & 1.0417 \end{pmatrix}.$$

The Hankel singular values are

$$\sigma_1 = 2.3081, \qquad \sigma_2 = .0669.$$

The right and left eigenvectors associated with σ_2 are

$$x_2 = y_2 = \begin{pmatrix} 0.6595 \\ -0.7517 \end{pmatrix}.$$

The Schmidt pair associated with σ_2 is

$$\hat{w}_2(\lambda) = \frac{0.24395\lambda - 0.0922}{(1 - 0.5\lambda)(1 - 0.2\lambda)}, \qquad \hat{v}_2(\lambda) = \frac{0.24395 - 0.0922\lambda}{(\lambda - 0.5)(\lambda - 0.2)}.$$

From this $\hat{X}$ can be computed:

$$\hat{X} = \sigma_2 \frac{\hat{w}_2(\lambda)}{\hat{v}_2(\lambda)} - \hat{G}$$

$$= 5.227 \frac{\lambda + 0.0013}{\lambda - 2.646}.$$

For this example $\hat{\tilde{G}}_2 = \hat{X}$.

4.6 FROM CONTINUOUS TO DISCRETE

The development we undertake in this book is restricted to discrete-time systems. As we have seen already, such systems are governed by difference equations with discrete inputs and outputs. In practice, however, a control system typically consists of a continuous-time process interconnected with a digital controller via a sample and hold interface. Such a system is known by the name of *sampled-data* or *hybrid system*. In this situation, the closed loop dynamics are hybrid in the sense that they combine both discrete-time and continuous-time dynamics. The exogenous inputs and regulated variables are in general continuous-time signals.

Traditionally, two independent approaches to design digital controllers are followed. The first is to design a continuous-time controller and then discretize it, and the second is to design the controller purely in discrete-time using the discrete-equivalence of the plant. Both of these procedures ignore the intersampling behavior of the system and may result in designs that do not meet the specifications. Of course, whenever a digital controller is connected with a continuous-time plant, issues of sampling frequency, aliasing, and choice of prefilters (anti-aliasing filters) arise and consequently limit the best achievable performance.

A direct approach in addressing sampled-data systems is to view them as periodic continuous-time systems, and design the controller directly to meet the performance objectives. Although a very powerful theory has been developed along these lines, it is beyond the scope of this book to develop it in detail. Instead, we show that the design of the controller can be performed using discrete-time techniques applied to a higher dimensional system, which takes the intersampling behavior into consideration.

First we need to introduce adequate notation. A superscript c will be used in all variables associated with a continuous-time model. Consider the following FDLTI continuous-time system in state-space form:

$$\dot{x}^c(t) = A^c x^c(t) + B_1^c w^c(t) + B_2^c u^c(t), \quad x^c(0) = x_0,$$

$$z^c(t) = C_1^c x^c(t) + D_{11}^c w^c(t) + D_{12}^c u^c(t), \tag{4.39}$$

$$y^c(t) = C_2^c x^c(t),$$

where $t \in \mathbb{R}_+$, or, in short,

$$G^c = \left[\begin{array}{c|cc} A^c & B_1^c & B_2^c \\ \hline C_1^c & D_{11}^c & D_{12}^c \\ C_2^c & 0 & 0 \end{array} \right].$$

The above corresponds to the standard setup described in the previous chapter, but for a continuous-time process. We will assume that the anti-aliasing filter needed to sample y^c has been absorbed into G^c. The fact that such filter is low-pass and thus strictly proper (i.e., rolls off at high frequencies), results in D_{21}^c and D_{22}^c being zero. We will also make the standard stabilizability and detectability assumptions on (A^c, B_2^c) and (C_2^c, A^c), respectively.

Definition 4.6.1. The sample operator with sampling period T, denoted by $\mathbf{S}_T$, is defined as follows:
$$\mathbf{S}_T : \mathcal{L}_{\infty,e}^n \longrightarrow \ell^n,$$
$$y^c \longrightarrow y(k) = y^c(kT), \ k = 0, 1, \dots \qquad \blacksquare$$

In this definition, it is assumed that $\mathbf{S}_T$ operates only on continuous signals (i.e., only a subset of $\mathcal{L}_{\infty,e}^n$ on which the action is well defined), and produces a sequence that corresponds to samples of that signal at multiples of T. Note that the sample operator is time-varying, i.e., time-shifted inputs do not produce index-shifted outputs in general, unless the input is time-shifted by an integer multiple of T. Hence, we conclude that $\mathbf{S}_T$ is time-periodic with period T.

The hold operation, on the other hand, takes signals from the discrete domain into the continuous domain. This can be done in several different ways, and here we adopt the mapping corresponding to a zero-order hold.

Definition 4.6.2. The hold operator with sampling period T, denoted by $\mathbf{H}_T$, is defined as follows:
$$\mathbf{H}_T: \ell^n \longrightarrow \mathcal{L}_{\infty,e}^n,$$
$$u \longrightarrow u^c(t) = u(k), \ kT \leq t < (k+1)T. \qquad \blacksquare$$

A zero-order hold takes a sequence and generates a piecewise constant continuous-time signal, where the jump discontinuities occur at multiples of T. It will be assumed throughout that the sampler is synchronized with the hold operator, that is, a jump discontinuity in u^c occurs at the same instant y^c is sampled. With this notation, the standard sampled-data disturbance rejection problem is depicted in Figure 4.5.

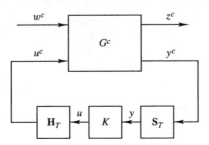

Figure 4.5 General sampled-data setup.

The following technical assumption, known as the nonpathological sampling condition, is required to guarantee the stabilizability of the sampled-data system of Figure 4.5:

Assumption: If λ is an eigenvalue of A^c with nonnegative real part, then A^c has no eigenvalues at $\lambda + j2\pi k/T$, for any nonzero integer k.   ∎

Note that this assumption is always satisfied for reasonable values of T, that is, values of the sampling frequency that are large enough to capture the relevant frequencies of the system. A typical choice in practical applications is between 5 to 20 times the expected bandwidth of the closed-loop system.

It is clear from Figure 4.5 and the previous discussion that the closed-loop map from w^c to z^c, denoted by Φ^c, is linear time-varying. In fact, it is time-periodic with period T. The standard performance problem can be posed as follows:

$$\inf_{K \text{ stab.}} \|\Phi^c\|_{\mathcal{L}_p-\text{ind}}. \tag{4.40}$$

Given a sampling period T, the above problem is an exact statement of the performance problem that one would like to solve. However, instead of presenting the solution to this problem, we develop an approximate and more intuitive approach. The main idea is simple namely, discretize the system by assuming that the input disturbance is a sequence passing through a hold operator to generate w^c, and that the output error is another sequence that results from sampling z^c. This situation is depicted in Figure 4.6, where the sampling period for the signals w^c and z^c is denoted by τ. Clearly, this representation is only an approximate one since, on one hand, the disturbances entering the real system may vary in between sampling instances, and on the other hand, the sequence of sampled errors may be missing important intersampling information.

Two different cases are of interest:

1. When $\tau = T$, which is the standard dicretization approach, and
2. When $\tau = T/N$ where $N \in \mathbf{Z}_+$ and $N > 1$.

The second method results in a class of discrete-time systems known as *multirate systems*, that is, systems where discrete-time signals run at different rates.

From the above definitions it is clear that, for any K stabilizing the sampled-data system,

$$\|\mathbf{S}_\tau \Phi^c \mathbf{H}_\tau\|_{\ell_p-\text{ind}} \leq \|\Phi^c\|_{\mathcal{L}_p-\text{ind}}$$

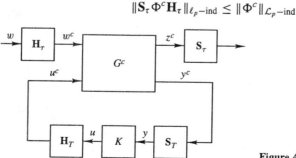

Figure 4.6 Multirate discrete-time setup.

since the class of input signals is restricted to piecewise constant signals and only samples of the output signals are considered. Furthermore, a closer look at the above inequality indicates that, given a real ordered sequence $\tau_0 \geq \tau_1 \geq \cdots \geq \tau_i \geq \cdots$, then

$$\|\mathbf{S}_{\tau_0}\mathbf{\Phi}^c\mathbf{H}_{\tau_0}\|_{\ell_p-\text{ind}} \leq \|\mathbf{S}_{\tau_1}\mathbf{\Phi}^c\mathbf{H}_{\tau_1}\|_{\ell_p-\text{ind}} \leq \cdots \leq \|\mathbf{S}_{\tau_i}\mathbf{\Phi}^c\mathbf{H}_{\tau_i}\|_{\ell_p-\text{ind}} \leq \cdots \leq \|\mathbf{\Phi}^c\|_{\mathcal{L}_p-\text{ind}}.$$

In other words, the faster w^c and z^c are sampled the better the approximation to the sampled-data system. In fact, it can be shown that for any stabilizing controller

$$\lim_{\tau \to 0} \|\mathbf{S}_\tau\mathbf{\Phi}^c\mathbf{H}_\tau\|_{\ell_p-\text{ind}} = \|\mathbf{\Phi}^c\|_{\mathcal{L}_p-\text{ind}}.$$

So the induced norm of the sampled-data system is recovered eventually. Furthermore, if τ is made sufficiently small, then a sub-optimal controller to Problem (4.40) can be computed to any degree of accuracy by solving the multirate discrete-time problem. In fact, it can be shown that there exist constants C_1, C_2 that depend on the plant G^c such that for every stabilizing controller

$$\|\mathbf{S}_\tau\mathbf{\Phi}^c\mathbf{H}_\tau\|_{\ell_p-\text{ind}} \leq \|\mathbf{\Phi}^c\|_{\mathcal{L}_p-\text{ind}} \leq C_1\tau + (1 + C_2\tau)\|\mathbf{S}_\tau\mathbf{\Phi}^c\mathbf{H}_\tau\|_{\ell_p-\text{ind}}. \tag{4.41}$$

4.6.1 Lifting of the Multirate System

It remains to find an effective way to design controllers for multirate systems. The fact that T and τ are commensurate, i.e., one is an integer multiple of the other, is exploited in the following transformations. Given a multirate system Φ as in Figure 4.6 with $T = N\tau$, define $\tilde{\Phi}$ as:

$$\tilde{\Phi} := W_{n_z}\Phi W_{n_w}^{-1}$$

where W_{n_z} is a "grouping" operator defined as follows:

$$W_{n_z} : \ell^{n_z} \longrightarrow \ell^{n_z N},$$

$$z = \{z(0), z(1), \ldots\} \longrightarrow W_{n_z}z = \left\{ \begin{pmatrix} z(0) \\ \vdots \\ z(N-1) \end{pmatrix}, \begin{pmatrix} z(N) \\ \vdots \\ z(2N-1) \end{pmatrix}, \ldots \right\},$$

and W_{n_w} is defined similarly. Notice that $\tilde{\Phi}$ has a larger number of inputs and outputs than Φ. Indeed, $\tilde{\Phi}$ is $n_z N \times n_w N$ and, most importantly, is no longer multirate. This can be seen more easily by defining a new input signal $\tilde{w} := W_{n_w}w$ and a new output signal $\tilde{z} := W_{n_z}z$. Then,

$$z = \Phi w = \Phi W_{n_w}^{-1}\tilde{w} \implies \tilde{z} = \tilde{\Phi}\tilde{w}$$

where the signals $\tilde{w}$ and $\tilde{z}$ run at the same rate as u and y, namely, at frequency of $2\pi/T$.

The above operation is called a *lifting* operation and is norm preserving. That is,

$$\sup_{0 \neq z \in \ell_p^{n_z}} \frac{\|\Phi z\|_p}{\|z\|_p} = \sup_{0 \neq z \in \ell_p^{n_z N}} \frac{\|\tilde{\Phi}z\|_p}{\|z\|_p}.$$

Therefore, controllers can be designed for the lifted system and then used in the multirate system without affecting performance. Moreover, since the lifted system is a standard discrete-time LTI system, all the techniques presented in this book can be applied.

4.6.2 State-Space Realization

We would like to find a state-space realization of the lifted system directly from the continuous-time model (4.39). Note that the evolution of the state $x(\cdot)$ in Equation (4.39) is given by

$$x^c(t) = e^{A^c t} x^c(0) + \int_0^t e^{A^c(t-\eta)} B_1^c w^c(\eta)\, d\eta + \int_0^t e^{A^c(t-\eta)} B_2^c u^c(\eta)\, d\eta.$$

Consider the propagation of the state from $t = kT$ to $t = (k+1)T$. Clearly, in such time period the control input remains constant and equal to $u(k)$. The input disturbance, on the other hand, may attain N different values corresponding to the expanded input $\tilde{w}(k)$. Therefore, the lifted discrete-time system state obeys the following difference equation:

$$x(k+1) = Ax(k) + B_1 \tilde{w}(k) + B_2 u(k), \quad x(0) = x_0$$

where

$$A = e^{A^c T}, \qquad B_2 = \int_0^T e^{A^c \eta} B_2^c\, d\eta,$$

$$B_1^T = \begin{pmatrix} \int_{kT}^{kT+\tau} e^{A^c[(k+1)T-\eta]} B_1^c\, d\eta \\[2mm] \int_{kT+\tau}^{kT+2\tau} e^{A^c[(k+1)T-\eta]} B_1^c\, d\eta \\[2mm] \vdots \\[2mm] \int_{kT+(N-1)\tau}^{(k+1)T} e^{A^c[(k+1)T-\eta]} B_1^c\, d\eta \end{pmatrix}.$$

Next, we compute the output equations. By definition,

$$\tilde{z}(k) = \begin{pmatrix} z(kN) \\ z(kN+1) \\ \vdots \\ z((k+1)N) \end{pmatrix}$$

where

$$z(kN + i) = z^c(kT + i\tau) = C_1^c x^c(kT + i\tau) + D_{11}^c w(kN + i) + D_{12}^c u(k)$$

for $i = 0, 1, \ldots, N - 1$. Therefore, by substituting the value of the state at $t = kT + i\tau$ and expanding, the equation for $\tilde{z}(k)$ can be written as

$$\tilde{z}(k) = C_1 x(k) + D_{11}\tilde{w}(k) + D_{12}u(k)$$

where

$$C_1 = \begin{pmatrix} C_1^c \\ C_1^c e^{A^c \tau} \\ \vdots \\ C_1^c e^{A^c (N-1)\tau} \end{pmatrix}, \quad D_{12} = \begin{pmatrix} D_{12}^c \\ D_{12}^c + \int_0^\tau C_1^c e^{A^c \eta} B_2^c \, d\eta \\ \vdots \\ D_{12}^c + \int_0^{(N-1)\tau} C_1^c e^{A^c \eta} B_2^c \, d\eta \end{pmatrix},$$

and

$$D_{11} = \begin{pmatrix} D_{11}^c & 0 & 0 & \cdots \\ \int_{kT}^{kT+\tau} C_1^c e^{A^c[(k+1)T-\eta]} B_1^c d\eta & D_{11}^c & 0 & \cdots \\ \vdots & & \ddots & \\ \int_{kT}^{kT+\tau} C_1^c e^{A^c[(k+1)T-\eta]} B_1^c d\eta & \cdots & \int_{kT+(N-2)\tau}^{kT+(N-1)\tau} C_1^c e^{A^c[(k+1)T-\eta]} B_1^c d\eta & D_{11}^c \end{pmatrix}.$$

Finally, the measured output equation is simply given by $y(k) = C_2 x(k)$, where $C_2 = C_2^c$.

Note that the standard discrete-time equations, i.e., without fast sampling w^c and z^c, can be obtained directly from the above by letting $N = 1$ and $\tau = T$. That is,

$$A = e^{A^c T}, \quad B_1 = \int_0^T e^{A^c \eta} B_1^c d\eta, \quad B_2 = \int_0^T e^{A^c \eta} B_2^c d\eta$$

$$C_1 = C_1^c, \qquad D_{11} = D_{11}^c \qquad, \qquad D_{12} = D_{12}^c$$

$$C_2 = C_2^c, \qquad D_{21} = 0 \qquad, \qquad D_{22} = 0$$

4.7 SUMMARY

This chapter has summarized various properties and results related to FDLTI systems. The polynomial matrix approach will be used in Chapter 6 to provide methods for computing zeros of MIMO systems using only matrix computations, thus avoiding the computation of the Smith-McMillan decomposition. Zeros will play an instrumental role in characterizing feasible closed-loop maps.

We have shown in this chapter that computing norms of FDLTI systems can be done using efficient algorithms. The Hankel operator associated with such systems is essentially a finite-dimensional operator and thus can be represented as a matrix using an appropriate basis. The norm of this operator can be computed exactly. The Hankel operator is shown to be intimately related to model reduction problems, and later on, will provide methods for computing the optimal $\mathcal{H}_\infty$ controller.

Theorem 4.3.1 provides a very interesting relationship between the ℓ_1 and the $\mathcal{H}_\infty$ norm. We have shown that the bound between the two norms is tight. It will be shown in Chapter 10 that this bound is tight in design problems as well. In other words, we will exhibit a plant such that if the controller is designed minimizing the $\mathcal{H}_\infty$ norm, then the ℓ_1 norm of the system is equal to the upper bound given by Theorem 4.3.1. This has implications on the utility of the $\mathcal{H}_\infty$ norm for rejecting bounded-amplitude disturbances by minimizing the amplitude of the output.

EXERCISES

4.1. Let $\mathbf{P}$, $\mathbf{Q}$ be the reachability and observability gramians. Verify:
   (a) The eigenvalues of $\mathbf{PQ}$ are the same as the eigenvalues of $\mathbf{R}^T\mathbf{QR}$, where $\mathbf{RR}^T = \mathbf{P}$.
   (b) Show that the singular values are well defined.
   (c) Show that the vectors x_i's can be chosen as in Equation (4.15).

4.2. Construct a similarity transformation such that both the reachability and observability gramians satisfy:

$$\mathbf{P} = \mathbf{Q} = diag(\sigma_1, \dots \sigma_n).$$

The resulting realization is a balanced realization.

4.3. Consider a stable system with a state-space description given by $(A, B, C, 0)$, both reachable and observable. We are interested in computing the norm of the Hankel operator directly from the definition. The precise quantity that we would like to compute is:

$$\alpha = \max_{u \in \ell_2(\mathbf{Z}_-)} \left\{ \sum_{t=0}^{\infty} y(t)^2 \,\Big|\, \sum_{t=-\infty}^{-1} u(t)^2 \leq 1 \right\}.$$

In words, we want to find the maximal achievable energy of the output after $t = 0$ which is the response to an input starting at $-\infty$ and ending at -1, with energy less than or equal to one.
   (a) Find an expression for $x(0)$ in terms of the above input u. Does the value $x(-\infty)$ enter your expression. Explain.
   (b) Can any $\xi \in \mathbb{R}^n$ be achieved by some choice of an input of the above form? If so, find an expression of the minimum energy input, u_{min} that achieves the value $x(0) = \xi$. Compute the square of the energy of u_{min}. Write this expression in terms of the Reachability Gramian $\mathbf{P}$, and denote it by $\alpha_1(\xi)$.
   (c) If some input u_{min} results in $x(0) = \xi$, write an expression of the output for $t \geq 0$. Compute the square of the energy of the output (for $t \geq 0$) as a function of ξ. Write this in an expression involving the Observability Gramian $\mathbf{Q}$, and denote it by $\alpha_2(\xi)$.
   (d) Argue that α is given by

$$\alpha = \max_{\xi \in \mathbb{R}^n}\{\alpha_2(\xi)|\alpha_1(\xi) \leq 1\}.$$

   (e) Prove that

$$\alpha = \lambda_{\max}\{\mathbf{QP}\}.$$

4.4. Prove the result in Theorem 4.3.1 for the MIMO case.

4.5. (The Caratheodory-Fejer interpolation problem) Given a finite pulse response filter $\hat{F}(\lambda) = f_0 + f_1\lambda + \ldots + f_n\lambda^n$, assume that the coefficients were chosen to meet certain specifications. To make the filter attenuate all bounded energy disturbances, we would like to find another filter $\hat{F}_1(\lambda)$ such that:

$$\hat{F}_1(\lambda) = \sum_{i=0}^{\infty} \tilde{f}_i\lambda^i,$$

$$\tilde{f}_k = f_k \qquad \text{for } k = 0, 1, \ldots, n,$$

$$\|\hat{F}_1\|_\infty < 1.$$

Use Nehari's theorem to show that we can find $\hat{F}_1(\lambda)$ with the above conditions if and only if the matrix M

$$\begin{pmatrix} f_n & f_{n-1} & \cdots & f_1 & f_0 \\ f_{n-1} & f_{n-2} & \cdots & f_0 & 0 \\ \vdots & \vdots & & 0 & 0 \\ f_0 & 0 & \cdots & 0 & 0 \end{pmatrix}$$

satisfies $\sigma_{max}(M) < 1$.

Hint: $\hat{F}_1(\lambda) = \hat{F}(\lambda) - \lambda^{n+1}\hat{Q}(\lambda)$. Notice that the $\mathcal{L}_\infty$ norm does not change if a function is multiplied by $\lambda^{-(n+1)}$. Also, notice that $\|\hat{F}\|_\infty = \|\hat{F}(\frac{1}{\lambda})\|_\infty$.

4.6. A class of uncertain systems is modeled as

$$P = P_0 + \Delta W$$

where P_0 is a time invariant nominal plant, W is a stable weight and Δ is an unknown time-invariant perturbation with $\|\hat{\Delta}\|_\infty < 1$. A designer wants to validate this model. She/he conducts an experiment of length N and records the set of data:

$$[u(k), \ y(k) \mid k = 0, 1, \ldots N - 1].$$

Let

$$\bar{y}(k) = y(k) - (P_0 u)(k) \quad \text{for all } k = 0, 1, \ldots N - 1,$$

and

$$\bar{u}(k) = (Wu)(k) \quad \text{for all } k = 0, 1, \ldots N - 1,$$

and assume that $\bar{u}(0) \neq 0$. Also, let $\mathcal{U}$ denote the Toeplitz matrix associated with $\bar{u}$, i.e.,

$$\mathcal{U} = \begin{pmatrix} \bar{u}(0) & 0 & \cdots & 0 \\ \bar{u}(1) & \bar{u}(0) & \cdots & 0 \\ & & \ddots & \\ \bar{u}(N-1) & \bar{u}(N-2) & \cdots & \bar{u}(0) \end{pmatrix}.$$

Similarly define the matrix $\mathcal{Y}$ associated with $\bar{y}$. Show that the designer can invalidate the model if and only if

$$\mathcal{Y}\mathcal{Y}^T \geq \mathcal{U}\mathcal{U}^T.$$

Hint: Remember Δ is time-invariant. You will need to use the Caratheodory-Fejer interpolation problem (see Exercise 4.5). Also, if $P > 0$ (i.e. positive definite) and $Q \geq 0$ (positive semi-definite), then $\lambda_{max}(QP^{-1}) < 1$ if and only if $Q < P$.

4.7. Consider the following system, G:

$$A = \begin{pmatrix} 0 & 1 \\ -.475 & 1.45 \end{pmatrix}, \quad b = \begin{pmatrix} 0 \\ 1 \end{pmatrix}, \quad c = (.475 \quad -.5).$$

(a) Compute the ℓ_1 norm of G with an accuracy within .001 by summing the first N components of the pulse response, and bounding its tail. (In other words, find N so that $\sum_{k>N} |cA^k b| < .001$).

(b) Compute the ℓ_1 norm of G with an accuracy within .001 using the Hankel singular values to estimate the error.

(c) Compare the length of the impulse response used in both of these schemes. If they are not comparable, explain why.

4.8. Let G be a SISO linear, causal, periodic operator of period N. Show that the lifted system

$$\tilde{G} = W_N G W_N^{-1}$$

is LTI, where W is the grouping operator. If $\tilde{G}$ has the pulse response $\tilde{G}(k)$, show that $\tilde{G}(0)$ is lower triangular. Verify that any LTI system of dimension $N \times N$ with a pulse response $\tilde{G}(k)$ such that $\tilde{G}(0)$ is lower triangular corresponds to a lifted periodic system of period N. This condition on $\tilde{G}$ is known as the causality condition.

4.9. Given the following system:

$$G: \quad x(k+1) = Ax(k) + B_1 w(k) + B_2 u(k),$$

where w is an unknown exogenous signal satifying $\|w\|_\infty \leq 1$. Consider the following problem: Find an input $u = Fx$ (a state feedback) such that

$$\|x\|_\infty \leq X_{max}, \quad \|u\|_\infty \leq U_{max}, \quad \text{for all } |x(0)|_\infty \leq X_{max} \text{ and } \|w\|_\infty \leq 1.$$

Show that all possible solutions for F are characterized by a set of linear constraints. Hint: The problem is really a static one.

NOTES AND REFERENCES

Finite dimensional LTI systems have been extensively covered in many textbooks, for example [Che84]. The algebraic view of such systems can be found in [Kai80]. For a complete study of zeros of multivariable systems, see [Kai80, GLR82, BGR88, BR90, DDV79].

The material on Hankel operators is taken from [Glo89] as well as [Fra87]. For more details on Hankel model reduction, and state-space solutions, see [Glo84, Glo89]. Hankel operators are discussed in a great detail in [Pow82]. The proof of Nehari's theorem is found in [AAK71, Neh57] and the detailed proof of Theorem 4.5.3 can be found in [Glo84].

The bounds on induced norms using Hankel singular values are generalizations of the SISO results in [BD87]. The algorithm for computing the $\mathcal{H}_\infty$ norm is simply the discrete-time version of the continuous-time algorithm in [BB91]; see also [IG91]. The algorithm for computing the ℓ_1 norm can be found in [BB92].

Analysis and design of sampled-data systems have both been addressed in many texts; see [Ack85, AW90, FPW90, RF58]. For a treatment of sampled-data systems as periodic systems by considering the exact hybrid problem, see [CF90, CF91a, CF91b, BP92, BDP93, DF92, DG92, DG93, KA92, SK91]. The general inequality in Equation 4.41 is proved in [BDP93].

CHAPTER 5 ——————————————————

Controller Parametrization

A central objective of this development is to capture the fundamental limitations of controller design in achieving specific performance objectives, and to provide nonconservative methods for design. For this reason, it is crucial that a parametrization of *all* stabilizing controllers be furnished, which in turn gives a parametrization of all achievable closed-loop maps. In the sequel, such a parametrization is presented with the following properties:

1. The representation of all stabilizing controllers is linear fractional in a free stable parameter.
2. All closed-loop maps are affine in the same parameter (i.e., forms a translated subspace in the space of stable systems).
3. The parametrization is complete, providing all possible nonlinear controllers that result in bounded gain stability.

We will see in later chapters that the second property is fundamental in developing synthesis techniques for robust control.

5.1 MOTIVATION: SISO PLANTS

Consider the standard closed-loop system in Figure 5.1 with real rational plant and controller. The closed-loop system is stable if and only if the map

$$H(\hat{P}, \hat{K}) = \begin{pmatrix} \dfrac{\hat{P}}{I - \hat{P}\hat{K}} & \dfrac{1}{I - \hat{P}\hat{K}} \\ \dfrac{1}{I - \hat{P}\hat{K}} & \dfrac{\hat{K}}{I - \hat{P}\hat{K}} \end{pmatrix} \tag{5.1}$$

is stable. In classical stability analysis, $\hat{P}$ and $\hat{K}$ are written as ratios of coprime polynomials:

$$\hat{P}(\lambda) = \frac{\hat{N}(\lambda)}{\hat{M}(\lambda)}, \qquad \hat{K}(\lambda) = \frac{\hat{Y}(\lambda)}{\hat{X}(\lambda)}. \tag{5.2}$$

It is well known from standard stability analysis that the closed-loop system in Figure 5.1 is stable if and only if the closed-loop return matrix $\hat{D}$, defined as

$$\hat{D}(\lambda) := \hat{M}(\lambda)\hat{X}(\lambda) - \hat{N}(\lambda)\hat{Y}(\lambda), \tag{5.3}$$

has all the roots outside the unit disc. This condition guarantees that all the closed-loop transfer functions are stable and that no unstable pole-zero cancellations occur between the plant and the controller. Rewrite the factors of the controller as

$$\hat{Y}_1(\lambda) = \frac{\hat{Y}(\lambda)}{\hat{D}(\lambda)}, \qquad \hat{X}_1(\lambda) = \frac{\hat{X}(\lambda)}{\hat{D}(\lambda)}.$$

$\hat{Y}_1$ and $\hat{X}_1$ are stable rational functions, and are not necessarily polynomials. The following equality holds:

$$\hat{M}(\lambda)\hat{X}_1(\lambda) - \hat{N}(\lambda)\hat{Y}_1(\lambda) = I. \tag{5.4}$$

This equation is known as the Bezout equation, in which all the elements are allowed to be stable rational systems. For any $\hat{M}$, $\hat{N}$, $\hat{X}_1$, $\hat{Y}_1$ that satisfy Equation (5.4), it follows that $\hat{M}$ and $\hat{N}$ have no common zeros inside the unit disc. Similarly, $\hat{X}_1$ and $\hat{Y}_1$. In view of this we make the following definition.

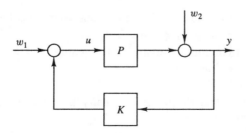

Figure 5.1 Standard diagram for SISO stability.

Definition 5.1.1.   Two stable rational SISO functions are coprime with respect to the space of stable systems if they do not have common zeros in the closed unit disc. ∎

Notice that two coprime polynomials are coprime with respect to the stable domain. The converse, however, is not true.

Example 5.1.1

Let

$$\hat{M}(\lambda) = (\lambda - 2)(\lambda - .5), \qquad \hat{N}(\lambda) = (\lambda - 2)(\lambda - .4).$$

It follows that $\hat{M}(\lambda)$ and $\hat{N}(\lambda)$ are not coprime over the space of polynomials; however, they are coprime over the space of stable systems.

In the sequel, coprimeness will always mean coprimeness over the stable domain. Lemma 5.1.1, which follows, summarizes the above discussion.

Lemma 5.1.1.   Let the nominal plant $\hat{P} = \hat{N}/\hat{M}$, where both $\hat{N}$ and $\hat{M}$ are stable rational functions in λ and coprime. Then every stabilizing controller $\hat{K}$ can be written as $\hat{K} = \hat{Y}/\hat{X}$, with $\hat{X}$ and $\hat{Y}$ coprime, such that

$$\hat{M}(\lambda)\hat{X}(\lambda) - \hat{N}(\lambda)\hat{Y}(\lambda) = I.$$ ∎

Using Lemma 5.1.1, it is evident that all stabilizing controllers are simply solutions of the above Bezout equation, and are independent of the particular choice of $\hat{M}$ and $\hat{N}$. The following theorem gives all such controllers.

Theorem 5.1.1.   Let the nominal plant $\hat{P} = \hat{N}/\hat{M}$, where both $\hat{N}$ and $\hat{M}$ are stable rational functions in λ and coprime. Assume that a controller $\hat{K}$ is known to stabilize $\hat{P}$, with $\hat{K} = \hat{Y}/\hat{X}$ and

$$\hat{M}(\lambda)\hat{X}(\lambda) - \hat{N}(\lambda)\hat{Y}(\lambda) = I.$$

Then all stabilizing controllers of $\hat{P}$ are given by

$$\hat{K}(\lambda) = \frac{\hat{Y}(\lambda) - \hat{M}(\lambda)\hat{Q}(\lambda)}{\hat{X}(\lambda) - \hat{N}(\lambda)\hat{Q}(\lambda)},$$

where $\hat{Q}$ is any arbitrary stable function.

Proof. This parametrization can be obtained by parametrizing all solutions to the Bezout equation. First, if $\hat{K}$ is given by the above formula, then

$$\hat{M}(\lambda)(\hat{X}(\lambda) - \hat{N}(\lambda)\hat{Q}(\lambda)) - \hat{N}(\lambda)(\hat{Y}(\lambda) - \hat{M}(\lambda)\hat{Q}(\lambda)) = I.$$

Thus $\hat{K}$ stabilizes $\hat{P}$. On the other hand, let $\hat{K} = \hat{Y}_1/\hat{X}_1$ be a stabilizing controller. Then $\hat{X}_1$ and $\hat{Y}_1$ satisfy

$$\hat{M}(\lambda)\hat{X}_1(\lambda) - \hat{N}(\lambda)\hat{Y}_1(\lambda) = \hat{D}(\lambda),$$

where $\hat{D}$ has all its zeros outside the unit disc. Define

$$\hat{Q}_1 := -(\hat{X}\hat{Y}_1 - \hat{Y}\hat{X}_1)\hat{D}^{-1}.$$

Then, with some algebra, the controller $\hat{K}$ is given by

$$\hat{K}(\lambda) = \frac{\hat{Y}_1(\lambda)}{\hat{X}_1(\lambda)} = \frac{\hat{Y}(\lambda) - \hat{M}(\lambda)\hat{Q}_1(\lambda)}{\hat{X}(\lambda) - \hat{N}(\lambda)\hat{Q}_1(\lambda)}.$$

This establishes the result.   ∎

The computations necessary for obtaining all stabilizing controllers involve computing one solution to Equation (5.4), which is equivalent to finding one stabilizing controller. Such a computation can be done systematically using state-space computations, as will be demonstrated later. A pictorial representation of this parametrization is shown in Figure 5.2.

Generality. For any rational plant $P = NM^{-1}$, the parametrization

$$K = (Y - MQ)(X - NQ)^{-1} \quad Q \text{ is stable} \tag{5.5}$$

is a complete parametrization in the following sense:

1. All LTI stabilizing controllers in the ℓ_∞ sense are parametrized by $Q \in \ell_1$. All FDLTI stabilizing controllers are parametrized by $Q \in \mathcal{RH}_\infty$.
2. All LTV stabilizing controllers in the ℓ_∞ sense are parametrized by Q in the set of all LTV ℓ_∞-stable operators.
3. All NLTV stabilizing controllers in the ℓ_∞ sense are parametrized by Q in the set of all NLTV ℓ_∞-stable operators.
4. Similarly for ℓ_2 stability; e.g., all LTI stabilizing controllers in the ℓ_2 sense are parametrized by $\hat{Q} \in \mathcal{H}_\infty$, and so on.

It is evident that our proofs only cover the LTI cases.

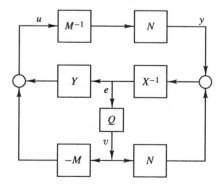

Figure 5.2 All stabilizing controllers.

Closed-loop maps.   The parametrization of all stabilizing controllers gives in turn a parametrization of all achievable closed-loop maps $H(P, K)$ in Equation (5.1):

$$H(P, K) = \begin{pmatrix} N(X - NQ) & M(X - NQ) \\ M(X - NQ) & M(Y - MQ) \end{pmatrix}. \tag{5.6}$$

The beauty of this parametrization is that all the admissible closed-loop maps are affine in the parameter Q, which belongs to a vector space (typically ℓ_1 or the subset $\mathcal{RH}_\infty$). This is a great simplification from the linear fractional dependence of the closed-loop maps on the set of stabilizing controllers. The latter set does not form a subspace and, in general, has quite a complicated structure.

Achievable closed-loop maps.   A general discussion of the characterization of achievable closed-loop maps is presented in Chapter 6. In here, however, we present a special case, namely the sensitivity function of a SISO system. Consider the sensitivity function as a function of the Q parameter:

$$S \triangleq (I - PK)^{-1} = M(X - NQ).$$

Suppose the $S_d \in \mathcal{RH}_\infty$ is a desired sensitivity function. We would like to check whether S_d can be exactly achieved by some controller, or consequently a $Q \in \mathcal{RH}_\infty$. We can solve directly for Q_d given S_d:

$$\hat{Q}_d = \frac{\hat{M}\hat{X} - \hat{S}_d}{\hat{M}\hat{N}}.$$

It is clear that the desired map can be attained if and only if $\hat{Q}_d$ is stable. Exact conditions guaranteeing this can be readily derived: Let $z_1, \ldots, z_m$ be the zeros of $\hat{N}$ inside the closed unit disc and $p_1, \ldots, p_n$ be the zeros of $\hat{M}$ inside the closed unit disc. For simplicity, assume that they are distinct and simple. Then $\hat{Q}_d$ is stable if and only if

$$\begin{aligned} S_d(z_i) &= 1 \quad \text{for all } i = 1, \ldots, m, \\ S_d(p_i) &= 0 \quad \text{for all } i = 1, \ldots, n. \end{aligned} \tag{5.7}$$

To state this differently: For an internally stabilizing controller, the sensitivity function evaluated at a pole of the plant inside the unit disc is equal to zero, and evaluated at a zero of the plant inside the unit disc is equal to one. If the plant is stable minimum-phase, then any closed-loop sensitivity function is achievable. This is consistent with classical stability conditions for SISO plants.

If S_d is not achievable while still desired as a closed-loop sensitivity function, then Q can be chosen to approximate S_d, say, as an operator on ℓ_∞. More precisely, Q can be chosen to minimize

$$\|S_d - M(X - NQ)\|_1. \tag{5.8}$$

This is known as the model matching problem and will be discussed in detail in future chapters.

Asymptotic tracking.   In many situations it is desirable to implement controllers that result in zero steady-state error for certain fixed commands, such as steps or sinusoids. A parametrization of such controllers can be easily achieved. Let $\hat{r}$ be the the λ-transform of the command to be tracked (which is assumed to be a rational function). For asymptotic tracking, the error function (see Figure 5.3)

$$\hat{e} = \hat{S}\hat{r} = \hat{M}(\hat{X} - \hat{N}\hat{Q})\hat{r} \qquad (5.9)$$

should have no poles inside the closed unit disc. To achieve this, the poles of $\hat{r}$, say a_i (typically on the unit circle, or inside the unit disc), have to be zeros of $\hat{S}$. This is reflected on $\hat{Q}$ through the following interpolation conditions (assume the poles are simple):

$$\hat{M}(\hat{X} - \hat{N}\hat{Q})(a_i) = 0. \qquad (5.10)$$

If $\hat{M}(a_i) = 0$, then the above condition is satisfied for any choice of $\hat{Q}$. If $\hat{M}(a_i) \neq 0$, the above condition translates into

$$(\hat{X} - \hat{N}\hat{Q})(a_i) = 0 \qquad \text{or} \qquad \hat{Q}(a_i) = \frac{\hat{X}(a_i)}{\hat{N}(a_i)}, \qquad (5.11)$$

which is equivalent to requiring the controller to have a pole at a_i. Clearly this condition cannot be satisfied if $\hat{N}(a_i) = 0$.

Another way of viewing this is as follows: A controller that results in asymptotic tracking must contain as poles all of the poles of the signal to be tracked. Hence, the controller has the form $\hat{K} = \hat{K}_1/\hat{r}$. To parametrize all possible K_1's, the factor $1/\hat{r}$ can be lumped in the plant, and the problem of asymptotic tracking is equivalent to stabilizing the plant $\hat{P}_1 = \hat{P}/\hat{r}$, as in Figure 5.4.

Coprime factorization.   If $\hat{P} = \hat{N}\hat{M}^{-1}$, $\hat{N}$ and $\hat{M}$ are stable and satisfy Equation (5.4) for some stable $\hat{X}$ and $\hat{Y}$, then $\hat{M}$, $\hat{N}$ are called coprime and Equation (5.4) is called the Bezout identity. From the previous analysis, it is evident that obtaining a coprime factorization is a first step in getting the parametrization of stabilizing controllers. On the other hand, given a stabilizing controller, a coprime factorization of the plant can be defined. This idea provides a general procedure for getting the parametrization for MIMO problems, as will be demonstrated next.

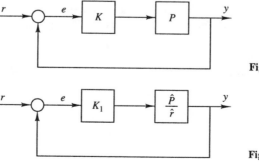

Figure 5.3   Tracking.

Figure 5.4   Stability vs. tracking.

5.2 GENERAL PROBLEMS

Consider the standard 2-input 2-output setup shown in Figure 5.5. There are several approaches for arriving to the parametrization of all stabilizing controllers, one of which is coprime factorization. In the sequel, this approach is presented, and then interpreted in terms of an augmented observer-based controller. This interpretation not only offers a deeper explanation but it also provides a way for calculating all necessary transfer functions. First, we make the following necessary assumption:

Assumption: The general plant G is stabilizable from u and detectable from y. Equivalently, all the unstable poles of the system are reachable and observable from u and y, respectively. This implies that a controller stabilizes G if and only if it stabilizes G_{22}. Hence, it suffices to parametrize all controllers for G_{22}.   ∎

5.2.1 Coprime Factorization

Definition 5.2.1.   A doubly-coprime factorization of G_{22} is a set of maps $N, M, \tilde{N}, \tilde{M}$, with $G_{22} = NM^{-1} = \tilde{M}^{-1}\tilde{N}$ satisfying

$$\begin{pmatrix} \tilde{X} & -\tilde{Y} \\ -\tilde{N} & \tilde{M} \end{pmatrix} \begin{pmatrix} M & Y \\ N & X \end{pmatrix} = I, \tag{5.12}$$

for some stable $X, Y, \tilde{X}$ and $\tilde{Y}$. Further, M and N are referred to as right coprime factors while $\tilde{M}$ and $\tilde{N}$ are referred to as left coprime factors of G_{22}.   ∎

It follows from the definition that if M and N are right coprime, then the system

$$\begin{pmatrix} \hat{M} \\ \hat{N} \end{pmatrix}$$

has no zeros in the closed unit disc and thus has a stable left inverse. Similarly, if $\tilde{N}$ and $\tilde{M}$ are left coprime, then the operator

$$\begin{pmatrix} \hat{\tilde{N}} & \hat{\tilde{M}} \end{pmatrix}$$

has no zeros in the closed unit disc and has a stable right inverse.

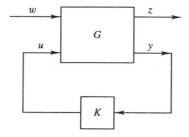

Figure 5.5   General setup.

Definition 5.2.2. A stable operator $U \in \ell_1^{m \times m}$ is a unit if $U^{-1} \in \ell_1^{m \times m}$. ∎

Hence, $\hat{U} \in \mathcal{RH}_\infty$ with full normal rank is a unit if and only if $\hat{U}(\lambda)$ has no zeros in the closed unit disc. Clearly the matrices

$$\begin{pmatrix} \tilde{X} & -\tilde{Y} \\ -\tilde{N} & \tilde{M} \end{pmatrix}, \qquad \begin{pmatrix} M & Y \\ N & X \end{pmatrix}$$

are units. We will see shortly that this property provides a characterization of stabilizing controllers, and is a direct extension of the SISO case.

Finally, if all the quantities in Definition 5.2.1 are polynomials, then the matrix

$$\begin{pmatrix} M & Y \\ N & X \end{pmatrix}$$

is unimodular (i.e., has a polynomial inverse). Restricting the factorization to polynomial matrices will have some interesting features, as we shall see in later chapters.

The following theorem establishes the existence of a doubly-coprime factorization.

Theorem 5.2.1. Every finite dimensional system G_{22} has a doubly-coprime factorization.

Proof. The proof of this theorem is obtained by exhibiting state-space formulae for the different factors in Definition 5.2.1. This will be done later in this section. ∎

5.2.2 Characterization of Stability

Lemma 5.2.1. Let a doubly-coprime factorization of G_{22} be given as in Definition 5.2.1. A controller K stabilizes G_{22} if and only if K has a right coprime factorization $K = Y_1 X_1^{-1}$ such that the map

$$\begin{pmatrix} M & Y_1 \\ N & X_1 \end{pmatrix}$$

is a unit in ℓ_1.

Proof. Consider Figure 5.6. The following equation can be easily verified:

$$\begin{pmatrix} M & -Y_1 \\ -N & X_1 \end{pmatrix} \begin{pmatrix} \xi \\ \eta \end{pmatrix} = \begin{pmatrix} v_1 \\ v_2 \end{pmatrix}.$$

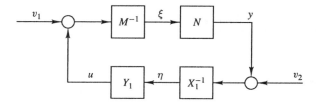

Figure 5.6 Internal stability.

If the inverse of the above matrix is stable, then the operator mapping v_1 and v_2 to ξ and η is stable. Since u and y are related to ξ and η in a stable way, the above implies that the closed-loop system is stable. On the other hand, suppose that the closed-loop system is stable. From the Bezout equation, we have

$$\tilde{X}M - \tilde{Y}N = I.$$

This implies that

$$\xi = \tilde{X}M\xi - \tilde{Y}N\xi.$$

However, $M\xi = v_1 + u$ and $N\xi = y$. Therefore,

$$\|\xi\| \le c_1\|v_1\| + c_2\|v_2\|.$$

A similar argument follows for η. Thus the map from v_1 and v_2 to ξ and η is stable. This verifies the assertion of the lemma.   ∎

In a similar fashion, if $G_{22} = \tilde{M}^{-1}\tilde{N}$, with $\tilde{M}$, $\tilde{N}$ left coprime, and a controller $K = \tilde{X}_1^{-1}\tilde{Y}_1$, with $\tilde{X}_1$, $\tilde{Y}_1$ left coprime, then K stabilizes G_{22} if and only if

$$\begin{pmatrix} \tilde{X}_1 & -\tilde{Y}_1 \\ -\tilde{N} & \tilde{M} \end{pmatrix}$$

is a unit in ℓ_1.

5.2.3 A Complete Parametrization

From Lemma 5.2.1, it is clear that a possible stabilizing controller is $K = YX^{-1}$, where X, Y are from Definition 5.2.1. Also, any other solution of the Bezout equation (5.12) gives a stabilizing controller. In fact, these are all possible stabilizing controllers.

Theorem 5.2.2.   Let a doubly-coprime factorization of G_{22} be given as in Definition 5.2.1. All stabilizing controllers are given by

$$K = (Y - MQ)(X - NQ)^{-1} = (\tilde{X} - Q\tilde{N})^{-1}(\tilde{Y} - Q\tilde{M}), \quad Q \text{ is stable.} \quad (5.13)$$

Proof. First we show that the parametrization in Equation (5.13) is stabilizing, and that the two sides of the equation are equal. Multiplying Equation (5.12) from the left by the matrix

$$\begin{pmatrix} I & Q \\ 0 & I \end{pmatrix}$$

and from the right by its inverse, we get

$$\begin{pmatrix} \tilde{X} - Q\tilde{N} & -\tilde{Y} + Q\tilde{M} \\ -\tilde{N} & \tilde{M} \end{pmatrix} \begin{pmatrix} M & Y - MQ \\ N & X - NQ \end{pmatrix} = I.$$

This implies that the parametrization is stabilizing (by Lemma 5.2.1). Also

$$(\tilde{X} - Q\tilde{N})(Y - MQ) = (\tilde{Y} - Q\tilde{M})(X - NQ).$$

On the other hand, assume that $K = Y_1 X_1^{-1}$ is a stabilizing controller with X_1, Y_1 coprime. By Lemma 5.2.1, the matrix

$$\begin{pmatrix} M & Y_1 \\ N & X_1 \end{pmatrix}$$

is a unit in ℓ_1, i.e., has a stable inverse. This implies that the following matrix

$$\begin{pmatrix} \tilde{X} & -\tilde{Y} \\ -\tilde{N} & \tilde{M} \end{pmatrix} \begin{pmatrix} M & Y_1 \\ N & X_1 \end{pmatrix} = \begin{pmatrix} I & \tilde{X}Y_1 - \tilde{Y}X_1 \\ 0 & -\tilde{N}Y_1 + \tilde{M}X_1 \end{pmatrix} \tag{5.14}$$

is a unit, and so the matrix $D := -\tilde{N}Y_1 + \tilde{M}X_1$ has a stable inverse. Define Q as follows:

$$Q := -(\tilde{X}Y_1 - \tilde{Y}X_1)D^{-1}.$$

Substituting this in Equation (5.14) and multiplying it from the left by the matrix

$$\begin{pmatrix} M & Y \\ N & X \end{pmatrix}$$

we get

$$\begin{pmatrix} M & Y_1 \\ N & X_1 \end{pmatrix} = \begin{pmatrix} M & (Y - MQ)D \\ N & (X - NQ)D \end{pmatrix}.$$

By equating entries in the above matrix, it follows that the controller K is in the parametrization.   ∎

Rational controllers.   In Equation 5.13, we did not specify the space in which Q lies. As explained earlier, if $Q \in \ell_1$, then Equation 5.13 gives a complete parametrization of all ℓ_∞ stablizing controllers. If $Q \in \mathcal{H}_\infty$, then the parametrization gives all ℓ_2 stabilizing controllers. All finite-dimensional controllers are parametrized by $Q \in \mathcal{RH}_\infty$.

5.2.4 Closed-Loop Maps

The closed-loop map from w to z is given by

$$\Phi = G_{11} + G_{12}K(I - G_{22}K)^{-1}G_{21}.$$

By direct substitution, it can be verified that

$$K(I - G_{22}K)^{-1} = (Y - MQ)\tilde{M}.$$

From this, it follows that

$$\Phi = H - UQV, \tag{5.15}$$

where

$$H = G_{11} + G_{12}Y\tilde{M}G_{21},$$

$$U = G_{12}M,$$

$$V = \tilde{M}G_{21}.$$

The closed-loop map is affine in the free parameter Q, exactly as in the SISO case.

5.2.5 Parametrization from a Given Controller

The above parametrization can be obtained from any nominal stabilizing controller. In the sequel, we consider the case where the nominal controller is observer based. Deriving the parametrization from this nominal controller provides state-space equations for computing all necessary factors in Definition 5.2.1.

Let the plant G have the state-space description

$$\left[\begin{array}{c|cc} A & B_1 & B_2 \\ \hline C_1 & D_{11} & D_{12} \\ C_2 & D_{21} & D_{22} \end{array}\right]. \tag{5.16}$$

We assume that (A, B_2) is stabilizable and (C_2, A) is detectable. This means that the unreachable and unobservable modes are stable. We also consider the general case where D_{22} may not be zero, or equivalently, the plant is not strictly causal. In this case, the parametrization will have an added constraint, namely the well-posedness of the closed-loop system.

Let F be a stabilizing state feedback matrix, i.e., $A + B_2F$ has all its eigenvalues strictly in the unit disc, and L be a stabilizing filter gain matrix, i.e., $A + LC_2$ has all its eigenvalues strictly in the unit disc. Consider the controller

$$\left[\begin{array}{c|c} A + B_2F + LC_2 + LD_{22}F & -L \\ \hline F & 0 \end{array}\right]. \tag{5.17}$$

This controller is known as the Observer-Based controller (or Model-Based controller), and is well known to be a stabilizing controller. Let x_K denote its states. Consider an external input v added to the output of the controller such that the input to the plant is given by $u = Fx_K + v$. The state-space realization of the controller becomes

$$\left[\begin{array}{c|cc} A + B_2F + LC_2 + LD_{22}F & -L & B_2 + LD_{22} \\ \hline F & 0 & I \end{array}\right], \tag{5.18}$$

i.e., v also drives the observer. Let

$$e = y - y_K = C_2x + D_{22}u - (C_2x_K + D_{22}u) = C_2\tilde{x}$$

where $\tilde{x} := x - x_K$. The error e is known as the innovations (which is the difference between the estimated output and the actual output). By direct substitution, and for any external input v, $\tilde{x}$ satisfies the difference equation

$$\tilde{x}(k+1) = (A + LC_2)\tilde{x}(k), \tag{5.19}$$

which implies that e depends only on the initial conditions of $\tilde{x}$ and is not driven by v. In other words, the transfer function from v to e is equal to zero. Consider the augmented controller constructed by connecting e to v through any stable system, i.e.,

$$v = Qe$$

as shown in Figure 5.7. The new augmented controller is a stabilizing controller since Q affects the system only in the forward loop. Define the maps:

$$
\begin{align}
H & \text{ Map from } w \text{ to } z, \notag \\
V & \text{ Map from } w \text{ to } e, \tag{5.20} \\
U & \text{ Map from } v \text{ to } z. \notag
\end{align}
$$

Because the map from v to e is identically equal to zero, the observer-based parametrization can be redrawn as a model matching problem with Q in a forward loop. Notice that the maps H, U, V are all stable maps since the nominal controller stabilizes the plant. This is shown in Figure 5.8. Thus, for every stable Q, the map from y to u gives a new stabilizing controller. The interesting fact is that these are all the stabilizing controllers for G. To prove this, we have to show that this controller can be described as in the configuration shown in Figure 5.2. This can be accomplished by reading off the variables on Figure 5.2 from Figure 5.7. First, observe that

$$
\begin{bmatrix} M & Y \\ N & X \end{bmatrix} \begin{bmatrix} -v \\ e \end{bmatrix} = \begin{bmatrix} u \\ y \end{bmatrix}.
$$

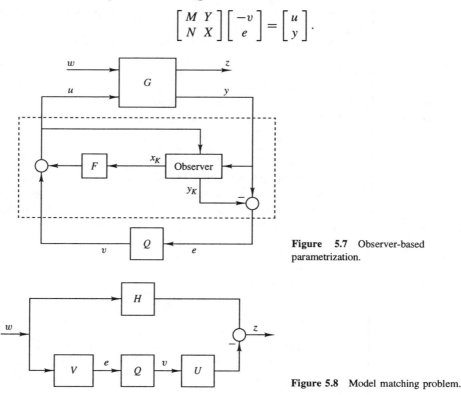

Figure 5.7 Observer-based parametrization.

Figure 5.8 Model matching problem.

A realization of this map can be obtained directly from the observer-based controller

$$
\left[
\begin{array}{c|cc}
A + B_2 F & -B_2 & -L \\
\hline
F & -I & 0 \\
C_2 + D_{22} F & -D_{22} & I
\end{array}
\right].
$$

In the same way,

$$
\begin{bmatrix} \tilde{X} & -\tilde{Y} \\ -\tilde{N} & \tilde{M} \end{bmatrix}
\begin{bmatrix} u \\ y \end{bmatrix}
= \begin{bmatrix} -v \\ e \end{bmatrix}.
$$

A realization of the above map can be obtained directly from the observer-based controller

$$
\left[
\begin{array}{c|cc}
A + LC_2 & (B_2 + LD_{22}) & -L \\
\hline
F & -I & 0 \\
-C_2 & -D_{22} & I
\end{array}
\right].
$$

The above realizations provide realizations for the various variables needed to parametrize all stabilizing controllers. In particular

$$
X = \left[
\begin{array}{c|c}
A + B_2 F & -L \\
\hline
C_2 + D_{22} F & I
\end{array}
\right], \qquad
Y = \left[
\begin{array}{c|c}
A + B_2 F & -L \\
\hline
F & 0
\end{array}
\right],
$$

$$
\tilde{X} = \left[
\begin{array}{c|c}
A + LC_2 & (B_2 + LD_{22}) \\
\hline
F & -I
\end{array}
\right], \quad
\tilde{Y} = \left[
\begin{array}{c|c}
A + LC_2 & L \\
\hline
F & 0
\end{array}
\right].
$$

$$\text{(5.21)}$$

The right coprime factors of the plant are given by

$$
M = \left[
\begin{array}{c|c}
A + B_2 F & -B_2 \\
\hline
F & -I
\end{array}
\right], \qquad
N = \left[
\begin{array}{c|c}
A + B_2 F & -B_2 \\
\hline
C_2 + D_{22} F & -D_{22}
\end{array}
\right],
\qquad \text{(5.22)}
$$

and the left coprime factors are given by

$$
\tilde{M} = \left[
\begin{array}{c|c}
A + LC_2 & -L \\
\hline
-C_2 & I
\end{array}
\right], \qquad
\tilde{N} = \left[
\begin{array}{c|c}
A + LC_2 & B_2 + LD_{22} \\
\hline
C_2 & D_{22}
\end{array}
\right].
\qquad \text{(5.23)}
$$

It can be verified by direct computation that the above quantities form a doubly-coprime factorization of G. Thus, the parametrization of stabilizing controllers obtained by adjusting the observer-based controller is complete.

The parametrization as an LFT.   It is evident that the observer-based parametrization can be expressed as a standard LFT. Let J be the system mapping y and v to u and e. It follows that J has the realization

$$
\left[
\begin{array}{c|cc}
A + B_2 F + LC_2 + LD_{22} F & -L & B_2 + LD_{22} \\
\hline
F & 0 & I \\
-(C_2 + D_{22} F) & I & -D_{22}
\end{array}
\right]
$$

and that all real rational stabilizing controllers are given by

$$
K(Q) = F_\ell(J, Q), \qquad Q \in \mathcal{RH}_\infty.
$$

This representation is not particular to the observer-based parametrization. It can be shown that for any doubly-coprime factorization, all controllers are given by

$$K(Q) = F_\ell(\bar{J}, Q), \qquad Q \in \mathcal{RH}_\infty, \tag{5.24}$$

where

$$\bar{J} = \begin{pmatrix} Y X^{-1} & \tilde{X}^{-1} \\ X^{-1} & -X^{-1}N \end{pmatrix}.$$

Computation.   To compute the parametrization of all stabilizing controllers, two quantities need to be furnished: the control gain F and the filter gain L. The poles of the closed-loop system with the nominal controller will be the union of the reciprocal of the eigenvalues of $A + B_2 F$ and $A + L C_2$. If the realization is reachable and observable, then these eigenvalues can be placed arbitrarily in the unit disc. The location of the poles is not important since the parametrization gives all stabilizing controllers for any choice of stabilizing F and L.

An important point to stress is that computing a doubly-coprime factorization and finding a stabilizing controller are somewhat equivalent (we have just demonstrated this for the observer-based controller). It is not very hard to show that starting from any stabilizing controller, a parametrization of all stabilizing controllers can be found.

If the doubly-coprime factorization is desired to be over polynomials, then the F and L should be chosen so that all the eigenvalues are at the origin. This is known as a deadbeat controller.

Stable plants.   The parametrization for stable plants is quite straightforward. A doubly-coprime factorization of the plant G_{22} is given by

$$\begin{pmatrix} I & 0 \\ -G_{22} & I \end{pmatrix} \begin{pmatrix} I & 0 \\ G_{22} & I \end{pmatrix} = I. \tag{5.25}$$

The parametrization is then given by

$$K = -Q(I - G_{22}Q)^{-1}, \tag{5.26}$$

which does not require any computation.

Example 5.2.1

Consider the following 1-input 2-output unstable plant:

$$\hat{G}_{22} = \begin{pmatrix} \dfrac{-\lambda(3.5\lambda - 1)}{\lambda^2 - 2.5\lambda + 1} \\ \dfrac{1.5\lambda^2}{\lambda^2 - 2.5\lambda + 1} \end{pmatrix}$$

with minimal state-space realization

$$G_{22} = \left[\begin{array}{cc|c} -1 & -3 & 1 \\ 1.5 & 3.5 & 0 \\ \hline 1 & 0 & 0 \\ 0 & 1 & 0 \end{array} \right].$$

We want to compute a polynomial doubly-coprime factorization of $\hat{G}_{22}$. This simple example can be solved by inspection: Choose the right and left coprime factors of $\hat{G}_{22}$ to be

$$\hat{M} = \lambda^2 - 2.5\lambda + 1, \qquad \hat{N} = \begin{pmatrix} -\lambda(3.5\lambda - 1) \\ 1.5\lambda^2 \end{pmatrix},$$

and

$$\hat{\tilde{M}} = \begin{pmatrix} \hat{M} & 0 \\ 0 & \hat{M} \end{pmatrix}, \qquad \hat{\tilde{N}} = \hat{N}.$$

Then, the rest of the factors can be obtained by solving Equation (5.12). A less obvious polynomial factorization can be obtained by a direct application of the state-space Equations (5.21)–(5.23), where the gain matrices F and L are chosen so that $A + B_2 F$ and $A + LC_2$ are nilpotent (note that this choice is not unique). In particular, let

$$F = -\frac{1}{6}(15 \quad 31) \qquad \text{and} \qquad L = \frac{1}{8}\begin{pmatrix} -13 & 6 \\ 12.5 & -7 \end{pmatrix},$$

then the right coprime factors of $\hat{G}_{22}$ are as before while the left coprime factors are given by

$$\hat{\tilde{M}} = \frac{1}{8}\begin{pmatrix} 6\lambda^2 - 13\lambda + 8 & 6\lambda \\ -\lambda(7\lambda - 12.5) & -7\lambda + 8 \end{pmatrix}, \qquad \hat{\tilde{N}} = \frac{1}{8}\begin{pmatrix} -\lambda(21\lambda - 8) \\ 24.5\lambda^2 \end{pmatrix}.$$

The corresponding right and left coprime factors of the observer-based controller are

$$\hat{X} = \frac{1}{8}\begin{pmatrix} \frac{679}{12}\lambda^2 + 13\lambda + 8 & -\lambda(\frac{217}{6})\lambda - 6) \\ -\lambda(24.25\lambda + 12.5) & 15.5\lambda^2 + 7\lambda + 8 \end{pmatrix}, \quad \hat{Y} = \frac{\lambda}{96}(-194\lambda + 385 \quad 124\lambda - 254),$$

and

$$\hat{\tilde{X}} = \frac{889}{96}\lambda^2 + \frac{5}{2}\lambda + 1, \qquad \hat{\tilde{Y}} = \frac{\lambda}{96}(-254\lambda + 385 - 254).$$

The results can be verified by direct substitution.

5.3 SEPARATION STRUCTURE OF STABILIZING CONTROLLERS

The purpose of this section is to show that the parametrization of closed-loop maps decomposes into two *natural* components:

State-feedback + Output Estimation.

This has become apparent from the previous development of the observer-based controller; however, the ideas are more general and there is a deeper algebraic structure. We explore this structure further. It is interesting to note that this separation structure carries through in certain optimization problems such as in the optimal $\mathcal{H}_2$ problem.

5.3.1 Special Problems

We would like to study the parametrization of all closed-loop maps of certain special problems, from which we derive the parametrization for the standard problem. We refer to the standard problem as the Output Feedback problem (OF). The associated plant is given by

$$G_{\mathrm{OF}} = \left[\begin{array}{c|cc} A & B_1 & B_2 \\ \hline C_1 & D_{11} & D_{12} \\ C_2 & D_{21} & 0 \end{array} \right].$$

In here we assumed that the plant G_{22} is strictly proper. For the generalization, see Exercise 5.12.

We define four special problems. Although we use the same notation for the state-space parameters, this will be a generic notation and will not necessarily refer to the same parameters of the OF problem.

FI: Full Information problem. The associated plant is given by

$$G_{\mathrm{FI}} = \left[\begin{array}{c|cc} A & B_1 & B_2 \\ \hline C_1 & D_{11} & D_{12} \\ I & 0 & 0 \\ 0 & I & 0 \end{array} \right].$$

In here, it is assumed that both the states and the disturbance are available for measurement.

FC: Full Control problem. The associated plant is given by

$$G_{\mathrm{FC}} = \left[\begin{array}{c|ccc} A & B_1 & I & 0 \\ \hline C_1 & D_{11} & 0 & I \\ C_2 & D_{21} & 0 & 0 \end{array} \right].$$

In here, it is assumed that there are two independent inputs directly affecting both the states and the regulated variables.

DF: Disturbance Feedforward problem. The associated plant is given by

$$G_{\mathrm{DF}} = \left[\begin{array}{c|cc} A & B_1 & B_2 \\ \hline C_1 & D_{11} & D_{12} \\ C_2 & I & 0 \end{array} \right].$$

with $(A - B_1 C_2)$ stable. In here, it is assumed that the disturbance enters the output directly. The assumption of stability is discussed later.

OE : Output Estimation problem. The associated plant is given by

$$G_{\mathrm{OE}} = \left[\begin{array}{c|cc} A & B_1 & B_2 \\ \hline C_1 & D_{11} & I \\ C_2 & D_{21} & 0 \end{array} \right].$$

with $(A - B_2 C_1)$ stable. In here, it is assumed that the input enters the regulated variable directly. The assumption of stability is discussed later.

The above four problems, although different in structure, have very strong relations. These relations are summarized below.

Duality.   One problem is the dual of another if the first is obtained through the transposition of the second. In this sense FI is the dual of FC and DF is the dual of OE and vise versa. For dual problems, we have the following proposition.

Proposition 5.3.1.   Suppose G_1 and G_2 are dual problems. If $\Phi_1 \in \mathcal{RH}_\infty$ is a feasible closed-loop map for G_1 then Φ_1^T is a feasible closed-loop map for G_2.   ■

As a consequence, if we parametrize all stabilizing controllers for FI problems, then we obtain all stabilizing controllers for FC. Similarly, with DF and OE.

Equivalence.   We say two problems are equivalent if every feasible closed-loop map of one is achievable by the other through some controller. We will show that FI problems are equivalent to DF problems and FC problems are equivalent to OE problems.

Proposition 5.3.2.   There exists a stabilizing controller K_{FI} for G_{FI} such that $F_\ell(G_{FI}, K_{FI}) = \Phi$ if and only if there exists a stabilizing controller K_{DF} for G_{DF} such that $F_\ell(G_{DF}, K_{DF}) = \Phi$.   ■

Proof. One direction is quite simple. If K_{DF} stabilizes G_{DF} such that $F_\ell(G_{DF}, K_{DF}) = \Phi$, we can define

$$K_{FI} = K_{DF}(C_2 \quad I).$$

Then K_{FI} is stabilizing, and $F_\ell(G_{FI}, K_{FI}) = F_\ell(G_{DF}, K_{DF}) = \Phi$.

On the other hand, let K_{FI} stabilize G_{FI} with $F_\ell(G_{FI}, K_{FI}) = \Phi$. We need to construct a stabilizing controller K_{DF} for G_{DF} such that $F_\ell(G_{DF}, K_{DF}) = \Phi$. Consider the system V shown below.

V is a mapping between u and y to x and w and has the realization

$$V = \left[\begin{array}{c|cc} A - B_1 C_2 & B_1 & B_2 \\ \hline I & 0 & 0 \\ -C_2 & I & 0 \end{array} \right].$$

From the assumptions on DF, V is stable and has no zeros (assume of course that B_2 has full column rank). It follows that G_{FI} is obtained from the composition of V and G_{DF}, as depicted in Figure 5.9. Define a controller K_{DF} by the relation

$$K_{DF} y = K_{FI} V \begin{pmatrix} y \\ u \end{pmatrix}.$$

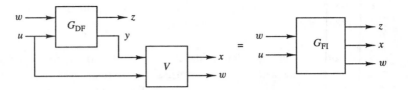

Figure 5.9   Composition of V and G_{DF} gives G_{FI}.

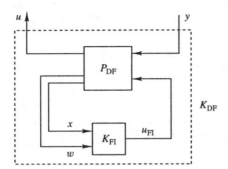

Figure 5.10 K_{DF} As a LFT.

Then K_{DF} stabilizes G_{DF} and results in a closed-loop map

$$F_\ell(G_{DF}, K_{DF}) = F_\ell(G_{FI}, K_{FI}) = \Phi,$$

which concludes the proof.   ∎

The controller K_{DF} in the previous theorem can be expressed in LFT form, as shown in Figure 5.10. This is stated in the next corollary, and follows by direct computation from the previous theorem.

Corollary 5.3.1.   The controller K_{DF} is given by

$$K_{DF} = F_\ell(P_{DF}, K_{FI}),$$

where

$$P_{DF} = \left[\begin{array}{c|cc} A - B_1 C_2 & B_1 & B_2 \\ \hline 0 & 0 & I \\ I & 0 & 0 \\ -C_2 & I & 0 \end{array} \right].$$
                                                                              ∎

Summary.   From the previous discussion, it follows that to parametrize all achievable closed-loop maps for these special problems, we only need to find the parametrization for FI problems. From that, we get the parametrization for FC problems by duality, and for DF and OE problems by equivalence. Figure 5.11 shows these relationships.

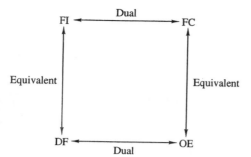

Figure 5.11 Duality and equivalence.

5.3.2 Parametrization of All Closed-Loop Maps for FI Problems

This parametrization can be summarized in the following theorem.

Theorem 5.3.1.   Let F be any matrix such that $A + B_2 F$ is stable. The class of controllers

$$K_{\mathrm{FI}}(Q) = (\,F \quad Q\,) \quad \text{for all } Q \in \mathcal{RH}_\infty$$

parametrizes all achievable closed-loop maps for G_{FI}.

Proof. The easiest way of showing this parametrization is by following Exercise 5.8. First, we stabilize the system by the constant matrix F, and then we parametrize all controllers stabilizing the closed-loop system. Let $u = Fx + v$. The system mapping w and v to z, x and w is given by

$$\tilde{G} = \left[\begin{array}{c|cc} A + B_2 F & B_1 & B_2 \\ \hline C_1 + D_{12} F & D_{11} & D_{12} \\ I & 0 & 0 \\ 0 & I & 0 \end{array}\right].$$

Thus, a controller $\tilde{K}$ stabilizes the plant $\tilde{G}$ if and only if the controller $\tilde{K} + (\,F \quad 0\,)$ stabilizes G. To parametrize all controllers $\tilde{K}$, we observe that the plant

$$\tilde{G}_{22} = \left[\begin{array}{c|c} A + B_2 F & B_2 \\ \hline I & 0 \\ 0 & 0 \end{array}\right]$$

is stable, and hence all stabilizing controllers are given by

$$\tilde{K} = -Q(I - \tilde{G}_{22}Q)^{-1}, \quad Q = (\,Q_1 \quad Q_2\,) \in \mathcal{RH}_\infty.$$

All closed-loop maps T_{zw} are given by

$$\begin{aligned} T_{zw} &= \tilde{G}_{11} + \tilde{G}_{12}\tilde{K}(I - \tilde{G}_{22}\tilde{K})^{-1}\tilde{G}_{21} \\ &= \tilde{G}_{11} - \tilde{G}_{12}(\,Q_1 \quad Q_2\,)\tilde{G}_{21}, \end{aligned}$$

where

$$\tilde{G}_{21} = \left[\begin{array}{c|c} A + B_2 F & B_1 \\ \hline I & 0 \\ 0 & I \end{array} \right].$$

It follows that $(Q_1 \quad Q_2) \tilde{G}_{21}$ can be any arbitrary stable system. Thus, we can pick $Q_1 = 0$, and Q_2 to be any arbitrary element in $\mathcal{RH}_\infty$, from which we get

$$K_{\mathrm{FI}} = (F \quad 0) - Q(I - \tilde{G}_{22}Q)^{-1} \text{ with } Q_1 = 0$$

$$= (F \quad Q_2) \quad Q_2 \in \mathcal{RH}_\infty,$$

which concludes the proof.   ∎

Notice that we have parametrized all closed-loop maps, but not all stabilizing controllers. It is possible that more than one controller can give the same closed-loop map. However, we are not concerned about this.

5.3.3 Parametrization of All Closed-Loop Maps for OE Problems

We have shown that OE problems are dual to DF problems, which are equivalent to FI problems. This enables us to parametrize all closed-loop maps for OE problems. This is given in the following theorem.

Theorem 5.3.2.   Let L be any matrix such that $A + LC_2$ is stable. The class of controllers given by

$$K_{\mathrm{OE}} = F_\ell \left(P_{\mathrm{OE}}, \begin{pmatrix} L \\ Q \end{pmatrix} \right) = F_\ell(J_{\mathrm{OE}}, Q) \quad \text{for all } Q \in \mathcal{RH}_\infty,$$

where

$$P_{\mathrm{OE}} = \left[\begin{array}{c|ccc} A - B_2 C_1 & 0 & I & -B_2 \\ \hline C_1 & 0 & 0 & I \\ C_2 & I & 0 & 0 \end{array} \right], \quad J_{\mathrm{OE}} = \left[\begin{array}{c|cc} A - B_2 C_1 + LC_2 & L & -B_2 \\ \hline C_1 & 0 & I \\ C_2 & I & 0 \end{array} \right],$$

parametrizes all achievable closed-loop maps for G_{OE}.

Proof. This result follows by using the duality between OE and DF, and then using the equivalence between DF and FI. The second formula for the controller follows by absorbing L into the system (see Exercise 5.11).   ∎

5.3.4 Output Feedback and Separation Structure

Let G_{OF} be given as before. Consider the change of variables

$$u = Fx + v.$$

Define G_{temp} to be the map from w and u to v and y. It has the realization

$$G_{\text{temp}} = \left[\begin{array}{c|cc} A & B_1 & B_2 \\ \hline -F & 0 & I \\ C_2 & D_{21} & 0 \end{array}\right].$$

Since $A + B_2 F$ is stable, G_{temp} is an OE problem . It can be seen that K stabilizes G_{OF} if and only if K stabilizes G_{temp} since both have the same G_{22} and the same A matrix. As shown in Figure 5.12, all closed-loop maps are given by

$$T_{zw} = G_c + U T_{vw}, \tag{5.27}$$

where

$$G_c = \left[\begin{array}{c|c} A + B_2 F & B_1 \\ \hline C_1 + D_{12} F & D_{11} \end{array}\right], \qquad U = \left[\begin{array}{c|c} A + B_2 F & B_2 \\ \hline C_1 + D_{12} F & D_{12} \end{array}\right].$$

To find all closed-loop maps T_{vw} we apply the parametrization from Theorem 5.3.2. The set of controllers is given by

$$K(Q) = F_\ell(J_{\text{OE}}, Q),$$

with

$$J_{\text{OE}} = \left[\begin{array}{c|cc} A + B_2 F + L C_2 & L & -B_2 \\ \hline -F & 0 & I \\ C_2 & I & 0 \end{array}\right].$$

All closed-loop maps are given by

$$\begin{aligned} T_{zw} &= F_\ell(G_{\text{OF}}, K(Q)) \\ &= G_c + U F_\ell\big(G_{\text{temp}}, F_\ell(J_{\text{OE}}, Q)\big). \end{aligned} \tag{5.28}$$

5.3.5 Comments

The parametrization in Equation 5.28 is precisely the one we arrived at through the observer-based parametrization. In fact, the only result used in this section from the previous development of coprime factorization is the parametrization of all closed-loop maps for the FI problem. However, since this problem has a lot of structure, it is possible to arrive at this parametrization without using any ideas from coprime factorization. If this is done, the development in this section gives a different approach to parametrization

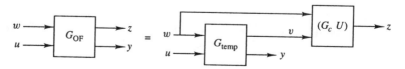

Figure 5.12   Separation structure.

that is more in tune with the theory of linear fractional transformations as opposed to coprime factorization.

The decomposition of all closed-loop maps as

$$T_{zw} = G_c + U T_{vw}$$

shows the algebraic separation structure of all stabilizing controllers. The map G_c is purely a function of the state feedback matrix F, while T_{vw} is an output estimation of the state-feedback.

A by-product of the development in this section is that there is an equivalence between FI problems and OE problems. In the case of designing controllers to minimize a certain objective function that is invariant under transposition, it is sufficient to know how to solve such problems for the FI problem. The OE problem as well as the rest of the special problems can be solved using either duality or equivalence. Of course, solving for the optimal FI problem and the optimal OE to construct a controller for the general OF problem does not necessarily mean that the controller is optimal. The question whether the algebraic separation extends to one of optimal solutions is a function of the specific optimization problem.

5.4 SUMMARY

The parametrization of stabilizing controllers has two attractive features. The first feature is that the free parameter is an arbitrary element in a linear space, and the second feature is that the closed-loop map is affine in this free parameter. Utilizing these two features, various optimization problems involving norms on the closed-loop system can be solved exactly. This is done in later chapters.

EXERCISES

5.1. Let

$$\hat{M} = \begin{pmatrix} \dfrac{\lambda - 1}{\lambda + 3} & 1 \end{pmatrix}, \qquad \hat{N} = \begin{pmatrix} \dfrac{2\lambda + 2}{(\lambda + 3)^2} & \dfrac{\lambda - 2}{\lambda + 3} \end{pmatrix}.$$

Determine if $\hat{M}$, $\hat{N}$ are right coprime, or left coprime.

5.2. Can the 2-input 2-output system given by

$$\hat{G} = \begin{bmatrix} \dfrac{1}{\lambda - .5} & 0 \\ 0 & \dfrac{1}{(\lambda - 1)(\lambda - 2)} \end{bmatrix}$$

be stabilized by a controller from y to u?

5.3. Let

$$\hat{G}_{22} = \dfrac{\lambda}{\lambda - .5}.$$

(a) Give all stabilizing controllers of any stabilizable G containing G_{22}.

(b) Write down a parametrization of all possible sensitivity functions,

$$S = (1 - G_{22}K)^{-1}.$$

(c) Show that

$$\inf_{K-\text{stabilizing}} ||\hat{S}||_\infty \geq 1.$$

5.4. (Regulators) Let $G_{22} = \frac{\lambda}{\lambda - .5}$. Find all compensators K such that:

(a) The closed-loop system is stable.

(b) If $w(t) = 1$ for all $t \geq 0$, then $\lim_{t \to \infty} z(t) = 0$. Assume that w is an output disturbance and $z = y$. This is a standard disturbance rejection problem.

5.5. (Internal model control) The diagram in Figure 5.13 depicts a controller configuration implemented on a process P assuming a model P_0 is available. These types of controllers are quite common in process control.

(a) Show that if the closed-loop system is stable, then necessarily $\hat{P}_0$ and $\hat{P}$ have no common unstable poles.

(b) If $P_0 = P$ and is stable, show that the system is stable for any stable Q. Show that this configuration gives a parametrization of all stabilizing controllers of P_0.

5.6. Given the SISO closed-loop system in Figure 5.14.

(a) Find the matrix $H(G_{22}, K)$ that maps w_1 and w_2 to u and y.

(b) Let $K = \frac{-Q}{1 - G_{22}Q}$, $Q \in \mathcal{RH}_\infty$. Find conditions on Q such that K stabilizes G_{22}. Is this a complete parametrization? Note that G_{22} may not be stable.

5.7. Given a plant $G_{22} = NM^{-1}$, where M, N are right coprime, and let $K_0 = UV^{-1}$ be a stabilizing controller, where U, V are right coprime. Show that *all* stabilizing controllers are given by

$$K = (U - MQ)(V - NQ)^{-1}, \quad Q \text{ stable}.$$

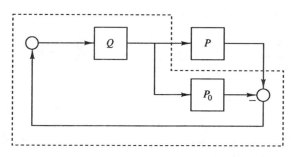

Figure 5.13 Internal model control.

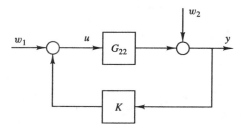

Figure 5.14 Standard diagram for SISO stability.

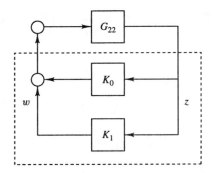

Figure 5.15 A new parametrization.

5.8. Let the nominal plant G_{22} be stabilized by a known controller K_0. A new procedure to parametrize all stabilizing controllers is described as follows (see Figure 5.15): If the first controller is implemented, then the map between the input w and the output z is stable. Considering this map as a new system, all stabilizing controllers of T_{zw} are given by

$$K_1(Q) = -Q(I - T_{zw}Q)^{-1} \qquad \text{for all } Q \in \ell_1.$$

Thus all stabilizing controllers are given by

$$K = K_0 + K_1(Q).$$

Is this argument correct? Prove or disprove. (Hint: Consider the cases when K_0 is stable and unstable.)

5.9. (Two-parameter compensators) Consider the two-parameter compensator given in Figure 5.16. Let P have the doubly-coprime factorization:

$$\begin{bmatrix} \tilde{X}_p & -\tilde{Y}_p \\ -\tilde{N}_p & \tilde{M}_p \end{bmatrix}\begin{bmatrix} M_p & Y_p \\ N_p & X_p \end{bmatrix} = I.$$

 (a) Find all stabilizing controllers $(C_1 \ C_2)$. Give exact formulae.
 (b) What is the relation between C_1 and C_2?
 (c) Calculate Φ_{vn}, Φ_{vr}, $\Phi_{(r-v)r}$.
 (d) Is it possible to minimize $||\hat{\Phi}_{vn}||_\infty$ and $||\hat{\Phi}_{(r-v)r}||_\infty$ independently? Why?
 (e) Let

$$z = \begin{pmatrix} r - v \\ v \end{pmatrix}, \quad w = \begin{pmatrix} r \\ n \end{pmatrix}.$$

 Calculate H_{zw}. Comment on the answer of part d.
 (f) Can C_1 be unstable? What went wrong? Explain. Can you redraw the block diagram to get rid of this problem? (Hint: Let $C = (C_1 \ C_2) = \tilde{V}^{-1}(\tilde{U}_1 \ \tilde{U}_2)$, a coprime factorization, where $\tilde{V}, \tilde{U}_1, \tilde{U}_2$ are stable. Put $\tilde{V}$ inside the loop only once!)

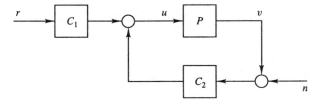

Figure 5.16 Two-parameter controller.

5.10. Show that all stabilizing controllers can be represented as an LFT of the form in Equation (5.24).

5.11. Verify the result in Theorem 5.3.2.

5.12. Suppose D_{22} in the OF problem is not equal to zero. Define the mapping on feasible controllers

$$\tilde{K} = K(I - D_{22}K)^{-1}.$$

Since the closed-loop system is well posed for any admissible controller, the above mapping is well defined. Show that

$$F_\ell(G_{OF}, K) = F_\ell(\tilde{G}_{OF}, \tilde{K}),$$

where

$$\tilde{G}_{OF} = \left[\begin{array}{c|cc} A & B_1 & B_2 \\ \hline C_1 & D_{11} & D_{12} \\ C_2 & D_{21} & 0 \end{array} \right].$$

Show that all stabilizing controllers are given by

$$F_\ell(J, Q), \qquad J = \left[\begin{array}{c|cc} A + B_2F + LC_2 + LD_{22}F & -L & B_2 + LD_{22} \\ \hline F & 0 & I \\ -(C_2 + D_{22}F) & I & -D_{22} \end{array} \right].$$

5.13. In this exercise, we show how to obtain a complete parametrization of periodic controllers (of period N) that stabilize an N-periodic plant, G_{22}. Let $\tilde{G}_{22}$ denote the lifted system (see Exercise 4.8).

(a) Does every stabilizing controller of $\tilde{G}_{22}$ define an admissible controller for G_{22} by applying the inverse of the lifting operation? Explain.

(b) Show that there exists a doubly-coprime factorization for $\tilde{G}_{22}$ in which all the elements satisfy the causality condition. (Hint: Think of the observer-based controller.)

(c) Show that the parametrization in Equation 5.13 with the above factorization provides a class of N-periodic stabilizing controllers if every Q satisfies the causality condition.

(d) Show that the above parametrization is complete. (Hint: Consider the construction of Q given a controller K.)

(e) Use the above to argue that if G_{22} is LTI, then all N-periodic stabilizing controllers are obtained by varying Q over all N-periodic stable systems.

NOTES AND REFERENCES

The parametrization of all stabilizing controllers presented explicitly for MIMO systems was derived in [YJB76a, YJB76b] and [Kuc79]. For SISO systems, the ideas appeared earlier; see [RF58]. A complete theory for stable coprime factorization of LTI systems and the application to parametrization of all stabilizing controllers is presented in [Vid85]. The proofs presented here follow in great detail the ones presented in [Fra87]. The observer-based parametrization is derived in [Doy84]. State-space formulae for the parametrization were derived in [KS82, NJB84, Net86]. The algebraic separation structure of stabilizing controllers is derived in [DGKF89] and more explicitly in [LZD91].

Internal model controllers (as in Exercise 5.5) commonly appear in process control and are discussed in detail in [MZ89]. Much of the material in this chapter has appeared in recent textbooks; see [BB91, DFT92, Fra87, Mac89, MZ89, Vid85].

Generalizations to nonlinear controllers and plants have been reported in [AD84, DL82, Ver88]. The parametrization shown in Exercise 5.13 is a special case of a more general result on multirate systems reported in [Mey90b]. See also [DVV92, VDVa].

CHAPTER 6 ———————————————

Achievable Closed-Loop Maps

In this chapter the parametrization of all stabilizing controllers is exploited to give a general characterization of all the achievable closed-loop maps. From the discussion in the previous chapter, all closed-loop maps are given by

$$\mathcal{H} = \{H - UQV \mid Q \text{ stable}\} \tag{6.1}$$

where H, U and V are stable maps of dimensions $n_z \times n_w$, $n_z \times n_u$ and $n_y \times n_w$, respectively. The set $\mathcal{H}$ forms an affine space inside the space of stable maps (typically ℓ_1). In here, a *dual* characterization in terms of interpolation conditions is given. In this representation, a map belongs to the set $\mathcal{H}$ if and only if it satisfies a set of (possibly infinite) equations. These equations in turn represent the limitations in achieving performance objectives. Some of these limitations are fundamental to the process, such as limitations due to unstable poles or non-minimum phase zeros. Others are functions of the performance objectives and are not necessarily inherent to the process. These different conditions are discussed in the sequel.

6.1 MOTIVATION

The parameter Q in Equation (6.1) is typically chosen so that the closed-loop map is equal to some desired map, i.e.,

$$H - UQ_dV =: \Phi_d. \tag{6.2}$$

121

A possible choice of Φ_d is the zero transfer function. In any such choice, Q has to invert both U and V to satisfy the equality. This may not be possible in general for two reasons:

1. U and/or V may not have stable inverses owing to zeros in the closed unit disc.
2. Q may not have enough degrees of freedom since, in general, U has at least as many rows as columns, and V has as many columns as rows.

To illustrate these ideas, consider the following example.

Example 6.1.1

Consider the problem in Figure 6.1, where P is stable and

$$z = \begin{pmatrix} y \\ u \end{pmatrix}, \qquad w = w_1. \tag{6.3}$$

The set of achievable closed-loop maps is given by

$$\mathcal{H} = \left\{ \begin{pmatrix} I \\ 0 \end{pmatrix} - \begin{pmatrix} P \\ I \end{pmatrix} Q \;\middle|\; Q \text{ stable} \right\}. \tag{6.4}$$

Clearly, not every choice of Φ_d can be achieved by an appropriate Q, even if the plant is minimum phase. This problem has an interesting trade-off: Suppose that it is desired to find the closed-loop map with the smallest possible norm, i.e.,

$$\min_{\Phi \in \mathcal{H}} \|\Phi\|_1.$$

The free parameter Q tries to invert the plant P; however, the gain of Q has to remain small. This optimization problem captures the trade-offs between sensitivity minimization and control effort.

It is easy to verify that a map Φ is an element of $\mathcal{H}$ only if

$$(I - P)\Phi = I, \quad \text{where} \quad \Phi = \begin{pmatrix} \Phi_1 \\ \Phi_2 \end{pmatrix}.$$

This equation carries quite a bit of information: It says that Φ_1 and Φ_2 are not decoupled. In fact, the choice of one may dictate exactly the choice of the other. Also, for any chosen Φ, $\hat{\Phi}_1$ evaluated at the non-minimum phase zeros of $\hat{P}$ is equal to one. Such constraints are fundamental to the process.

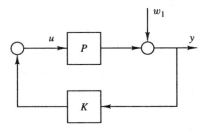

Figure 6.1   Mixed objectives.

On the other hand, consider the weighted problem where $z_1 = Wy$, for some stable weight with a stable inverse. The set of achievable closed-loop maps is given by

$$\mathcal{H}_W = \left\{ \begin{pmatrix} W \\ 0 \end{pmatrix} - \begin{pmatrix} WP \\ I \end{pmatrix} Q \ \middle| \ Q \text{ stable} \right\}. \tag{6.5}$$

A map Φ is in $\mathcal{H}_W$ only if

$$(I - WP)\Phi = W.$$

This new equation changes the relation between Φ_1 and Φ_2. However, $\hat{\Phi}_1$ still interpolates the zeros of $\hat{P}$ at one. The zeros seem to limit the possibilities of closed-loop maps in a fundamental way, whereas the relations between Φ_1 and Φ_2 depend on the objective function.

In the next few sections we will characterize the feasibility of the closed-loop map in terms of interpolation conditions derived from the zeros of the matrices $\hat{U}(\lambda)$ and $\hat{V}(\lambda)$, which lie in the unit disc. Later on, methods to compute these conditions are derived based on matrix computations.

6.2 INTERPOLATION CONDITIONS

Consider the composition $R := UQV$. The set of achievable closed-loop maps can be written as

$$\mathcal{H} = \{ H - R \mid R = UQV \text{ for some stable } Q \}.$$

We would like to find necessary and sufficient conditions on R such that it can be written as UQV for some stable Q. These conditions are generically known as *interpolation conditions*.

The notion of interpolation conditions can be viewed in at least two ways: as algebraic conditions on the matrix function $\hat{R}(\lambda)$, or as conditions on the left and right null-spaces of the operator R. In the sequel we are going to exploit the algebraic notion; however, later on for the purpose of computations, we will view the interpolation conditions as a null-space matching problem.

The algebraic notion of interpolation is better understood through a simple SISO example. Consider a scalar transfer function $\hat{r}(\lambda) = \hat{u}(\lambda)\hat{q}(\lambda)$ where $\hat{u}(\lambda)$ is a scalar, real rational and stable function ($\hat{u}(\lambda) \in \mathcal{RH}_\infty$), and $\hat{q}(\lambda)$ is any stable transfer function ($q \in \ell_1$). Assume that $\hat{u}(\lambda_0) = 0$ for $\lambda_0 \in \mathcal{D}$ (i.e., in the open unit disc). We would like to find conditions a stable function $\hat{r}(\lambda)$ should satisfy so that $\hat{q}(\lambda) = \hat{r}(\lambda)/\hat{u}(\lambda)$ is stable. Clearly, the condition is given by $\hat{r}(\lambda_0) = 0$. This is a consequence of a result from complex variable theory, tied to Wiener's theorem. The notation $(\cdot)^{(k)}(\lambda_0)$ denotes the kth order derivative with respect to λ, evaluated at λ_0.

Theorem 6.2.1.   Given a function $\hat{f}(\cdot)$ of the complex variable λ analytic in $\mathcal{D}$, then $(\hat{f})^{(k)}(\lambda_0) = 0$ for $k = 0, 1, \ldots, (\sigma - 1)$ and $\lambda_0 \in \mathcal{D}$ if and only if $\hat{f}(\lambda) = (\lambda - \lambda_0)^\sigma \hat{g}(\lambda)$ where $\hat{g}(\cdot)$ is analytic in $\mathcal{D}$.   ∎

Notice that this theorem is valid only if λ_0 does not lie on the boundary of the disc. Consider the following example.

Example 6.2.1

Let the function $\hat{f}$ be given by

$$\hat{f}(\lambda) = (1 - \lambda)^{\frac{1}{2}}.$$

This function is single-valued with the branch-cut taken as the interval $[1, \infty)$. It can be verified that the Taylor series expansion of $\hat{f}$ is absolutely summable. Obviously, $\hat{f}(\lambda)$ has a zero at $\lambda_0 = 1$; however, there is no function $\hat{g} \in \ell_1$ such that $\hat{f}(\lambda) = (\lambda - 1)\hat{g}(\lambda)$.

If the function $\hat{f}$ is real rational, i.e., $\hat{f} \in \mathcal{RH}_\infty$, then the conclusion of Theorem 6.2.1 is valid for any zero in the closed unit disc.

Corollary 6.2.1. Given a function $\hat{f} \in \mathcal{RH}_\infty$, then $(\hat{f})^{(k)}(\lambda_0) = 0$ for $k = 0, 1, \ldots, (\sigma - 1)$ and $\lambda_0 \in \overline{\mathcal{D}}$ if and only if $\hat{f}(\lambda) = (\lambda - \lambda_0)^\sigma \hat{g}(\lambda)$ where $\hat{g} \in \mathcal{RH}_\infty$. ∎

In the sequel it will be assumed, without loss of generality, that $\hat{U}(\lambda)$ has full column normal rank (i.e., rank of n_u for almost all λ) and $\hat{V}(\lambda)$ has full row normal rank (i.e., rank of n_y for almost all λ). Violation of these assumptions implies that there are redundancies in the controls and/or the measurements that can be easily removed.

Consider Smith-McMillan decompositions of the rational matrices $\hat{U}$ and $\hat{V}$ (Note: To simplify notation, the complex variable argument will be omitted in most expressions);

$$\hat{U} = \hat{L}_U \hat{M}_U \hat{R}_U, \tag{6.6}$$

$$\hat{V} = \hat{L}_V \hat{M}_V \hat{R}_V, \tag{6.7}$$

where, according to the rank assumption and Theorem 4.1.1,

$$\hat{M}_U = \begin{pmatrix} \frac{\hat{\epsilon}_1}{\hat{\psi}_1} & & \\ & \ddots & \\ & & \frac{\hat{\epsilon}_{n_u}}{\hat{\psi}_{n_u}} \\ 0 & \cdots & 0 \\ \vdots & \ddots & \vdots \\ 0 & \cdots & 0 \end{pmatrix}, \tag{6.8}$$

$$\hat{M}_V = \begin{pmatrix} \frac{\hat{\epsilon}'_1}{\hat{\psi}'_1} & & & 0 & \cdots & 0 \\ & \ddots & & \vdots & \ddots & \vdots \\ & & \frac{\hat{\epsilon}'_{n_y}}{\hat{\psi}'_{n_y}} & 0 & \cdots & 0 \end{pmatrix}. \tag{6.9}$$

Let Λ_{UV} denote the set of zeros of $\hat{U}$ and $\hat{V}$ in $\overline{\mathcal{D}}$. Following the notation of Theorem 4.1.1, let $\{\sigma_{U_i}(\lambda_0)\}_{i=1}^{n_u}$ denote the sequence of structural indices corresponding to $\hat{U}$ for each $\lambda_0 \in \Lambda_{UV}$. Similarly, define $\{\sigma_{V_j}(\lambda_0)\}_{j=1}^{n_y}$ for $\hat{V}$.

Next, consider the unimodular matrices in Equation (6.6). Since their inverses are polynomials, one can define the following polynomial row and column vectors:

$$
\begin{aligned}
\hat{\alpha}_i(\lambda) &= (\hat{L}_U^{-1})_i(\lambda) & i = 1, 2, \ldots, n_z, \\
\hat{\beta}_j(\lambda) &= (\hat{R}_V^{-1})^j(\lambda) & j = 1, 2 \ldots, n_w,
\end{aligned}
\tag{6.10}
$$

where $(M)_i$ denotes the ith row of the matrix M and $(M)^j$ denotes the jth column of matrix M. Now we are ready to present the main interpolation theorem.

Theorem 6.2.2.   Assume that $\Lambda_{UV} \subset \mathcal{D}$. Given $R \in \ell_1^{n_z \times n_w}$, there exists a $Q \in \ell_1^{n_u \times n_y}$ such that $R = UQV$ if and only if for all $\lambda_0 \in \Lambda_{UV} \subset \mathcal{D}$ the following conditions are satisfied:

$$
i) \ (\hat{\alpha}_i \hat{R} \hat{\beta}_j)^{(k)}(\lambda_0) = 0 \ \text{for} \
\begin{cases}
i = 1, \ldots, n_u \\
j = 1, \ldots, n_y \\
k = 0, \ldots, \sigma_{U_i}(\lambda_0) + \sigma_{V_j}(\lambda_0) - 1
\end{cases}.
$$

$$
ii) \
\begin{cases}
(\hat{\alpha}_i \hat{R})(\lambda) \equiv 0 \ for \ i = n_u + 1, \ldots, n_z \\
(\hat{R} \hat{\beta}_j)(\lambda) \equiv 0 \ for \ j = n_y + 1, \ldots, n_w
\end{cases}.
$$

Proof. Consider the following factorization of $\hat{M}_U$ and $\hat{M}_V$ (where $\mathbf{0}$ denotes a block of zeros of appropriate dimensions)

$$
\hat{M}_U =: \begin{pmatrix} \hat{\hat{\mathcal{E}}}_U \hat{\hat{\Psi}}_U^{-1} \\ \mathbf{0} \end{pmatrix}, \qquad \hat{M}_V =: \begin{pmatrix} \hat{\hat{\Psi}}_V^{-1} \hat{\hat{\mathcal{E}}}_V & \mathbf{0} \end{pmatrix},
$$

where $\hat{\hat{\mathcal{E}}}_U$ and $\hat{\hat{\mathcal{E}}}_V$ retain the zeros in Λ_{UV} while $\hat{\hat{\Psi}}_U$ and $\hat{\hat{\Psi}}_V$ capture the stable (i.e., minimum phase) zeros of $\hat{U}$ and $\hat{V}$ along with their (stable) poles. Thus, both $\bar{\Psi}_U$ and $\bar{\Psi}_V$ are units in ℓ_1. Then

$$
\hat{R} = \hat{L}_U \begin{pmatrix} \hat{\hat{\mathcal{E}}}_U \hat{\hat{Q}} \hat{\hat{\mathcal{E}}}_V & \mathbf{0} \\ \mathbf{0} & \mathbf{0} \end{pmatrix} \hat{R}_V,
$$

where $\hat{\hat{Q}} := \hat{\hat{\Psi}}_U^{-1} \hat{R}_U \hat{Q} \hat{L}_V \hat{\hat{\Psi}}_V^{-1}$. Clearly, $\bar{Q} \in \ell_1^{n_u \times n_y}$ if and only if $Q \in \ell_1^{n_u \times n_y}$. Next, define the following partitions of $\hat{L}_U$ and $\hat{R}_V$:

$$
\hat{L}_U = \begin{pmatrix} \hat{L}_{U,1} & \hat{L}_{U,2} \end{pmatrix}, \qquad \hat{R}_V = \begin{pmatrix} \hat{R}_{V,1} \\ \hat{R}_{V,2} \end{pmatrix},
\tag{6.11}
$$

where $\hat{L}_{U,1}$ has n_u columns and $\hat{R}_{V,1}$ has n_y rows. Then, given $R \in \ell_1^{n_z \times n_w}$,

$$
\exists Q \in \ell_1^{n_u \times n_y} \ \text{such that} \ R = UQV
$$
$$
\Updownarrow
$$
$$
\exists \bar{Q} \in \ell_1^{n_u \times n_y} \ \text{such that} \ R = L_{U,1} \bar{\mathcal{E}}_U \bar{Q} \bar{\mathcal{E}}_V R_{V,1}.
$$

Necessity of condition i) follows immediately. Take any $i \in \{1, \ldots, n_u\}$ and $j \in \{1, \ldots, n_y\}$, then

$$(\hat{\alpha}_i \hat{R} \hat{\beta}_j)(\lambda) = \prod_{\lambda_0 \in \Lambda_{UV}} (\lambda - \lambda_0)^{\sigma_{U_i}(\lambda_0)} \, \hat{\bar{q}}_{ij}(\lambda) \prod_{\lambda_0 \in \Lambda_{UV}} (\lambda - \lambda_0)^{\sigma_{V_j}(\lambda_0)}$$

which implies condition i) by Theorem 6.2.1 and the fact that $\bar{q}_{ij}$ is in ℓ_1.

Necessity of condition ii) results from the following: Take any $i \in \{n_u + 1, \ldots, n_z\}$ and $j \in \{n_y + 1, \ldots, n_w\}$, then $(\hat{\alpha}_i \hat{R})(\lambda) \equiv 0$ and $(\hat{R} \hat{\beta}_j)(\lambda) \equiv 0$ since $(\hat{\alpha}_i \hat{L}_{U,1})(\lambda) \equiv 0$ and $(\hat{R}_{V,1} \hat{\beta}_j)(\lambda) \equiv 0$.

To show that conditions i) and ii) are sufficient, we proceed by backwards construction; by Theorem 6.2.1,

$$i) \Longrightarrow \begin{pmatrix} \hat{\alpha}_1 \\ \vdots \\ \hat{\alpha}_{n_u} \end{pmatrix} \hat{R} \left(\hat{\beta}_1 \cdots \hat{\beta}_{n_y} \right) = \hat{\bar{\mathcal{E}}}_U \hat{W} \hat{\bar{\mathcal{E}}}_V$$

for some $W \in \ell_1^{n_u \times n_y}$, since $R \in \ell_1^{n_z \times n_w}$. Moreover,

$$ii) \Longrightarrow \begin{pmatrix} \hat{\alpha}_{n_u+1} \\ \vdots \\ \hat{\alpha}_{n_z} \end{pmatrix} \hat{R} \equiv \mathbf{0} \quad \text{and} \quad \hat{R} \left(\hat{\beta}_{n_y+1} \cdots \hat{\beta}_{n_w} \right) \equiv \mathbf{0}.$$

Therefore, combining these equations into one, we get

$$\hat{L}_U^{-1} \hat{R} \hat{R}_V^{-1} = \begin{pmatrix} \hat{\bar{\mathcal{E}}}_U \hat{W} \hat{\bar{\mathcal{E}}}_V & \mathbf{0} \\ \mathbf{0} & \mathbf{0} \end{pmatrix},$$

which implies that $W = \bar{Q}$ is the solution.   ∎

In the case where $\hat{U}$ or $\hat{V}$ has zeros on the unit circle, the above result does not follow for arbitrary $R \in \ell_1$. However, the result does hold if $R \in \mathcal{RH}_\infty$.

Corollary 6.2.2.   The result in Theorem 6.2.2 holds for $\Lambda_{UV} \subset \overline{\mathcal{D}}$ if $\hat{R} \in \mathcal{RH}_\infty$.   ∎

In words, Theorem 6.2.2 provides a set of algebraic conditions that are necessary and sufficient for R to be feasible (i.e., equivalent to UQV for some stable Q). The conditions in i) ensure that the unstable zero structure of the composition UQV is preserved. This is captured by the polynomial vectors $\hat{\alpha}_i$, $i \leq n_u$ and $\hat{\beta}_j$, $j \leq n_y$, which carry the directional information of the zeros. The conditions in ii) impose the correct rank conditions. In fact, it is possible to view the collection of $\hat{\alpha}_i$'s and $\hat{\beta}_j$'s for $i > n_u$ and $j > n_y$, as two polynomial bases (not necessarily of minimal degree) for the left and right null-spaces of $\hat{R}(\lambda)$. By virtue of the Smith-McMillan decomposition (Theorem 4.1.1) these sets of polynomial vectors are linearly independent (over the field of rational functions), so they generate a minimal set of constraints on $\hat{R}$.

In the sequel, we will refer to the conditions in *i*) as the *zero interpolation conditions*, and to the conditions in *ii*) as the *rank interpolation conditions*. Rank interpolation conditions are also known by the names of *relations* and *convolution* conditions.

6.3 PROBLEM CLASSIFICATION

Here we adopt the following classification of problems of the form of Equation (6.1):

- **One-Block Problems:** problems with as many exogenous disturbances as measurements and as many controls as regulated outputs, i.e., $n_w = n_y$ and $n_z = n_u$. These are also known as square or good rank problems.
- **Two-Block Column Problems:** problems with as many exogenous disturbances as measurements, but with more regulated outputs than controls, i.e., $n_w = n_y$ and $n_z > n_u$.
- **Two-Block Row Problems:** problems with as many controls as regulated outputs, but with more exogenous disturbances than measurements, i.e., $n_w > n_y$ and $n_z = n_u$.
- **Four-Block Problems:** problems with more exogenous disturbances than measurements and with more regulated outputs than controls, i.e., $n_w > n_y$ and $n_z > n_u$.

A problem is labeled multiblock when it is not one-block. Multiblock problems are also known as bad rank or nonsquare problems.

Clearly, one-block problems only require zero interpolation conditions and have no rank interpolation conditions, whereas multiblock problems require both zero and rank interpolation conditions.

6.4 COMPUTATIONAL ASPECTS

The problem of finding the Smith-McMillan decomposition of rational matrices is at the heart of the interpolation problem. This decomposition has been studied thoroughly owing to its strong connections with several important notions in system theory (e.g., multivariable zeros and poles), although mostly from an algebraic point of view. The standard algebraic algorithm to compute such objects is based on the Euclidean division algorithm, which is known to be numerically sensitive. In general, it is desirable to have algorithms based on matrix computations that are easily implementable in digital computers.

Here we present a matrix approach to the problem of finding the interpolation conditions. Such approach avoids the explicit computation of the Smith-McMillan decomposition. Furthermore, it is computationally attractive since it is mostly based on finding the null-spaces of certain Toeplitz-like matrices.

6.5 COMPUTATION OF ZERO DIRECTIONS

In Theorem 6.2.2 we have shown how the internal stability of the closed-loop system is assured if the left zero structure of the unstable zeros of $\hat{U}$ and the right zero structure of the unstable zeros of $\hat{V}$ are preserved in $\hat{R}$. Such structure is characterized by the zero frequency and its directional properties. The algebraic and geometric multiplicity of the zero is defined in Definition 4.1.5, and its directional properties are given by the corresponding polynomial vector $\hat{\alpha}_i$ (left zero structure) and $\hat{\beta}_j$ (right zero structure). Despite its numerical problems, the Smith-McMillan decomposition provides the most natural way of characterizing the zero and pole structure of a rational matrix. To circumvent the formal Smith-McMillan decomposition of $\hat{U}(\lambda)$ and $\hat{V}(\lambda)$, it is necessary to find an alternative set of conditions that unequivocally defines the zero structure of a rational matrix. Such a set is presented in this section.

It is well known that a zero of a square system given in state-space form (A, B, C, D), is characterized by the solution of a generalized eigenvalue problem of the form:

$$\begin{pmatrix} A - z_0 I & B \\ C & D \end{pmatrix} \begin{pmatrix} x_0 \\ u_0 \end{pmatrix} = 0$$

where $z_0 := \lambda_0^{-1}$, x_0 is known as the state zero direction and u_0 is known as the zero input direction. However, the numerical stability of such eigenvalue problem deteriorates quickly when there are zeros with algebraic multiplicity greater than one. Indeed, such difficulty is equivalent to finding the Jordan decomposition of a defective matrix (i.e., a nondiagonalizable matrix), which is known to be a hard numerical problem.

Although it is difficult to obtain the full zero structure directly from the state-space description of a system, the location or frequency of the zeros can be reliably computed. In the sequel, we will assume that the locations of the unstable zeros of the rational matrices $\hat{U}(\lambda)$ and $\hat{V}(\lambda)$ are available.

In what follows, a method for computing the structure of zeros of a rational matrix is presented. Such computations provide all the necessary quantities for constructing the interpolation conditions of one-block problems.

To begin with, we introduce some useful definitions along with some notation.

Definition 6.5.1. Given a rational matrix $\hat{H}(\lambda)$ analytic at λ_0 and a positive integer σ, define the following block-lower-triangular Toeplitz matrix:

$$T_{\lambda_0,\sigma}(\hat{H}) = \begin{pmatrix} H_0 & 0 & 0 & \cdots & 0 \\ H_1 & H_0 & 0 & \cdots & 0 \\ \vdots & & & \ddots & \vdots \\ H_{\sigma-1} & H_{\sigma-2} & H_{\sigma-3} & \cdots & H_o \end{pmatrix}, \tag{6.12}$$

where the H_i's are given by the Taylor expansion of $\hat{H}(\lambda)$ at λ_0, that is,

$$\hat{H}(\lambda) = H_0 + (\lambda - \lambda_0) H_1 + (\lambda - \lambda_0)^2 H_2 + (\lambda - \lambda_0)^3 H_3 + \cdots$$

and $H_i = \frac{1}{i!} (\hat{H})^{(i)} (\lambda_0)$.   ∎

Definition 6.5.2.   Given an $m \times n$ (real) rational matrix $\hat{H}(\lambda)$ analytic at λ_0, a right null chain of order σ at λ_0 is an ordered set of column vectors in $\mathbb{R}^n$, $\{x_1, x_2, \ldots, x_\sigma\}$, such that $x_1 \neq 0$ and

$$T_{\lambda_0, \sigma}(\hat{H}) \begin{pmatrix} x_1 \\ x_2 \\ \vdots \\ x_\sigma \end{pmatrix} = \mathbf{0}.$$

Similarly, a left null chain of order σ at λ_0 is an ordered set of column vectors in $\mathbb{R}^m$, $\{y_1, y_2, \ldots, y_\sigma\}$, such that $y_1 \neq 0$ and

$$T_{\lambda_0, \sigma}(\hat{H}^T) \begin{pmatrix} y_1 \\ y_2 \\ \vdots \\ y_\sigma \end{pmatrix} = \mathbf{0}. \qquad \blacksquare$$

The next result shows that, if $\hat{H}$ is square, the existence of a right (left) null chain of order σ at λ_0 is equivalent to the existence of a zero at λ_0 of algebraic multiplicity σ. Afterwards, we establish a complete equivalence between the structure of zeros and the null chains associated with them. But first we give a complementary lemma.

Lemma 6.5.1.   Given an $n \times 1$ rational vector $\hat{z}(\lambda)$ analytic at λ_0, and an $n \times n$ unimodular matrix $\hat{W}(\lambda)$, then

$$\hat{z}^{(k)}(\lambda_0) = 0 \quad \text{for} \quad k = 0, 1, \ldots, \sigma - 1$$

$$\Updownarrow$$

$$(\hat{W}\hat{z})^{(k)}(\lambda_0) = 0 \quad \text{for} \quad k = 0, 1, \ldots, \sigma - 1.$$

Proof. The only-if part follows from expanding the derivative of the product $\hat{W}\hat{z}$, that is

$$(\hat{W}\hat{z})^{(k)}(\lambda_0) = \sum_{j=0}^{k} \binom{k}{j} \hat{W}^{(k-j)}(\lambda_0) \hat{z}^{(j)}(\lambda_0) = 0$$

since $\hat{z}^{(j)}(\lambda_0) = 0$ for $j = 0, 1, \ldots, \sigma - 1$.

The if part is shown by induction. For $k = 0$, $(\hat{W}\hat{z})(\lambda_0) = 0$ implies that $\hat{z}(\lambda_0) = 0$ since $\hat{W}(\lambda_0)$ is full rank. Next, assume that $\hat{z}^{(k)}(\lambda_0) = 0$ for $k = 0, 1, \ldots, i - 1 \leq \sigma - 2$, then

$$0 = (\hat{W}\hat{z})^{(i)}(\lambda_0) = \sum_{j=0}^{i} \binom{i}{j} \hat{W}^{(i-j)}(\lambda_0) \hat{z}^{(j)}(\lambda_0) = \hat{W}(\lambda_0) \hat{z}^{(i)}(\lambda_0),$$

which implies that $\hat{z}^{(i)}(\lambda_0) = 0$. $\qquad \blacksquare$

Theorem 6.5.1. A full rank, $n \times n$, rational matrix $\hat{H}(\lambda)$, analytic at λ_0, has a zero at λ_0 of geometric multiplicity l and a sequence of structural indices equal to, at least, $\sigma_{n-l+1}, \ldots, \sigma_n$ ($\sigma_1 = \cdots = \sigma_{n-l} = 0$) if and only if the following conditions hold:

1. There exist l polynomial vectors, $\hat{u}_1, \ldots, \hat{u}_l$, such that

$$(\hat{H}\hat{u}_j)^{(k)}(\lambda_0) = 0 \quad for \quad k = 0, \ldots, \sigma_{n-l+j} - 1, \quad j = 1, \ldots, l.$$

2. The vectors $\{\hat{u}_1(\lambda_0), \ldots, \hat{u}_l(\lambda_0)\}$ are linearly independent and

$$\mathrm{span}\{\hat{u}_1(\lambda_0), \ldots, \hat{u}_l(\lambda_0)\} = \mathcal{N}[\hat{H}(\lambda_0)].$$

Proof. Necessity follows directly from the Smith-McMillan decomposition of $\hat{H}(\lambda)$:

$$\hat{H}(\lambda) = \hat{L}(\lambda)\hat{M}(\lambda)\hat{R}(\lambda).$$

Say that the jth entry ($j \geq n - l + 1$) on the diagonal of $\hat{M}$ has a factor $(\lambda - \lambda_0)^{\sigma_j}$. Then, pick $\hat{u}_{j-n+l}$ to be the jth column of $\hat{R}^{-1}$. With this choice

$$\hat{H}\hat{u}_{j-n+l} = \hat{H}(\hat{R}^{-1})^j = (\lambda - \lambda_0)^{\sigma_j}\hat{L}\hat{p}_{j-n+l} \quad \text{for } j = n - l + 1, \ldots, n,$$

where $\hat{p}_{j-n+l}(\lambda)$ is a rational vector analytic at λ_0. Clearly, this implies that $(\hat{H}\hat{u}_{j-n+l})^{(k)}(\lambda_0) = 0$ for $k = 0, \ldots, \sigma_j - 1$, and further the set $\{\hat{u}_1(\lambda_0), \ldots, \hat{u}_l(\lambda_0)\}$ is linearly independent (since $\hat{R}$ is unimodular) and spans the null space of $\hat{H}(\lambda_0)$.

The proof of sufficiency is not as straightforward. Let $\hat{z}_j := \hat{H}\hat{u}_j$ for $j = 1, \ldots, l$ and define the following auxiliary rational vectors:

$$\hat{y}_j(\lambda) := (\hat{L}^{-1}\hat{z}_j)(\lambda), \qquad \hat{v}_j(\lambda) := (\hat{R}\hat{u}_j)(\lambda) \quad \text{for } j = 1, \ldots, l.$$

Then, we have that $\hat{y}_j(\lambda) = \hat{M}(\lambda)\hat{v}_j(\lambda)$. Note that $\hat{u}_1(\lambda_0), \cdots, \hat{u}_l(\lambda_0)$ are linearly independent if and only if $\hat{v}_1(\lambda_0), \cdots, \hat{v}_l(\lambda_0)$ are linearly independent since $\hat{R}$ is unimodular. Furthermore, since multiplication by a unimodular matrix preserves the zero structure, this direction of the proof can be restated as follows: for all $j = 1, \ldots, l$

$$\exists \, \hat{v}_j(\lambda) \text{ such that } \hat{v}_1(\lambda_0) \cdots \hat{v}_l(\lambda_0) \text{ are linearly independent}$$

$$\text{and } \hat{y}_j^{(k)}(\lambda_0) = 0 \text{ for } k = 0, \ldots, \sigma_{n-l+j-1}$$

$$\Downarrow$$

$$\exists \text{ a factor } (\lambda - \lambda_0)^{\sigma_{n-l+j}} \text{ in the } n - l + j \text{ diagonal entry of } \hat{M}(\lambda).$$

Now, it follows from above that

$$\hat{y}_j(\lambda) = (\lambda - \lambda_0)^{\sigma_{n-l+j}}\hat{p}_j(\lambda).$$

Let $\hat{\epsilon}_j(\lambda)$, for $j = 1, \ldots, n$ be the diagonal entries of the matrix $\hat{M}$. Then,

$$\begin{pmatrix} \hat{\epsilon}_1 & & \\ & \ddots & \\ & & \hat{\epsilon}_n \end{pmatrix} (\hat{v}_1 \quad \cdots \quad \hat{v}_l) = (\hat{p}_1 \quad \cdots \quad \hat{p}_l) \begin{pmatrix} (\lambda - \lambda_0)^{\sigma_{n-l+1}} & & \\ & \ddots & \\ & & (\lambda - \lambda_0)^{\sigma_n} \end{pmatrix}.$$

$$(6.13)$$

First, we show that the matrix $(\hat{v}_1(\lambda_0) \cdots \hat{v}_l(\lambda_0))$ has the structure

$$\begin{pmatrix} 0 \\ \hat{V}(\lambda_0) \end{pmatrix} \tag{6.14}$$

where $\hat{V}(\lambda_0)$ is a $l \times l$ nonsingular matrix. The top zero block results from the fact that the matrix $\hat{M}(\lambda_0)$ has a null-space of dimension l (otherwise there would be more linearly independent vectors than l); hence, $\hat{\epsilon}_1, \ldots, \hat{\epsilon}_{n-l}$ do not have zeros at λ_0. The matrix $\hat{V}(\lambda_0)$ is nonsingular since the vectors $(\hat{v}_1(\lambda_0) \cdots \hat{v}_l(\lambda_0))$ are linearly independent.

From Equation (6.13), it follows that for all λ

$$\begin{pmatrix} \hat{\epsilon}_{n-l+1} & & \\ & \ddots & \\ & & \hat{\epsilon}_n \end{pmatrix} \hat{V} = \hat{P} \begin{pmatrix} (\lambda - \lambda_0)^{\sigma_{n-l+1}} & & \\ & \ddots & \\ & & (\lambda - \lambda_0)^{\sigma_n} \end{pmatrix}, \tag{6.15}$$

where the matrices $\hat{V}$ and $\hat{P}$ are obtained from the decompositions

$$(\hat{v}_1(\lambda) \quad \ldots \quad \hat{v}_l(\lambda)) = \begin{pmatrix} \tilde{\hat{V}} \\ \hat{V} \end{pmatrix}$$

and

$$(\hat{p}_1(\lambda) \quad \ldots \quad \hat{p}_l(\lambda)) = \begin{pmatrix} \tilde{\hat{P}} \\ \hat{P} \end{pmatrix}.$$

Let

$$\hat{E} = \begin{pmatrix} \hat{\epsilon}_{n-l+1} & & \\ & \ddots & \\ & & \hat{\epsilon}_n \end{pmatrix}, \qquad \hat{D} = \begin{pmatrix} (\lambda - \lambda_0)^{\sigma_{n-l+1}} & & \\ & \ddots & \\ & & (\lambda - \lambda_0)^{\sigma_n} \end{pmatrix}.$$

Let $\hat{R}_1$ and $\hat{R}_2$ be unimodular matrices such that

$$\hat{V}\hat{R}_1 = \hat{L} \text{ where } \hat{L} \text{ is lower triangular,}$$

and

$$\hat{R}_2\hat{P} = \hat{U} \text{ where } \hat{U} \text{ is upper triangular .}$$

Then, Equation (6.15) can be factored as follows

$$\hat{E}\hat{L} = \hat{R}_2^{-1}\hat{U}\hat{D}\hat{R}_1$$

Clearly, the matrix $\hat{E}\hat{L}$ has the same zero structure as the matrix $\hat{U}\hat{D}$. By direct computation of the Smith-McMillan matrix of $\hat{U}\hat{D}$, it follows that $(\lambda - \lambda_0)^{\sigma_{n-l+j}}$ is a factor of jth diagonal element. Since $\hat{L}$ has full rank at λ_0, it follows that $(\lambda - \lambda_0)^{\sigma_{n-l+j}}$ is a factor of ϵ_{n-l+j}. This completes the proof.   ∎

Note that a similar result holds showing the existence of left vectors associated with a zero. The proof follows from constructing right vectors for the matrix $\hat{H}^T$.

As pointed out in the proof of Theorem 6.5.1, if $\hat{H}$ has a zero of geometric multiplicity greater than one, say l, then there are l different vectors that pick up the structure of the zero. To relate these vectors to a null chain, we make the following definition.

Definition 6.5.3. A canonical set of right null chains of $\hat{H}(\lambda)$ at λ_0 is an ordered set of right null chains, i.e., $x^i = (x_1^i \; \ldots \; x_{\sigma_i}^i)$ for $i = 1, \ldots, l$, such that

1. $\{x_1^1, x_1^2, \ldots, x_1^l\}$ are linearly independent,
2. $\mathrm{span}\{x_1^1, x_1^2, \ldots, x_1^l\} = \mathcal{N}[\hat{H}(\lambda_0)]$, and
3. $\sigma_1 \geq \sigma_2 \geq \cdots \geq \sigma_l$.

A canonical set of left null chains is defined similarly.   ∎

The following corollary restates the result of Theorem 6.5.1 in terms of null chains.

Corollary 6.5.1. A full rank, $n \times n$, rational matrix $\hat{H}(\lambda)$, analytic at λ_0, has a zero at λ_0 of geometric multiplicity l and a sequence of structural indices equal to, at least, $\sigma_{n-l+1}, \ldots, \sigma_n$ ($\sigma_1 = \sigma_{n-l} = 0$) if and only if there exits a canonical set of right null chains with the same structural indices at λ_0.

Proof. Both directions follow immediately from Theorem 6.5.1 since there is a direct correspondence between null chains and right vector polynomials associated with λ_0. For a given $\hat{u}_j$, a null chain is defined as

$$\hat{u}_j(\lambda) = x_1^j + (\lambda - \lambda_0)x_2^j + \cdots + (\lambda - \lambda_0)^{\sigma_{n-l+j}-1}x_{\sigma_{n-l+j}}^j.$$

On the other hand, every null chain defines a vector in the same way. It follows that

$$(\hat{H}\hat{u}_j)^{(k)}(\lambda_0) = 0 \quad \text{for} \quad k = 0, \ldots, \sigma_{n-l+j} - 1$$

$$\Updownarrow$$

$$\sum_{i=0}^{k} H_{k-i}x_{i+1}^j = 0 \quad \text{for} \quad k = 0, \ldots, \sigma_{n-l+j} - 1$$

$$\Updownarrow$$

$$T_{\lambda_0,\sigma}(\hat{H})(\; x_1^j \; \cdots \; x_{\sigma_{n-l+j}}^j \;)^T = 0$$

Clearly

$$\mathrm{span}\{x_1^1, x_1^2, \ldots, x_1^l\} = \mathcal{N}[\hat{H}(\lambda_0)]$$

if and only if

$$\mathrm{span}\{\hat{u}_1(\lambda_0), \hat{u}_2(\lambda_0), \ldots, \hat{u}_l(\lambda_0)\} = \mathcal{N}[\hat{H}(\lambda_0)].$$ ∎

A similar result follows showing the equivalence of left zero directions with left null chains. Next, we show that the zero interpolation conditions of Theorem 6.2.2 can be stated in terms of the canonical set of right null chains of $\hat{V}$ and the canonical set of left null chains of $\hat{U}$ at each $\lambda_0 \in \Lambda_{UV}$. To facilitate the presentation, we introduce an extension of Definition 6.5.3.

Definition 6.5.4. An extended set of right null chains of a full rank $n \times n$ rational matrix $\hat{H}(\lambda)$ at λ_0, is a canonical set of right null chains augmented with $n - l$ vectors in

$\mathbb{R}^n$, i.e., $\{x_1^{l+1}, \ldots, x_1^n\}$, such that $\mathrm{span}\{x_1^1, x_1^2, \ldots, x_1^n\} = \mathbb{R}^n$. The order associated with these added chains is zero. ∎

From the above definition, if a square rational matrix has no zeros at λ_0, then the corresponding canonical set of null chains is empty and the extended set is a basis for $\mathbb{R}^n$, e.g., the columns of an $n \times n$ identity matrix.

6.6 ZERO INTERPOLATION CONDITIONS

Next we apply the above results and definitions to the zero interpolation conditions of a one-block problem. In the context of Theorem 6.2.2 we have the following equivalence: For $j = 1, \ldots, n_y$ and $k = 0, \ldots, \sigma_{V_j}$,

$$(\hat{V}\hat{\beta}_j)^{(k)}(\lambda_0) = 0 \iff T_{\lambda_0, \sigma_{V_j}}(\hat{V})x^{n_y-j+1} = 0,$$

where x^i is an extended set of right null chains for $\hat{V}$ at λ_0. The sequence of x^i's has to be reversed in the above equation due to the fact that σ_{V_j} is a nondecreasing sequence of algebraic multiplicities while an extended set of null chains is defined with the opposite ordering (see Definitions 4.1.1 and 6.5.3). Note that if $\sigma_{V_j} = 0$ then both conditions are satisfied trivially (i.e., there are no conditions). Similarly, for $i = 1, \ldots, n_u$ and $k = 0, \ldots, \sigma_{U_i}$,

$$(\hat{\alpha}_i\hat{U})^{(k)}(\lambda_0) = 0 \iff T_{\lambda_0, \sigma_{U_i}}(\hat{U}^T)y^{n_u-i+1} = 0.$$

In other words, the extended set of left and right null chains are locally (i.e., for each λ_0) equivalent to the polynomial vectors $\hat{\alpha}_i$'s and $\hat{\beta}_j$'s. Having made this observation, we are ready to present an alternative set of zero interpolation conditions.

6.6.1 Conditions Using Null Chains

Given an element of an extended set of right null chains at λ_0, x^j, of order σ_j, define the following polynomial vector:

$$\hat{x}_{\lambda_0}^j(\lambda) := x_1^j + (\lambda - \lambda_0)x_2^j + \cdots + (\lambda - \lambda_0)^{\sigma_j - 1}x_{\sigma_j}^j$$

if $\sigma_j > 0$, and $\hat{x}_{\lambda_0}^j(\lambda) := x_1^j$ if $\sigma_j = 0$. Similarly, define $\hat{y}_{\lambda_0}^i(\lambda)$ for an element of an extended set of left null chains, y^i, of order σ_i. With this notation we have the following corollary.

Corollary 6.6.1.   Given a one-block problem, the zero interpolation conditions of Theorem 6.2.2 are equivalent to the following: For all $\lambda_0 \in \Lambda_{UV}$,

$$(\hat{y}_{\lambda_0}^i \hat{R} \hat{x}_{\lambda_0}^j)^{(k)}(\lambda_0) = 0 \text{ for } \begin{cases} i = 1, \ldots, n_u \\ j = 1, \ldots, n_y \\ k = 0, \ldots, \sigma_{U_i}(\lambda_0) + \sigma_{V_j}(\lambda_0) - 1 \end{cases}$$

where y^i and x^j are elements of the extended sets of left and right null chains of $\hat{U}$ and $\hat{V}$, respectively, and σ_{U_i}, σ_{V_j} are the corresponding orders (i.e., algebraic multiplicities).

Proof. Follows directly from Theorems 6.2.2 and 6.5.1, and from the above definitions. ∎

6.6.2 Computation of Null Chains

This subsection discusses a simple algorithm to compute the extended set of null chains at λ_0, of a full normal rank square rational matrix analytic at λ_0. Let $\hat{H}(\lambda)$ denote an $n \times n$ rational matrix and assume that λ_0 is given; then the algorithm is based on the computation of a basis for the null-space of $T_{\lambda_0,\sigma}(\hat{H})$ for increasing values of σ.

Consider the construction of an extended set of right null chains (the left null chains are obtained simply by transposing $\hat{H}$). Then from Definition 6.5.2, given some positive integer σ, any vector in the kernel of $T_{\lambda_0,\sigma}(\hat{H})$ such that $x_1 \neq 0$ is a potential member of the set. Let B_σ denote a matrix whose columns form a basis for the right null-space of $T_{\lambda_0,\sigma}(\hat{H})$; then the following algorithm generates an extended set of right null chains:

Step 1:   Compute B_σ for $\sigma = 1, 2, \ldots$ until the top n rows are filled with zeros (no more null chains can be extracted at this point). Then the maximum order of any chain, σ_1, is given by the current value of the counter (σ) minus one. Note that this iteration process is guaranteed to stop since the rational matrix $\hat{H}$ is finite dimensional (i.e., its zeros have finite algebraic multiplicity).

Step 2:   Let b_i for $i = 1, \ldots, r$ denote each column of B_{σ_1}. Reduce the dimension of the b_i's by removing all sets of n contiguous zeros at the top of each vector. The result is a collection of r vectors (possibly of different dimensions) such that the top n entries of each one define a non-zero vector in $\mathbb{R}^n$. (Note that at least one will have dimension $n\sigma_1$.)

Step 3:   Sort the resulting vectors in decreasing order of dimension. Let l be the rank of the $n \times r$ matrix that results from collecting the first n rows of each vector. Then, select the first l vectors such that the reduced matrix that results from collecting the first n rows of each vector has rank l. Such collection forms a canonical set of right null chains.

Step 4:   Extend the set by augmenting the collection with $n - l$ vectors such that the set of n vectors formed with the first n rows define a basis in $\mathbb{R}^n$.

If the system $\hat{H}(\lambda)$ is given in state-space form, say (A, B, C, D), then the Toeplitz matrices $T_{\lambda_0,\sigma}(\hat{H})$ can be easily computed using the following equation (see Definition 6.5.1):

$$H_k = \begin{cases} \lambda_0 C(I - \lambda_0 A)^{-1} B + D & \text{for } k = 0 \\ C(I - \lambda_0 A)^{-k-1} A^{k-1} B & \text{for } k = 1, 2, \ldots \end{cases}$$

Note that $(I - \lambda_0 A)^{-1}$ always exists since λ_0 is in the unit disc and $\hat{H}$ is stable (i.e., analytic in the closed unit disc). A word of warning is necessary, however, when λ_0 is close to the unit circle and A has a stable eigenvalue that is also close to the unit circle and next to λ_0. Such cases may give rise to numerical difficulties. Besides this fact, the rest of the algorithm only involves the computation of null-spaces that can be done efficiently through the well-known QR decomposition.

Finally, it is worth pointing out that the algorithm presented in this section is applicable to one-block problems only. Later on, we will see that this algorithm will be utilized in the computation of the zero structure of multiblock problems as well.

Example 6.6.1

To illustrate the workings of the algorithm introduced in the previous section, a simple example is offered. Let $\hat{H}(\lambda)$ be a 3×3 polynomial matrix given by

$$\hat{H}(\lambda) = \begin{pmatrix} (\lambda - 0.5)^2 & \lambda(\lambda + 2)(\lambda - 0.5) & 0 \\ (\lambda - 0.5)^3 & \lambda(\lambda - 0.5) & 0 \\ 0 & 0 & \lambda^2 \end{pmatrix}.$$

We have chosen a polynomial matrix just to make the example tractable without the aid of a computer. Let us construct an extended set of right null chains for the zero at $\lambda_0 = 0.5$. The first step is to compute the null-space of $T_{\lambda_0,\sigma}(\hat{H})$ for $\sigma = 1, 2, \ldots$. In particular, for $\sigma = 3$ we have

$$T_{0.5,3}(\hat{H}) = \begin{pmatrix} 0 & 0 & 0 & 0 & 0 & 0 & 0 & 0 & 0 \\ 0 & 0 & 0 & 0 & 0 & 0 & 0 & 0 & 0 \\ 0 & 0 & .25 & 0 & 0 & 0 & 0 & 0 & 0 \\ 0 & .5 & 0 & 0 & 0 & 0 & 0 & 0 & 0 \\ 0 & .5 & 0 & 0 & 0 & 0 & 0 & 0 & 0 \\ 0 & 0 & 1 & 0 & 0 & .25 & 0 & 0 & 0 \\ 1 & 1.5 & 0 & 0 & .5 & 0 & 0 & 0 & 0 \\ 0 & 1 & 0 & 0 & .5 & 0 & 0 & 0 & 0 \\ 0 & 0 & 1 & 0 & 0 & 1 & 0 & 0 & .25 \end{pmatrix}, \quad B_3 = \begin{pmatrix} 0 & 0 & 0 \\ 0 & 0 & 0 \\ 0 & 0 & 0 \\ 0 & 0 & 1 \\ 0 & 0 & 0 \\ 0 & 0 & 0 \\ 1 & 0 & 0 \\ 0 & 1 & 0 \\ 0 & 0 & 0 \end{pmatrix}.$$

Clearly, the first three rows of B_3 are zero, so we stop increasing σ. Then, the maximum algebraic multiplicity of $\lambda_0 = 0.5$ is two, i.e., $\sigma_1 = 2$. Next (Step 2), reduce each column of B_3 by eliminating the leading blocks of zeros to get

$$b_1 = \begin{pmatrix} 1 \\ 0 \\ 0 \end{pmatrix}, \quad b_2 = \begin{pmatrix} 0 \\ 1 \\ 0 \end{pmatrix}, \quad b_3 = \begin{pmatrix} 1 \\ 0 \\ 0 \\ 0 \\ 0 \\ 0 \end{pmatrix}.$$

Then (Step 3), reorder the set of vectors in decreasing dimension, i.e., $\{b_3, b_1, b_2\}$, and compute the rank of the matrix formed with the first three rows:

$$l = \text{rank} \left[\begin{pmatrix} 1 & 1 & 0 \\ 0 & 0 & 1 \\ 0 & 0 & 0 \end{pmatrix} \right] = 2.$$

The canonical set of right null chains is then given by $\{x^1, x^2\}$ where

$$
x^1 = \begin{pmatrix} 1 \\ 0 \\ 0 \\ 0 \\ 0 \\ 0 \end{pmatrix} \quad \text{and} \quad x^2 = \begin{pmatrix} 0 \\ 1 \\ 0 \end{pmatrix}
$$

with their corresponding orders (i.e., algebraic multiplicity) being $\sigma_1 = 2$ and $\sigma_2 = 1$. This indicates that the geometric multiplicity of λ_0 is two. Finally (Step 4), to get an extended set of right null chains we augment the collection with $x^3 = (0\ 0\ 1)^T$ having order $\sigma_3 = 0$ (by definition).

6.7 EFFECTS OF POLES AND ZEROS

The previous sections showed that poles and zeros in the unit disc play a major role in characterizing the achievable closed-loop maps. Although this implies that performance will degrade in the presence of such poles and zeros (by the mere fact that there are more constraints on the achievable closed-loop maps), the degree of degradation is not immediately evident. Clearly, the best achievable performance can be accomplished by directly optimizing the performance over the set of admissible closed-loop maps. On the other hand, it is desirable to obtain some quantitative and qualitative insight in the trade-offs without solving any optimization problem. This is hard. In general, however, there are some interesting special problems. We will discuss an example of such problems.

6.7.1 Frequency Trade-offs

The example presented here illustrates a concept known as the *water-bed effect*. Let S denote the standard sensitivity function. Qualitatively, this property indicates that if $\hat{S}(e^{i\theta})$ is made small for a range of θ, then it will peak somewhere outside this range. The amount of peaking is dependent on the pole/zero locations (in the disc). The results presented are for SISO problems.

Definitions.   The sensitivity function

$$
S = (I - PK)^{-1} = (I - L)^{-1},
$$

where $\hat{L} = \hat{P}\hat{K}$ has n-poles $p_1, \ldots, p_n$ and m-zeros $z_1, \ldots, z_m$ in the disc. Assume for simplicity that all such poles and zeros are simple and lie in the open unit disc (i.e., no interpolations on the boundary). Notice that the sensitivity function satisfies

$$
\hat{S}(p_k) = 0 \quad \text{for } k = 1, \ldots, n,
$$
$$
\hat{S}(z_k) = 1 \quad \text{for } k = 1, \ldots, m.
$$

More so, the only non-minimum phase zeros of $\hat{S}(\lambda)$ are located at p_i. Define the allpass function

$$\hat{B}(\lambda) = \prod_{i=1}^{n} \frac{\lambda - p_i}{1 - \bar{p}_i \lambda}.$$

The function $\hat{B}$ is stable and has magnitude equal to 1 for all $\theta \in [0, 2\pi)$. Since the only zeros of $\hat{S}$ are the p_i's, we can write

$$\hat{S} = \hat{S}_{\mathrm{mp}} \hat{B},$$

where $\hat{S}_{\mathrm{mp}}$ is stable and minimum phase. Notice also that

$$|\hat{S}(e^{i\theta})| = |\hat{S}_{\mathrm{mp}}(e^{i\theta})| \quad \text{for all } \theta \in [0, 2\pi).$$

Given any complex number $a = re^{ix}$, define the real valued function of θ:

$$\begin{aligned} w(a, \theta) &= \frac{d}{d\theta} \arg \frac{e^{i\theta} - a}{1 - \bar{a}e^{i\theta}} \\ &= \frac{1 - r^2}{1 - 2r\cos(\theta - x) + r^2}. \end{aligned}$$

Also, define the function $v(a, \theta)$ as:

$$v(a, \theta) = \frac{1}{2}(w(a, \theta) + w(a, -\theta)).$$

Norm minimization.   To motivate the discussion, consider the problem of minimizing the weighted sensitivity function:

$$\min_{\substack{K \text{ stabilizing}}} \|SW\|_{\ell_p - ind}.$$

For any stabilizing controller, we have

$$\begin{aligned} \|SW\|_{\ell_p - ind} &\geq \|\hat{S}\hat{W}\|_{\infty} \\ &\geq \max_{1 \leq i \leq m} |\hat{S}(z_i)\hat{W}(z_i)| \\ &\geq \max_{1 \leq i \leq m} |\hat{W}(z_i)|. \end{aligned}$$

The second inequality follows from the maximum modulous principle which indicates that the maximum magnitude of a function analytic in a region is always achieved at the boundary of the region. This implies that zeros in the unit disc limit the achievable performance of the system. While this highlights some of the limitations, it does not show the role of the unstable poles, nor the trade-offs at different frequencies. This is done next.

Poisson integral formula.   Let $\hat{F}$ be any function analytic in a region containing the unit disc, and $a = re^{ix}$ be any complex number in the disc, i.e., $|r| < 1$. The Poisson Integral Formula states that

$$\hat{F}(a) = \frac{1}{2\pi} \int_{-\pi}^{\pi} w(a, \theta) \hat{F}(e^{i\theta}) d\theta.$$

This formula is a variation of Cauchy's Integral Formula and simply shows that it is sufficient to know an analytic function on the boundary of the set in which it is analytic in order to evaluate it at any point in the set.

Application to the sensitivity function.   Since the function $\hat{S}_{mp}$ is analytic in the disc, and has no zeros in the unit disc, it follows that $\log \hat{S}_{mp}$ is also analytic in the disc. Applying Poisson Integral Formula on this function we get

$$\log \hat{S}_{mp}(a) = \frac{1}{2\pi} \int_{-\pi}^{\pi} w(a, \theta) \log \hat{S}_{mp}(e^{i\theta}) d\theta.$$

Evaluating this integral at z_i (a zero of $\hat{L}$ inside the disc), and taking the real part of both sides of the equation, we get

$$\log |\hat{S}_{mp}(z_i)| = \frac{1}{2\pi} \int_{-\pi}^{\pi} w(z_i, \theta) \log |\hat{S}_{mp}(e^{i\theta})| d\theta.$$

By substituting the values in the integral in terms of $\hat{S}$ and $\hat{B}$, we get

$$\log |\hat{B}^{-1}(z_i)| = \frac{1}{2\pi} \int_{-\pi}^{\pi} w(z_i, \theta) \log |\hat{S}(e^{i\theta})| d\theta.$$

Finally, since $\hat{S}(e^{i\theta})$ is real rational, $|\hat{S}(e^{i\theta})|$ is symmetric about $\theta = 0$. By splitting the integral, we get

$$\log |\hat{B}^{-1}(z_i)| = \frac{1}{2\pi} \int_{-\pi}^{0} w(z_i, \theta) \log |\hat{S}(e^{i\theta})| d\theta + \frac{1}{2\pi} \int_{0}^{\pi} w(z_i, \theta) \log |\hat{S}(e^{i\theta})| d\theta$$

$$= \frac{1}{\pi} \int_{0}^{\pi} \frac{1}{2} (w(z_i, \theta) + w(z_i, -\theta)) \log |\hat{S}(e^{i\theta})| d\theta$$

$$= \frac{1}{\pi} \int_{0}^{\pi} v(z_i, \theta) \log |\hat{S}(e^{i\theta})| d\theta.$$

We first make two observations:

1. $|\hat{B}^{-1}(z_i)| > 1$ for all z_i,
2. $v(z_i, \theta) \geq 0$ for all $\theta \in [0, \pi)$.

With these observations we see that the weighted $\log|\hat{S}|$ has a constant positive total integral that depends only on the poles and zeros of $\hat{L}$ in the unit disc. We conclude the following:

1. If $\hat{L}$ is designed to push $\hat{S}$ down at a specified frequency range, $\hat{S}$ will peak at another frequency outside this range, to keep the total integral constant; see Figure 6.2(a).
2. If $\hat{L}$ is designed to make $\hat{S}$ small for a larger frequency range, $\hat{S}$ will have a larger peak at a frequency outside that range, see Figure 6.2(b).
3. The amount of peaking depends on the poles and zeros. The worst situations occur when a pole is close to a zero (both inside the disc).

A lower bound on $\|\hat{S}\|_\infty$.   Assume that $\hat{L}$ is designed so that

$$|\hat{S}(e^{i\theta})| \le \gamma < 1 \quad \text{for all } \theta \in [0, \theta_0].$$

We would like to furnish a lower bound on the $\mathcal{H}_\infty$-norm of the sensitivity function. Define $\Theta(z_i, \theta_0)$ as

$$\Theta(z_i, \theta_0) = \int_0^{\theta_0} v(z_i, \theta)d\theta.$$

Note that $\Theta(z_i, \theta_0) \le \pi$ and is equal to π if $\theta_0 = \pi$. Using the integral formula, we have

$$\pi \log|\hat{B}^{-1}(z_i)| = \int_0^{\theta_0} v(z_i, \theta) \log|\hat{S}(e^{i\theta})|d\theta + \int_{\theta_0}^{\pi} v(z_i, \theta) \log|\hat{S}(e^{i\theta})|d\theta$$

$$\le \Theta(z_i, \theta_0) \log \gamma + (\pi - \Theta(z_i, \theta_0)) \log \|\hat{S}\|_\infty.$$

This yields the inequality

$$\|\hat{S}\|_\infty \ge \left(\frac{1}{\gamma}\right)^{\frac{\Theta(z_i, \theta_0)}{\pi - \Theta(z_i, \theta_0)}} |B^{-1}(z_i)|^{\frac{\pi}{\pi - \Theta(z_i, \theta_0)}}.$$

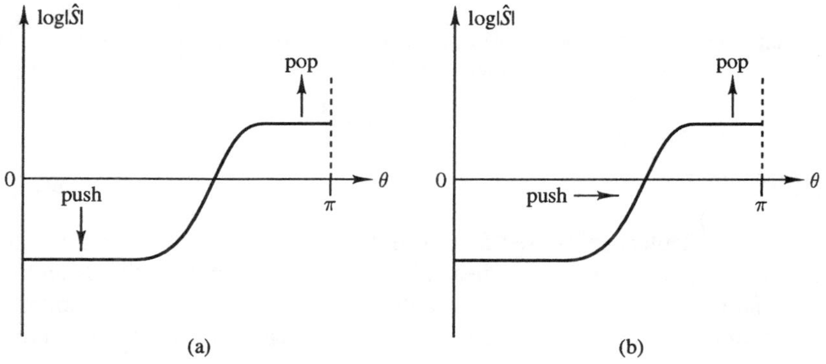

Figure 6.2   The water-bed effect.

This formula gives the trade-offs in a more quantitative fashion:

1. If γ is small, then $\|\hat{S}\|_\infty$ is large.
2. If θ_0 is close to π, then $\|\hat{S}\|_\infty$ is large.
3. If a zero is close to a pole, then $\|\hat{S}\|_\infty$ is large.

Finally, the complementary sensitivity function $PK(I-PK)^{-1}$ behaves in a similar fashion, with the roles of poles and zeros reversed.

6.8 A GENERAL DESIGN PROBLEM

If the performance objective is more elaborate than shaping the sensitivity function (or any scalar function), it is quite difficult to capture the trade-offs without performing some optimization. Theorem 6.2.2 shows that a closed-loop map is feasible if it satisfies both the zero interpolation conditions and the rank interpolation conditions. Each of these conditions is equivalent to a set of linear constraints on the closed-loop map, which can be represented by some linear operator. Thus, the interpolation conditions can be summarized as

$$\mathcal{A}_{\text{feas}}\Phi = b_{\text{feas}}, \tag{6.16}$$

where $\mathcal{A}_{\text{feas}}$ is a bounded operator on ℓ_1, and b_{feas} is a fixed element. An explicit construction of this operator will be presented in Chapter 11.

Following the discussion in Chapter 3, we formulate a general controller synthesis problem for achieving nominal performance. This problem will be referred to in future chapters:

$$\begin{aligned} &\min \|\Phi\|_1 \\ \text{Subject to} \quad & \\ &\mathcal{A}_{\text{feas}}\Phi = b_{\text{feas}} \\ &\mathcal{A}_{\text{perf}}\Phi \leq b_{\text{perf}} \end{aligned} \tag{6.17}$$

The operator $\mathcal{A}_{\text{perf}}$ summarizes the performance constraints that are linear, or that are approximated by linear constraints.

6.9 SUMMARY

In this chapter we offered a complete treatment of interpolation conditions that characterize closed-loop maps. These are zero interpolation conditions and rank interpolation conditions. Both of these are completely characterized using the Smith-McMillan decompositions of the systems $\hat{U}$ and $\hat{V}$. For the purpose of computation, a matrix approach for computing the zero interpolation conditions is proposed. While this approach can be generalized to compute the rank interpolation conditions, it turns out that this is not necessary in future computations since nonsquare problems will be converted to square problems by adding delays to the system (see Exercise 6.5). The characterization of closed-loop maps given in this chapter is exploited later to solve various synthesis problems.

Interpolation conditions give algebraic constraints on the achievable closed-loop maps. Analytic constraints are harder to obtain in general, although some special cases can be studied. Such constraints give a qualitative insight into the trade-offs involved in controller design.

EXERCISES

6.1. Let the set of closed loop maps be parametrized as

$$(H_1 \quad H_2) - Q(\tilde{V}_1 \quad \tilde{V}_2)$$

where $\tilde{V}_1$ is a square matrix with a causal possibly unstable inverse (equivalently has no zeros at the origin), and the pair $\tilde{V}_1$ and $\tilde{V}_2$ is left coprime. Let

$$\tilde{V}_1^{-1}\tilde{V}_2 = V_2 V_1^{-1}$$

where the pair V_1 and V_2 is right coprime. Show that $\Phi \in \mathcal{RH}_\infty$ is feasible if and only if

$$(\Phi_1 \quad \Phi_2)\begin{pmatrix} -V_2 \\ V_1 \end{pmatrix} = (H_1 \quad H_2)\begin{pmatrix} -V_2 \\ V_1 \end{pmatrix}.$$

6.2. Apply the above result to characterize all admissible controllers that guarantee the stability robustness of a class of systems described by perturbing the coprime factors of a given nominal plant. See Equation (3.16).

6.3. Consider the following decompositions of the matrices $\hat{U}$, $\hat{V}$:

$$\hat{U} = \begin{pmatrix} \hat{U}_1 \\ \hat{U}_2 \end{pmatrix}, \quad \hat{V} = (\hat{V}_1 \quad \hat{V}_2).$$

Assume that $\hat{U}_1$ and $\hat{V}_1$ are square matrices. Let $\hat{U}_2\hat{U}_1^{-1} = \hat{\tilde{U}}_1^{-1}\hat{\tilde{U}}_2$ where $\hat{\tilde{U}}_1$ and $\hat{\tilde{U}}_2$ are left coprime. Similarly, let $\hat{V}_1^{-1}\hat{V}_2 = \hat{\tilde{V}}_2\hat{\tilde{V}}_1^{-1}$ where $\hat{\tilde{V}}_1$ and $\hat{\tilde{V}}_2$ are right coprime. Prove that a given closed-loop map, $\hat{\Phi} = \hat{H} - \hat{R}$, is feasible if and only if the following conditions hold:
(a) R_{11} satisfies the standard zero interpolation conditions associated with $\hat{U}_1$ and $\hat{V}_1$.
(b) The following matrix equations hold:

$$(\tilde{U}_2 \quad -\tilde{U}_1)R = 0,$$

$$(R_{11} \quad R_{12})\begin{pmatrix} \tilde{V}_2 \\ -\tilde{V}_1 \end{pmatrix} = 0.$$

6.4. In the previous problem, show that there exists a feasible Φ with the property that $\Phi(k) = 0 \; \forall k \geq N$ (FIR) if and only if

$$(\tilde{U}_2 \quad -\tilde{U}_1)H \text{ is FIR}$$

and

$$(H_{11} \quad H_{12})\begin{pmatrix} \tilde{V}_2 \\ -\tilde{V}_1 \end{pmatrix} \text{ is FIR.}$$

6.5. Consider the matrix $\hat{U}$ given by

$$\begin{pmatrix} \hat{u}_1 \\ \hat{u}_2 \end{pmatrix}.$$

To compute the zero and rank interpolation conditions, the following procedure is suggested:

(a) Augment the matrix with delays of order N

$$\hat{U}_N = \begin{pmatrix} \hat{u}_1 & 0 \\ \hat{u}_2 & \lambda^N \end{pmatrix}$$

(b) Calculate the zeros and their directions of $\hat{U}_N$.

Explain how the zeros and the associated directions of $\hat{U}_N$, for $N = 1, 2, \ldots$, provide all the necessary information needed for the zero and rank interpolation conditions associated with $\hat{U}$.

6.6. In this exercise we would like to verify that periodic controllers do not improve the performance for LTI systems when measured by some induced norm, even though they enlarge the set of all achievable closed-loop maps. Consider the following two problems.

$$v_{TI} = \inf_{\substack{Q \text{ stable, time invariant}}} \| H - UQV \|_{\ell_p - ind},$$

$$v_P = \inf_{\substack{Q \text{ stable, } N\text{-Periodic}}} \| H - UQV \|_{\ell_p - ind}.$$

Show that

$$v_{TI} = v_P.$$

Hint: Show that for any periodic Q there exists a time-invariant Q such that its performance is at least as good as the periodic one. Use the average-delayed operator in Exercise 2.12

6.7. Consider the parametrization of all stabilizing controllers for a periodic plant (see Exercise 5.13)

(a) After lifting all systems, show that the closed-loop maps are parametrized as follows:

$$\tilde{\Phi} = \tilde{H} - \tilde{U}\tilde{Q}\tilde{V}$$

where $\tilde{H}, \tilde{U}$ and $\tilde{V}$ are known LTI systems that satisfy the causality condition, and $\tilde{Q}$ is an arbitrary stable system that also satisfies the causality condition.

(b) Suppose $\tilde{U}$ and $\tilde{V}$ are square systems. Show that a given $\tilde{\Phi}$ is feasible if and only if $\tilde{\Phi}$ satisfies the zero interpolation conditions of $\hat{U}$ and $\hat{V}$, as well as extra linear constraints to guarantee that $\tilde{Q}$ satisfies the causality condition. Show how to obtain these linear constraints.

6.8. Let a plant $\hat{P}$ have a zero in the unit disc at $\lambda = a$ (assume a is real). Let $z = (I - PK)^{-1}w_0$ be the output due to a unit step disturbance w_0 starting at $t = 0$. Suppose we wanted the output z to stay in the following time-template

$$|z(t)| \leq \begin{cases} M & 0 \leq t \leq T - 1 \\ m & t \geq T \end{cases}$$

where $M > 1 > m > 0$, and $T > 0$. Show that for a stabilizing controller to meet these specifications, we must have

$$T \log \frac{1}{|a|} \geq \log \left(\frac{M - m}{M - \gamma} \right), \quad \gamma = \frac{1 - |a|}{1 - a}.$$

Explain the trade-offs in this problem.

6.9. It is standard by now that the SISO model matching problem can be posed as an interpolation problem. Let $\hat{U}$ have N zeros in the disc, $a_1, \ldots a_N$, and $\hat{H}(a_i) = b_i, i = 1, \ldots N$ (assume the zeros are simple). Then there exists a $\hat{Q} \in \mathcal{RH}_\infty$ such that $\|\hat{H} - \hat{U}\hat{Q}\|_\infty \leq 1$ if and only if there exists a $\hat{\Phi} \in \mathcal{RH}_\infty$ such that

$$\|\hat{\Phi}\|_\infty \leq 1 \quad \text{and} \quad \hat{\Phi}(a_i) = b_i, \quad i = 1, \ldots N.$$

The last problem is called the Nevanlinna-Pick interpolation problem (NP). Define a matrix F as follows:

$$(F)_{ij} = \frac{1 - b_i \bar{b}_j}{1 - a_i \bar{a}_j}, \quad i = 1, \ldots N, \ j = 1, \ldots N,$$

where $\bar{a}$ denotes the complex conjugate of the complex number a. The following result holds: The NP problem has a solution if and only if F is positive semidefinite.

We would like to prove that if NP has a solution, then $F \geq 0$. We will not prove the other direction.

(a) Consider the output signal $y = \Phi u$. Show that if $u \in \ell_2(\mathbf{Z})$ and $\|\Phi\|_\infty \leq 1$, then

$$\sum_{t=-\infty}^{0} |y(t)|^2 \leq \sum_{t=-\infty}^{0} |u(t)|^2.$$

(b) If $|a| < 1$, define

$$u(t) = a^{-t}, \quad -\infty < t \leq 0,$$

and zero otherwise. Show that $y = \Phi u$ satisfies:

$$y(t) = \hat{\Phi}(a)a^{-t}, \quad -\infty < t \leq 0.$$

(c) Use the above parts to prove that if NP has a solution, then $F \geq 0$, i.e., positive semi-definite. (Hint: Let $u = \sum_i c_i a_i^{-t}$ with $c_i \in \mathbf{C}$. Use part 2 and rearrange.)

NOTES AND REFERENCES

The equivalence between an affine subspace in ℓ_1 and a subspace characterized by interpolation conditions for SISO systems is a consequence of Wiener's theorem [Con85, Rud73a]. This equivalence was heavily used in the $\mathcal{H}_\infty$ theory (see [Vid85] and references therein), and later in the ℓ_1 theory (see [DP87a, DP88a, MP91, Sta90, Vid91]). The characterization of zero and rank interpolation conditions follow [DBD93]. Rank interpolation conditions were also called convolution conditions in [Sta90, Sta91]. The terminology *one-block, four-block problems* is commonly used in the $\mathcal{H}_\infty$ literature [Fra87].

Results on Rational Matrices can be found in [Kai80] and [Bar84]. The theory of zeros of MIMO systems has been studied extensively, both from algebraic and state-space perspectives [DS74, MK76, SS89]. Symbolic methods for computing the Smith-McMillan decomposition were proposed in [BMM90] from which zeros and their directions could be computed. Computing zeros using generalized eigenvalue problems was discussed in [MK76]; see also [END82]. For numerical difficulty analysis associated with rank determination (and the QR algorithm) see [GV83].

A numerically stable method for computing zeros was proposed in [DDV79] to find the structural indices associated with poles and zeros of a stable rational matrix $\hat{H}$, by looking at the rank of $T_{\lambda_0,\sigma}(\hat{H})$ as σ increases. Such an approach, however, does not provide the directional information necessary to construct the interpolation conditions. In [DBD93] an extension of the ideas in [DDV79] was proposed by looking at the structure of the null-space of $T_{\lambda_0,\sigma}(\hat{H})$ for increasing values of σ. This approach has strong connections with the general interpolation theory of rational matrix functions [BGR88, BR90]. In particular, it exploits the analyticity of the matrices $\hat{U}$ and $\hat{V}$ in the disc.

The frequency domain trade-offs presented are the discrete-time version of the results presented in [FL88]. It is evident that there is no analog of the Bode integral formula for discrete-time systems. Proofs of the problems in Exercises 6.1, 6.2, 6.3, 6.4, and 6.5 can be found in [Dah92, DBD93, MP91]. The proof of Exercise 6.6 and the generalization to arbitrary time-varying operators can be found in [SD91, CD92]. For more details on periodic and multirate systems, see [DVV92, Mey90b, VDVa]. The Nevanlinna-Pick interpolation problem (Exercise 6.9) is discussed in [DFT92].

CHAPTER 7 ——————————

Stability and Performance Robustness

In general, controller design is performed on a mathematical model of the actual process. This model rarely describes the process accurately, and may behave quite differently for certain classes of inputs. In general, delays, nonlinearities, time variations, and other kinds of dynamics that are difficult to model are ignored. The mismatch between the model and the actual process should be represented in a certain way to avoid exciting unmodeled dynamics that could cause serious deterioration in the stability and performance of the system. This can be accomplished by introducing plant perturbations in various locations of the closed-loop system and then designing the controller to guard against their worst-case behavior.

In this chapter we discuss the stability robustness and performance robustness problems. Plant perturbations are all assumed to be norm-bounded but possibly structured. The aim is to capture the exact necessary and sufficient conditions on the closed-loop map that guarantee the stability and performance of the system in the presence of such plant perturbations. The cases where the perturbations are allowed to be arbitrary nonlinear time varying, nonlinear time invariant, linear time varying, or linear time invariant are all analyzed. Real parameter perturbations are discussed in special cases.

Certain performance robustness problems can be made equivalent to stability robustness problems by adding an extra block of perturbations to capture performance. This procedure addresses very specialized classes of problems, and not every performance specification can be incorporated in this fashion. We discuss these classes of problems in this chapter.

7.1 THE GENERAL PROBLEM

Before we start discussing the general robustness problem, we present a general framework that helps in setting up a wide range of problems in a unified fashion. This framework uses linear fractional transformations, which were defined in Chapter 3.

7.1.1 The General Setup Through LFT's

Consider a 3×3 block system matrix G as shown in Figure 7.1, mapping

$$\begin{pmatrix} v \\ w \\ u \end{pmatrix} \longrightarrow \begin{pmatrix} r \\ z \\ y \end{pmatrix},$$

where $v = \Delta r$, for some Δ, and $u = Ky$. The signals w and z denote the exogenous inputs and regulated outputs, respectively.

The transfer function matrix mapping

$$\begin{pmatrix} v \\ w \end{pmatrix} \longrightarrow \begin{pmatrix} r \\ z \end{pmatrix}$$

with $u = Ky$ (with the upper loop open) can be written as the lower LFT, $F_\ell(G, K)$, corresponding to G partitioned conformally with

$$\begin{pmatrix} v \\ w \\ \cdots \\ u \end{pmatrix} \longrightarrow \begin{pmatrix} r \\ z \\ \cdots \\ y \end{pmatrix}.$$

Similarly, the transfer matrix mapping

$$\begin{pmatrix} w \\ u \end{pmatrix} \longrightarrow \begin{pmatrix} z \\ y \end{pmatrix}$$

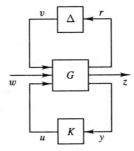

Figure 7.1 Perturbations through LFT.

with $v = \Delta r$ (with the lower loop open) can be written as the upper LFT, $F_u(G, \Delta)$, corresponding to G partitioned conformally with

$$\begin{pmatrix} v \\ \cdots \\ w \\ u \end{pmatrix} \longrightarrow \begin{pmatrix} r \\ \cdots \\ z \\ y \end{pmatrix}.$$

Then, the map from w to z that results from closing both loops (upper and lower), denoted by T_{zw}, can be expressed in terms of the corresponding LFT's, as follows

$$T_{zw} = F_\ell(F_u(G, \Delta), K) = F_u(F_\ell(G, K), \Delta).$$

For convenience, we define

$$M := F_\ell(G, K).$$

Example 7.1.1

Consider the case where the system has both input uncertainty and output uncertainty, as shown in Figure 7.2. Define new variables v_1, v_2, r_1 and r_2, where v_i is the output of Δ_i and r_i is the input to Δ_i. Let G be the transfer matrix from

$$\begin{pmatrix} v_1 \\ v_2 \\ w \\ u \end{pmatrix} \longrightarrow \begin{pmatrix} r_1 \\ r_2 \\ z \\ y \end{pmatrix}$$

and is given by

$$G = \begin{pmatrix} 0 & 0 & 0 & I \\ P_0 W_1 & 0 & 0 & P_0 \\ P_0 W_1 & W_2 & I & P_0 \\ P_0 W_1 & W_2 & I & P_0 \end{pmatrix}.$$

Then, the set of allowable perturbations has the form

$$\begin{pmatrix} \Delta_1 & 0 \\ 0 & \Delta_2 \end{pmatrix},$$

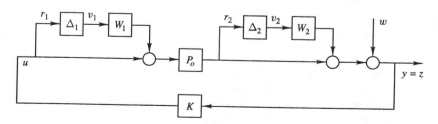

Figure 7.2   Input and output perturbations.

and the set of all admissible plants has the form

$$P = F_u(G, \Delta).$$

The closed-loop map from w to z is given by

$$T_{zw} = F_\ell(F_u(G, \Delta), K) = F_u(F_\ell(G, K), \Delta),$$

where $F_\ell(G, K)$ is given by

$$F_\ell(G, K) = \begin{pmatrix} 0 & 0 & 0 \\ P_0 W_1 & 0 & 0 \\ P_0 W_1 & W_2 & I \end{pmatrix} + \begin{pmatrix} I \\ P_0 \\ P_0 \end{pmatrix} K(I - P_0 K)^{-1} (P_0 W_1 \quad W_2 \quad I).$$

It is not hard to see that any class of perturbations occurring at different locations of the closed loop, in feedforward or feedback form, can always be put in linear fractional form, with block diagonal perturbations. The above example illustrates the procedure for doing this:

1. Define the signal r_i and v_i as the input and output of the perturbation block Δ_i.
2. Define G as the map between the inputs v, w, u and the outputs r, z, y.
3. Admissible plants have the form

$$P = F_u(G, \Delta)$$

   with $v = \Delta r$.
4. Admissible perturbations have the form

$$\Delta = \text{diag}(\Delta_1, \dots, \Delta_n).$$

5. The closed-loop map T_{zw} is given by

$$T_{zw} = F_\ell(F_u(G, \Delta), K) = F_u(F_\ell(G, K), \Delta)$$

   with $u = Ky$.

7.1.2 Stability and Performance Robustness

To address the problem of stability and performance robustness, we need to define how the perturbations connect to the rest of the nominal system. The notation just introduced becomes handy at this point. Let the set Δ denote the class of allowable perturbations. This set in general carries the structure information of the perturbations. For example, Δ may be the set of all diagonal perturbations of the form

$$\begin{pmatrix} \Delta_1 & 0 \\ 0 & \Delta_2 \end{pmatrix},$$

where Δ_1 is linear-time invariant ℓ_∞-stable, and Δ_2 is time-varying ℓ_∞-stable. Other combinations of parametric, linear, or nonlinear perturbations are also possible. In general, Δ will have a block diagonal structure. The subset of Δ containing elements with norm less than 1 is denoted by $\mathbf{B}_{\Delta, p}$, i.e.,

$$\mathbf{B}_{\Delta, p} = \left\{ \Delta \in \Delta \,\middle|\, \|\Delta\|_{\ell_p-\text{ind}} < 1 \right\}. \tag{7.1}$$

Consider the class of plants $\Omega(G, \Delta)$ described as

$$\Omega(G, \Delta) = \{P | P = F_u(G, \Delta) \text{ for some } \Delta \in \mathbf{B}_{\Delta, p}\}.$$

As it was shown in Example 7.1.1 and the procedure that followed, the set $\Omega(G, \Delta)$ represents a wide class of plant uncertainty. Consider the systems in Figures 7.3 and 7.4.

Stability robustness problem. Find necessary and sufficient conditions on M in Figure 7.3 such that the system is ℓ_p stable for all $\Delta \in \mathbf{B}_{\Delta, p}$.

Performance robustness problem. Find necessary and sufficient conditions on M in Figure 7.4 such that the system is ℓ_p stable for all $\Delta \in \mathbf{B}_{\Delta, p}$ and

$$\|T_{zw}\|_{\ell_p-\text{ind}} \leq 1 \quad \text{for all } \Delta \in \mathbf{B}_{\Delta, p}.$$

Notice that the signals w and z are eliminated from the stability robustness problem, and hence M denotes a different transfer function in each of the above problems.

We start with the stability robustness problem. The conditions on M will eventually depend on two things: the class of perturbations Δ and the notion of stability. Afterwards, we discuss the performance robustness problem.

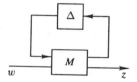

Figure 7.3 Stability robustness problem.   **Figure 7.4** Performance robustness.

7.2 STRUCTURED SMALL GAIN THEOREM

In this section we discuss the general stability robustness problem in the presence of structured uncertainty. This is accomplished by defining a function that gives the exact necessary and sufficient conditions for robust stability. This function is, in some sense, a restatement of the definition of robust stability and will be denoted by SN (Structured Norm).

Definition 7.2.1. The Structured Norm, SN, is a map from the space of stable systems to the nonnegative reals defined as

$$SN_{\Delta, p}(M) = \frac{1}{\inf_\Delta \left\{\|\Delta\|_{\ell_p-\text{ind}} \left| (I - M\Delta)^{-1} \text{is not } \ell_p\text{-stable}\right.\right\}},$$

if for every $\Delta \in \Delta$, $(I - M\Delta)^{-1}$ is ℓ_p-stable, then $SN_{\Delta, p}(M) := 0$.   ∎

The structured norm is not a norm. This is clear since there are nonzero matrices for which $SN_{\Delta, p}(M) = 0$. Also, it does not satisfy the triangle inequality.

Notice that the function $SN_{\Delta,p}(M)$ depends on the perturbation structure, and on the sense of stability. It is straightforward to verify directly from the definition that

$$(I - M\Delta)^{-1} \text{ is } \ell_p\text{-stable for all } \Delta \in \mathbf{B}_{\Delta,p} \quad \Leftrightarrow \quad SN_{\Delta,p}(M) \le 1.$$

This, we denote as the *Structured Small Gain Theorem*. A natural question to ask at this point is the following: When is the function $SN_{\Delta,p}(M)$ computable? It is not hard to guess that obtaining an algorithm to compute this function for any given class of perturbations may not be possible. Nevertheless, computing $SN_{\Delta,p}(M)$ may be an ambitious objective and one may very well be satisfied with approximations of it. In some special cases that cover a wide range of applications, we show how to compute $SN_{\Delta,p}(M)$ exactly.

An upper bound of $SN_{\Delta,p}(M)$.   An immediate application of the small gain theorem shows that

$$SN_{\Delta,p}(M) \le \|M\|_{\ell_p-\text{ind}}.$$

To see this, notice that if $\|\Delta\|_{\ell_p-\text{ind}} < 1/\|M\|_{\ell_p-\text{ind}}$, then $(I - M\Delta)^{-1}$ is ℓ_p-stable.

Clearly, this bound does not take into consideration any information on the structure of the perturbations, captured in the set Δ. We can exploit some of this information in the following way. Define the set $\mathbf{D}$ as follows:

$$\mathbf{D} = \big\{ D \,\big|\, D,\ D^{-1} \text{ are } \ell_p\text{-stable},\ D^{-1}\Delta D \in \Delta,$$

$$\text{and } \|D^{-1}\Delta D\|_{\ell_p-\text{ind}} = \|\Delta\|_{\ell_p-\text{ind}} \text{ for all } \Delta \in \Delta \big\}.$$

This set includes systems that are stable, with a stable inverse such that if $\Delta \in \mathbf{B}_{\Delta,p}$ then $D\Delta D^{-1} \in \mathbf{B}_{\Delta,p}$ for every $D \in \mathbf{D}$. From the definition of $SN_{\Delta,p}(M)$, it follows that

$$SN_{\Delta,p}(M) = SN_{\Delta,p}(D^{-1}MD).$$

This result has an easy interpretation. Consider the closed-loop system in Figure 7.5. The stability of this system is equivalent to the stability of the system in Figure 7.3; after all, it is the same system. But $D^{-1}\Delta D \in \mathbf{B}_{\Delta,p}$, implies that the stability robustness condition for M and $D^{-1}MD$ is the same, for any $D \in \mathbf{D}$. It follows from the small gain theorem, applied on Figure 7.5, that

$$SN_{\Delta,p}(M) \le \|D^{-1}MD\|_{\ell_p-\text{ind}} \quad \text{for all } D \in \mathbf{D}.$$

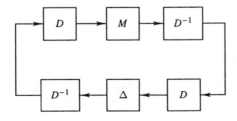

Figure 7.5   Scaled system.

Since D is arbitrary in $\mathbf{D}$ and the left-hand side does not depend on D, it follows that

$$SN_{\Delta, p}(M) \leq \inf_{D \in \mathbf{D}} \| D^{-1} M D \|_{\ell_p - \text{ind}}. \tag{7.2}$$

The quantity on the right is a scaled norm and gives a tighter upper bound on $SN_{\Delta, p}(M)$ that exploits the structure of Δ. However, it is not clear how tight the scaled norm is and whether or not it is equal to $SN_{\Delta, p}(M)$. This problem is discussed in later sections. In the sequel, we will be concerned with three classes of perturbations: The first containing linear time-varying perturbations, the second containing nonlinear perturbations, and the third containing linear time-invariant perturbations.

Time-varying versus time-invariant perturbations. As presented in Example 7.1.1, perturbation blocks are introduced to accommodate unmodeled dynamics, such as actuator and sensor dynamics, as well as modeling errors in the actual process. Some of these blocks may have to include time-varying and nonlinear systems to faithfully represent the ignored dynamics. These dynamics may not be as sinister as an arbitrary block of norm-bounded time-varying or nonlinear perturbations. However, it is quite difficult to represent them in any other way that is mathematically tractable. On the other hand, time-invariant perturbations may not capture the real environment, since nonlinearities and time variations always exist in a system. By studying the three models of perturbations, more insight into the trade-offs can be obtained under various modeling assumptions.

7.3 STABILITY ROBUSTNESS WITH LTV PERTURBATIONS

The space of all time-varying causal perturbations was discussed earlier and is denoted by $\mathcal{L}$. Consider the following class of perturbations:

$$\Delta_{LTV} = \left\{ \text{diag}(\Delta_1, \Delta_2, \ldots \Delta_n) \,\middle|\, \Delta_i \in \mathcal{L}^{p_i \times p_i} \right\}. \tag{7.3}$$

This represents the class of block diagonal, linear time-varying (LTV), causal perturbations. We have assumed that each block is a square matrix of dimension $p_i \times p_i$. The extension of the following development for nonsquare matrices is immediate. The various blocks in the perturbations are not related, nor dependent. Of course, there are interesting situations in which some blocks may be repeated, but they are not discussed here. At this point, it is not necessary to assume that this set contains only stable perturbations. The unstable elements do not alter the definition of the structured norm. Finally, the set $\mathbf{B}_{\Delta_{LTV}, p}$ is defined for a given ℓ_p space as in Equation (7.1).

Associated with this class is a set of scalings that commute with the set of perturbations. A reasonable and simple set to pick is

$$\mathbf{D} = \left\{ \text{diag}(d_1 I_{p_1}, d_2 I_{p_2} \ldots d_n I_{p_n}) \,\middle|\, d_i \in \mathbb{R}, d_i > 0 \right\}. \tag{7.4}$$

For this class of scales, we have

$$D^{-1} \Delta D = \Delta \quad \text{for all } \Delta \in \Delta_{LTV}.$$

The set $\mathbf{D}$ contains elements that commute with each individual perturbation.

Consider the interconnection of a stable linear time-invariant system M, with a structured perturbation $\Delta \in \mathbf{B}_{\Delta_{LTV}, p}$ as in Figure 7.3. We seek necessary and sufficient conditions for the stability robustness of such a closed-loop system. Since, M and Δ are both stable, the internal stability of the system is equivalent to the map $(I - M\Delta)$ having a ℓ_p-stable inverse.

7.3.1 ℓ_∞ Stability Robustness Conditions

We begin with the case of n SISO perturbation blocks, i.e., $p_i = 1$ for all $i = 1, \ldots n$. The structured norm for the space ℓ_∞ for time-invariant M satisfies

$$SN_{\Delta_{LTV}, \infty}(M) \leq \inf_{D \in \mathbf{D}} \|D^{-1}MD\|_1. \tag{7.5}$$

Before we present the main result, a certain nonnegative matrix, $\widehat{M}$, which depends solely on M, will be defined. M has n inputs and n outputs (corresponding to n perturbation blocks), and can be partitioned as follows:

$$M = \begin{pmatrix} M_{11} & \cdots & M_{1n} \\ \vdots & & \vdots \\ M_{n1} & \cdots & M_{nn} \end{pmatrix}.$$

Each M_{ij} is linear, time invariant, and stable, and thus $M_{ij} \in \ell_1$. We can now define $\widehat{M}$ as follows:

$$\widehat{M} = \begin{pmatrix} \|M_{11}\|_1 & \cdots & \|M_{1n}\|_1 \\ \vdots & & \vdots \\ \|M_{n1}\|_1 & \cdots & \|M_{nn}\|_1 \end{pmatrix}.$$

The following proposition relates the spectral radius of this matrix to the upper bound in Equation (7.5). The proof is a consequence of Perron-Frobenius theorem for positive matrices, which is stated first.

Theorem 7.3.1.   Let A be a nonnegative, primitive matrix, i.e., A^k has positive elements for some $k > 0$.

1. The spectral radius of A, $\rho(A)$, is a simple eigenvalue and the associated eigenvector contains only positive elements.
2. If there exists a positive vector x and a positive real number λ such that $Ax = \lambda x$, then $\rho(A) = \lambda$.   ∎

Proposition 7.3.1.   The following conditions are equivalent:

1. $\rho(\widehat{M}) \leq 1$, where $\rho(\cdot)$ denotes the spectral radius.
2. The system of inequalities $x < \widehat{M}x$ and $x \geq 0$ has no solutions, where the vector inequalities are to be interpreted componentwise.
3. $\inf_{D \in \mathbf{D}} \|D^{-1}MD\|_1 \leq 1$, where $\mathbf{D} := \{\operatorname{diag}(d_1, \ldots, d_n) : d_i > 0\}$.

Proof. We will show that $1 \Leftrightarrow 2$, and that $1 \Leftrightarrow 3$. For simplicity, we do this for an irreducible $\widehat{M}$ (has positive elements). So suppose that $\rho(\widehat{M}) \leq 1$. It follows that $(I - \gamma\widehat{M})^{-1}$ exists for any $\gamma < 1$. Since $(I - \gamma\widehat{M})^{-1} = I + \gamma\widehat{M} + \gamma^2\widehat{M}^2 + \dots$, all of its entries will be positive. Now if $x \geq 0$ is such that $x < \widehat{M}x$, then $(I - \gamma\widehat{M})x < 0$ for some $\gamma < 1$. Multiplying the latter by $(I - \gamma\widehat{M})^{-1}$ implies that $x < 0$. Then the condition in 2 holds. To show that 2 implies 1, suppose 1 does not hold, i.e., that $\rho(\widehat{M}) > 1$. Theorem 7.3.1 states that $\rho(\widehat{M})$ is itself a simple, real, and positive eigenvalue of $\widehat{M}$. Moreover, associated with $\rho(\widehat{M})$ we can find an eigenvector x' with strictly positive components. This implies that $\rho(\widehat{M})x' = \widehat{M}x'$, which in turn implies that 2 does not hold. Thus, we have demonstrated that $1 \Leftrightarrow 2$.

We now show $1 \Leftrightarrow 3$ by showing that $\rho(\widehat{M}) = \inf_{D \in \mathbf{D}} \|D^{-1}MD\|_1$. By definition,

$$\|D^{-1}MD\|_1 = \max_i \sum_{j=1}^n \|(D^{-1}MD)_{ij}\|_1 = \max_i \sum_{j=1}^n \frac{d_j}{d_i}\|M_{ij}\|_1.$$

The expression on the right is also equal to the induced norm of the matrix $D^{-1}\widehat{M}D$ as a map from $(\mathbb{R}^n, |\cdot|_\infty)$ to itself. Referring to this norm by $|\cdot|_1$, we therefore have $\|D^{-1}MD\|_1 = |D^{-1}\widehat{M}D|_1$. Since any matrix norm bounds from above the spectral radius of that matrix we have

$$\inf_{D \in \mathbf{D}} \|D^{-1}MD\|_1 = \inf_{D \in \mathbf{D}} |D^{-1}\widehat{M}D|_1 \geq \inf_{D \in \mathbf{D}} \rho(D^{-1}\widehat{M}D) = \rho(\widehat{M}).$$

But if we choose $D = \mathrm{diag}(d_1', \dots, d_n')$, where $(d_1', \dots, d_n')^T$ is the positive eigenvector corresponding to the eigenvalue $\rho(\widehat{M})$, the inequality becomes an equality and the equivalence between 1 and 3 is established. ∎

It is interesting to note that for the optimum scalings $D = \mathrm{diag}(d_1', \dots, d_n')$, all the rows of $D^{-1}MD$ have the same norm. As will be demonstrated shortly, this fact is used to show why condition 3 in the above theorem is necessary for system robustness.

A corollary of this proposition is a method for computing the optimal scaling D. This is summarized in the following.

Corollary 7.3.1.   Assume that $\widehat{M}$ is irreducible (has positive elements). Let x be the positive eigenvector associated with the maximum eigenvalue. Then the optimal scaling D_o defined as

$$\min_{D \in \mathbf{D}} \|D^{-1}MD\|_1 =: \|D_o^{-1}MD_o\|_1$$

is given by

$$D_o = \mathrm{diag}(x_1, \dots x_n).$$

Also, at the optimal scaling, all the rows of $D_o^{-1}MD_o$ have equal norms. ∎

Combining Proposition 7.3.1 with the upper bound on the structured norm, we get:

$$SN_{\Delta_{LTV}, \infty}(M) \leq \min_{D \in \mathbf{D}} \|D^{-1}MD\|_1 = \rho(\widehat{M}).$$

The next theorem shows that the upper bound is in fact equal to $SN_{\Delta_{LTV}, \infty}(M)$.

Theorem 7.3.2.   The system in Figure 7.3 achieves robust stability for all $\Delta \in \mathbf{B}_{\Delta_{LTV}, \infty}$ if and only if

$$\min_{D \in \mathbf{D}} \|D^{-1}MD\|_1 \leq 1.$$

Equivalently, the Structured Norm can be computed exactly and is given by

$$SN_{\Delta_{LTV}, \infty}(M) = \min_{D \in \mathbf{D}} \|D^{-1}MD\|_1 = \rho(\widehat{M}).$$

Proof. The fact that this condition is sufficient has been shown earlier.

We now demonstrate that $\inf_{D \in \mathbf{D}} \|D^{-1}MD\|_1 \leq 1$ is necessary for robust stability. To simplify the exposition, we show the proof for the case $n = 2$. The approach is to show how we can construct a destabilizing perturbation $\Delta \in \Delta$ whenever $\inf_{D \in \mathbf{D}} \|D^{-1}MD\|_1 > 1$. So suppose that $\inf_{D \in \mathbf{D}} \|D^{-1}MD\|_1 \geq \gamma > 1$. We have previously shown that this infimum is in fact a minimum, and it is achieved by an optimum scaling, D, obtained from the eigenvector corresponding to $\rho(\widehat{M})$. It was also indicated that the two rows of $D^{-1}MD$ will have equal norms. This can be expressed as follows:

$$\|(D^{-1}MD)_1\|_1 = \|(D^{-1}MD)_2\|_1 = \|D^{-1}MD\|_1 \geq \gamma > 1,$$

where $(D^{-1}MD)_i$ denotes the ith row of $D^{-1}MD$. The proof is divided into two parts: The first is a construction of an unbounded signal that gets amplified componentwise by $\|D^{-1}MD\|_1$ at the optimum D, and the second is a construction of a destabilizing perturbation using this signal.

Construction of unbounded signals. $D^{-1}MD$ appears in Figure 7.6 and has $\xi = (\xi_1, \xi_2)$ as its input and $z = (z_1, z_2)$ as its output. In the figure, $y = (y_1, y_2)$ consists of the output $z = (z_1, z_2)$ after a bounded signal, the output of a sign function, has been added to it. This bounded signal is interpreted as an external signal

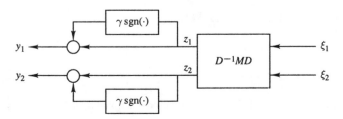

Figure 7.6 Signal construction.

injected for stability analysis. The strategy taken is to construct ξ satisfying the two requirements:

1. ξ is unbounded.
2. ξ results in a signal y, which satisfies $\| P_k \xi_i \|_\infty \leq \frac{1}{\gamma} \| P_k y_i \|_\infty$ for $i = 1, 2$, where P_k is the truncation operator.

The first requirement on ξ guarantees that if an admissible perturbation Δ were to map y to ξ, it would be a destabilizing one because the bounded external signal would have produced an unbounded internal signal ξ. The second requirement guarantees that such an admissible perturbation exists. In other words, if ξ and y satisfy the second condition, then it is always possible to find Δ_i, for $i = 1, 2$, so that Δ_i is causal, has induced norm less than one, and satisfies $\Delta_i y_i = \xi_i$. If the first requirement is also met, this Δ will be the destabilizing perturbation.

For simplicity we assume that all M_{ij}'s have finite pulse response of length N. The construction of ξ proceeds as follows. While maintaining $|\xi_i(k)| \leq 1$ for $k = 0, \ldots, N - 1$, the first N components of ξ can be constructed so as to achieve $\|(D^{-1}MD)_1\|_1$. Since $\|(D^{-1}MD)_1\|_1 \geq \gamma$, this implies that $\| P_{N-1} z_1 \|_\infty \geq \gamma$, which in turn implies that $\| P_{N-1} y_1 \|_\infty \geq 2\gamma$. Next, while still maintaining $|\xi_i(k)| \leq 1$, we pick the next N components of ξ so as to achieve the second row norm, $\|(D^{-1}MD)_2\|_1$. As a result we have $\| P_{2N-1} z_2 \|_\infty \geq \gamma$, which implies that $\| P_{2N-1} y_2 \|_\infty \geq 2\gamma$. Note that the second requirement on ξ has been met for $k = 0, \ldots, 2N - 1$. In addition, because of the way the first $2N$ terms of ξ have been constructed, we have

$$\| P_{2N-1} y_i \|_\infty \geq \gamma \| P_{2N-1} \xi_i \|_\infty + \gamma, \qquad i = 1, 2.$$

This allows us to relax the restriction on $|\xi_i(k)|$ for $k > 2N - 1$ without violating the second requirement on ξ. Specifically, we now allow $|\xi_i(k)|$ to be as large as 2 for $k = 2N, \ldots, 4N - 1$. In the same way we have proceeded before, we can pick $\xi(k)$ for this range of k so as to satisfy

$$\| P_{4N-1} y_i \|_\infty \geq \gamma \| P_{4N-1} \xi_i \|_\infty + \gamma, \qquad i = 1, 2,$$

which allows us to increase $|\xi_i(k)|$ by 1 for the next $2N$ components of ξ, and repeat the whole procedure again. From this construction, it is clear that when ξ is completely specified it will be unbounded and hence meets the first requirement. The second requirement is also met since all along $\xi_i(k)$ was chosen carefully so as not to become too large too soon.

Construction of the destabilizing perturbation.
Now we show the construction for Δ_1. Moreover, Δ_2 follows the same way. Suppose we are given $\xi_1 = \{\xi_1(i)\}_{i=0}^\infty \in \ell$ and $y_1 = \{y_1(i)\}_{i=0}^\infty \in \ell$ such that $\| P_k \xi_1 \|_\infty \leq \frac{1}{\gamma} \| P_k y_1 \|_\infty$ for all k. The construction of Δ_1 is trivial if $y_1 = 0$: Just pick Δ_1 itself to be zero. So assume $y_1 \neq 0$. We start the construction of Δ_1 by identifying a subsequence of the y_1, say

$(y_1(i_1), y_1(i_2), \ldots)$. This subsequence may be defined recursively in the following manner: Let i_1 be the smallest integer such that $y_1(i_1) \neq 0$. Given $y_1(i_n)$, let i_{n+1} be the smallest integer greater than i_n such that $|y_1(i_{n+1})| \geq |y_1(i_n)|$. Using the $\xi_1(i)$'s and $y_1(i_j)$'s we are now ready to construct Δ_1 through specifying its matrix representation as follows:

$$
\Delta_1 = \begin{pmatrix}
\ddots & & & & & & & \\
 & \frac{\xi_1(i_1)}{y_1(i_1)} & & & & & & \\
 & \vdots & 0 & & & & & \\
 & \vdots & \vdots & \ddots & & & & \\
 & \frac{\xi_1(i_2-1)}{y_1(i_1)} & 0 & \cdots & 0 & & & \\
 & & & & & \frac{\xi_1(i_2)}{y_1(i_2)} & & \\
 & & & & & \vdots & 0 & \\
 & & & & & \vdots & \vdots & \ddots \\
 & 0 & & & & \frac{\xi_1(i_3-1)}{y_1(i_2)} & 0 & \cdots & 0 \\
 & & & & & & & & \frac{\xi_1(i_3)}{y_1(i_3)} \\
 & & & & & & & & \vdots & \ddots
\end{pmatrix},
$$

where the first nonzero column is i_1th, the second is the i_2th, and so on. Notice that each row of the above matrix has at most one nonzero element, which, by the choice of the $y_1(i_j)$'s, will have its absolute value less than or equal to $1/\gamma$. This implies that $\|\Delta_1\|_{\ell_\infty-\text{ind}} \leq 1/\gamma < 1$. Moreover, Δ_1 is clearly causal and it can be easily checked that $\Delta_1 y_1 = \xi_1$, which is what we wanted to show.

Finally, it is clear that the above construction can be repeated so that Δ is strictly causal. To do this, construct the sequences y and ξ such that $\xi(0) = 0$, $y(0) \neq 0$, and

$$
\|P_k \xi_i\| \leq \frac{1}{\gamma} \|P_{k-1} y_i\|_\infty, \quad i = 1, 2, \text{ and } k \geq 1.
$$

(See Exercise 7.1.) This guarantees the well posedness of the closed-loop system, implying that $(I - \Delta M)^{-1}$ exists, and is by construction, unstable. ∎

Conservatism of the small gain condition.   Clearly, the small gain condition $\|M\|_1 \leq 1$ is always sufficient for robust stability, regardless of the structure of Δ (as long as the perturbation belongs to the admissible class). The degree of conservatism introduced by such a condition is very much problem dependent, going from nonconservative for unstructured perturbations to arbitrarily conservative for problems with at least two perturbation blocks. To illustrate this point, consider the following simple example.

Example 7.3.1

Consider a first-order SISO plant with perturbations at the plant input and output, as shown in Figure 7.2, where

$$
P_0 = \frac{a}{1 - 0.5\lambda}.
$$

For simplicity, no weights will be included. Assume that the loop is closed through a static gain, $K = k$. We will analyze the stability robustness properties of such a system, for different values of a and k.

The transfer function matrix seen by the perturbations is given by

$$M = \begin{pmatrix} (1 - P_0 K)^{-1} P_0 K & (1 - P_0 K)^{-1} K \\ (1 - P_0 K)^{-1} P_0 & (1 - P_0 K)^{-1} P_0 K \end{pmatrix}.$$

It is a simple exercise to show that the matrix of norms, $\widehat{M}$, can be computed exactly as

$$\widehat{M} = \begin{pmatrix} \dfrac{2ak}{1 - 2ak} & \dfrac{k}{1 - 2ak} \\ \dfrac{2a}{1 - 2ak} & \dfrac{2ak}{1 - 2ak} \end{pmatrix},$$

with a, $k > 0$ and $ak < 0.5$. Let us fix the feedback gain to one (i.e., $k = 1$), and vary the value of the parameter a while preserving nominal stability, that is, $0 \leq a < 0.5$. Table 7.1 shows the result of applying the small gain theorem and the spectral radius condition for different values of a.

Clearly, as $a \to 0$ all entries of M vanish except the upper right entry, which tends to one. This strong directional properties of M are unnoticed by the small gain theorem test, which measures the gain of M regardless of input-output directions. As a result, even when the loop is opened at the plant (i.e., $a = 0$), this test points to a maximum tolerable perturbation of induced norm equal to one. In contrast, the spectral radius test captures such directional information, and provides a tight margin (i.e., infinite margin for the extreme case $a = 0$).

Computational aspects.   One of the main contributions of Theorem 7.3.2 is that it provides simple and exact conditions for testing the system's stability robustness regardless of the number of perturbation blocks, n. This theorem, combined with Proposition 7.3.1, provides three equivalent conditions for robust stability. Each of these conditions provides a different perspective and has certain advantages over the

TABLE 7.1 COMPARISON OF
MAXIMUM TOLERABLE PERTURBATIONS

a	Maximum tolerable perturbation $\|\Delta\|_{\ell_\infty -\mathrm{ind}}$	
	Unstructured $1/\|M\|_1$	Structured $1/\rho(\widehat{M})$
0.49	0.0102	0.0180
0.1	0.8098	1.3401
0.01	0.9802	6.1255
0.001	0.9980	21.381
0.0001	0.9998	69.718
$\vdots$	$\vdots$	$\vdots$
0	1	∞

others. For example, the spectral radius condition is in general the easiest to compute. It is particularly useful when $\widehat{M}$ is large since it can be computed efficiently using power methods. Specifically, given an $\widehat{M}$ that is assumed primitive (i.e., $\widehat{M}^k > 0$ for some integer k), then for any x it satisfies

$$\min_i \frac{(\widehat{M}^{k+1}x)_i}{(\widehat{M}^k x)_i} \leq \rho(\widehat{M}) \leq \max_i \frac{(\widehat{M}^{k+1}x)_i}{(\widehat{M}^k x)_i}.$$

Furthermore, the upper and lower bounds both converge to $\rho(\widehat{M})$ as k goes to infinity. If $\widehat{M}$ is not primitive, it can be perturbed slightly to become primitive. Whereas the spectral radius test provides a yes or no answer concerning system robustness, the second test involving the Linear Matrix Inequality (LMI) is most useful for providing information about the effect of the individual entries of $\widehat{M}$ on the overall robustness of the system. This is achieved by translating the LMI condition into n algebraic conditions stated explicitly in terms of the entries of $\widehat{M}$. This is best demonstrated by an example. Suppose $\widehat{M}$ is a 2×2 matrix corresponding to a certain robustness problem with $n = 2$. The LMI condition states that robust stability is attained if and only if the system of inequalities:

$$\gamma x_1 \leq \|M_{11}\|_1 x_1 + \|M_{12}\|_1 x_2,$$

$$\gamma x_2 \leq \|M_{21}\|_1 x_1 + \|M_{22}\|_1 x_2,$$

for any $\gamma > 1$ has no solution whenever x_1 and x_2 are strictly positive. Among other things, this implies that $\|M_{11}\|_1 \leq 1$; otherwise $x = (1, \epsilon)$ (for some small ϵ) would be a solution for the two inequalities. The first inequality can be rewritten as

$$x_1 \leq \frac{\|M_{12}\|_1}{\gamma - \|M_{11}\|_1} x_2.$$

When combined with the second inequality, we have that

$$\gamma x_2 \leq \left(\|M_{21}\|_1 \frac{\|M_{12}\|_1}{\gamma - \|M_{11}\|_1} + \|M_{22}\|_1 \right) x_2$$

has no solution in $(0, \infty)$, which is equivalent to

$$\|M_{21}\|_1 \frac{\|M_{12}\|_1}{\gamma - \|M_{11}\|_1} + \|M_{22}\|_1 \leq 1.$$

This last condition, together with the condition that $\|M_{11}\|_1 \leq 1$, is therefore necessary for the inequality robustness conditions to hold. By retracing our steps backwards, it becomes clear that they are also sufficient. This procedure of constructing explicit norm conditions from the second robustness conditions can be repeated in the same way for any n. Notice that this form of stability robustness conditions can be useful in pointing out the interplay of the different gains in the system. If for a given problem the test fails, then a close look at these conditions can reveal which elements in the system are most critical. For instance, if either M_{12} or M_{21} is negligible, then the conditions reduce to the decoupled case, i.e., $\|M_{11}\|_1 \leq 1$ and $\|M_{22}\|_1 \leq 1$, although the system may not be diagonal. Such insight can be useful in guiding the design process.

Finally, the third robustness condition is useful for robust controller synthesis. This is discussed in more detail later on.

MIMO stability robustness conditions.   Here, stability robustness conditions are developed for the general MIMO problem. The results are simple generalizations of the SISO case.

Let $p = \sum_i p_i$. The matrix M is a $p \times p$ matrix, and can be partitioned as follows:

$$
M = \begin{pmatrix} M_{11} & \cdots & M_{1n} \\ \vdots & & \vdots \\ M_{n1} & \cdots & M_{nn} \end{pmatrix},
$$

where M_{ij} is $p_i \times p_i$ system. Also, let the set J be an index set for all possible collections of rows from the row blocks, and for each $j = (j_1 \ldots j_n) \in J$, define the submatrix $\widehat{M}_j$ as follows:

$$
\widehat{M}_j = \begin{pmatrix} \|(M_{11})_{j_1}\|_1 & \cdots & \|(M_{1n})_{j_1}\|_1 \\ \vdots & & \vdots \\ \|(M_{n1})_{j_n}\|_1 & \cdots & \|(M_{nn})_{j_n}\|_1 \end{pmatrix},
$$

where $(M_{ik})_{j_r}$ is the j_rth row of the block matrix M_{ik}. The following theorem shows how the stability robustness conditions are generalized for the MIMO case.

Theorem 7.3.3.   The system in Figure 7.3 is stable for all $\Delta \in \mathbf{B}_{\mathbf{\Delta}_{LTV}, \infty}$ if and only if one of the following holds:

1. $\displaystyle\inf_{D \in \mathbf{D}} \|DMD^{-1}\|_1 \leq 1$
2. $\| [d_1 M_{i1} \; \ldots \; d_i M_{ii} \; \ldots \; d_n M_{in}] \|_1 \leq d_i$ for all $i = 1, \ldots n$ and $d_i > 0$. These conditions are referred to as LMI (linear matrix inequalities).
3. $\rho(\widehat{M}_j) \leq 1$ for all $j \in J$.

Proof. The fact that $1 \Leftrightarrow 2$ is straightforward. To show that $1 \Leftrightarrow 3$ notice that

$$
\inf_{D \in \mathbf{D}} \|DMD^{-1}\|_1 = \inf_{d_1 \ldots d_n} \max_{j \in J} |\tilde{D}\widehat{M}_j \tilde{D}^{-1}|_1 \geq \max_{j \in J} \inf_{d_1 \ldots d_n} |\tilde{D}\widehat{M}_j \tilde{D}^{-1}|_1,
$$

where $\tilde{D} = \operatorname{diag}(d_1 \ldots d_n)$. If D is a minimizing solution of the left-hand side, then it can be shown that we can pick a $j^* \in J$ such that all the rows corresponding to the indices in j^* have norms equal to $\|DMD^{-1}\|_1$ (see Exercise 7.6). This implies that

$$
\inf_{D \in \mathbf{D}} \|DMD^{-1}\|_1 = \max_{j \in J} \inf_{d_1 \ldots d_n} |\tilde{D}\widehat{M}_j \tilde{D}^{-1}|_1,
$$

from which the equivalence follows.

Condition 1 is clearly sufficient for stability. To prove necessity, a destabilizing perturbation needs to be constructed. It follows that $\inf_{d_1 \ldots d_n} |\tilde{D}\widehat{M}_{j^*} \tilde{D}^{-1}|_1$ results in a system $\tilde{D}M_{j^*}\tilde{D}^{-1}$ with rows of equal norms (assuming $\widehat{M}_{j^*}$ is primitive). This is the same situation as the SISO problem, and a destabilizing perturbation can be constructed based on these rows only. The details of the construction are omitted. ∎

7.3.2 ℓ_2 Stability Robustness Conditions

In this section we develop the parallel theory for ℓ_2 stability. We will only discuss the case of SISO blocks. The structured norm for the space ℓ_2 satisfies

$$SN_{\Delta_{LTV},2}(M) \leq \inf_{D \in \mathbf{D}} \|D^{-1}\hat{M}D\|_\infty. \tag{7.6}$$

As in the ℓ_∞ case, the upper bound is equal to the structured norm. This is stated in the next theorem.

Theorem 7.3.4. The system in Figure 7.3 achieves robust stability for all $\Delta \in$ $\mathbf{B}_{\Delta_{LTV},2}$ if and only if

$$\inf_{D \in \mathbf{D}} \|D^{-1}\hat{M}D\|_\infty \leq 1.$$

Equivalently, the Structured Norm is given by

$$SN_{\Delta_{LTV},2}(M) = \inf_{D \in \mathbf{D}} \|D^{-1}\hat{M}D\|_\infty.$$

To prove this result, we first show the following two lemmas. For simplicity, we assume that M is FIR of length N_1. The results generalize in a straightforward way. Define

$$\kappa_i(f) := \|(Mf)_i\|_2^2 - \|f_i\|_2^2 \quad \text{for } i = 1, 2, \ldots n.$$

As usual, $(Mf)_i$ denotes the ith row of Mf.

Lemma 7.3.1. There exists a destabilizing perturbation $\Delta \in \mathbf{B}_{\Delta_{LTV},2}$ of the system in Figure 7.3, if there exists a function $f \in \ell_2^n$ such that

$$\kappa_i(f) > 0 \quad \text{for } i = 1, 2, \ldots n.$$

These conditions are called *Integral Quadratic Constraints*.

Proof. For simplicity, we will show all the results for the two block case, i.e., $n = 2$. Since $\kappa_i(f) > 0$, for $i = 1, 2$ then there exists an $N_2 \geq N_1$ and a $\gamma > 1$ such that

$$\|P_{N_2-1}(Mf)_i\|_2^2 \geq \gamma \|P_{N_2-1}f_i\|_2^2 \quad \text{for } i = 1, 2. \tag{7.7}$$

Without loss of generality, we can assume that f has a finite length N_2, i.e., $f(k) = 0$ for all $k \geq N_2$. The proof can be divided into two steps. The first step is the construction of a signal $\xi \in \ell^2 \setminus \ell_2^2$ (i.e., has infinite ℓ_2 norm) such that the output of each row is amplified by the amount γ. The next step is to use this signal to construct a destabilizing perturbation.

Construction of unbounded signals. Define the signal ξ as follows:

$$\xi = \sum_{k=0}^{\infty} S_{k(N_1+N_2)}f.$$

This signal can be visualized as a signal made up from the nonzero components of f (which we denote by f) by shifting it, and adding zeros in between, i.e.,

$$\xi = \{ \underbrace{f}_{N_2}, \overbrace{0}^{N_1}, f, 0, \ldots\}.$$

The action of M can be decomposed as follows:

$$y = M\xi = \sum_{k=o}^{\infty} S_{k(N_1+N_2)} Mf$$

$$= \sum_{k=o}^{\infty} S_{k(N_1+N_2)} (P_{N_2-1} Mf + (P_{N_1+N_2-1} - P_{N_2-1})Mf).$$

Define $M_0 \in \mathbb{R}^{2N_2 \times 2N_2}$ and $M_1 \in \mathbb{R}^{2N_1 \times 2N_2}$ as follows:

$$M_0 := P_{N_2-1} M P_{N_2-1}, \qquad M_1 := S_{-N_2}(P_{N_1+N_2-1} - P_{N_2-1})M P_{N_2-1}.$$

Then y can be written as

$$y = \{M_0 f, M_1 f, M_0 f, \ldots\} \tag{7.8}$$

decomposed conformally with ξ. It follows From Equations (7.7) and (7.8) that for any $k \geq 0$

$$\| P_{k(N_1+N_2-1)} y_i \|_2^2 \geq \gamma \| P_{k(N_1+N_2-1)} \xi_i \|_2^2 \quad \text{for } i = 1, 2.$$

Construction of a destabilizing perturbation.   Construct the matrix $\tilde{\Delta} \in \mathbb{R}^{2N_2 \times 2N_2}$ to be a block diagonal matrix, with $\tilde{\Delta}_i$ satisfying:

$$f_i = \tilde{\Delta}_i (M_0 f)_i, \quad \sigma_{\max}[\tilde{\Delta}_i] \leq \frac{1}{\gamma}, \quad \text{for } i = 1, 2.$$

Such a construction is straightforward and is given by

$$\tilde{\Delta}_i = \frac{f_i (M_0 f)_i^T}{|(M_0 f)_i|_2^2} \quad \text{for } i = 1, 2.$$

Define the perturbation

$$\Delta = \begin{pmatrix} 0 & \cdots & & & & \\ 0 & 0 & \cdots & & & \\ \tilde{\Delta} & 0 & 0 & \cdots & & \\ 0 & 0 & 0 & 0 & \cdots & \\ 0 & 0 & \tilde{\Delta} & 0 & \ddots & \\ \vdots & \vdots & \vdots & \vdots & & \ddots \end{pmatrix}, \quad \text{with} \quad \tilde{\Delta} = \begin{pmatrix} \tilde{\Delta}_1 & 0 \\ 0 & \tilde{\Delta}_2 \end{pmatrix}.$$

With this perturbation, the output in each channel for the input ξ is given by

$$(f_i, 0, f_i - \tilde{\Delta}_i(M_0 f)_i, 0, f_i - \tilde{\Delta}_i(M_0 f)_i, \ldots) = (f_i, 0, 0, \ldots) \in \ell_2.$$

This immediately implies that $(I - \Delta M)^{-1}$ is not ℓ_2-stable since it maps a signal in ℓ_2^2 to a signal in $\ell^2 \setminus \ell_2^2$. Notice also that $(I - \Delta M)^{-1}$ is well defined since Δ is strictly proper. This concludes the proof.   ∎

The next lemma establishes the equivalence between the conditions on $\kappa_i(f)$ and diagonally scaled norms. Indeed, if there exists an f such that $\kappa_i(f) > 0$, for $i = 1, 2$, then, for any positive constants d_1^2, d_2^2 we have

$$\sum_{i=1}^{2} d_i^2 \kappa_i(f) > 0 \iff \|DMf\|_2^2 > \|Df\|_2^2, \quad D = \begin{pmatrix} |d_1| & 0 \\ 0 & |d_2| \end{pmatrix}$$

$$\iff \|D\hat{M}D^{-1}\|_\infty > 1.$$

The converse is shown in the following lemma.

Lemma 7.3.2. Suppose that $\hat{M}$ is such that $\inf_{D \in \mathbf{D}} \|D\hat{M}D^{-1}\|_\infty$ has a finite nonzero minimizer. If

$$\inf_{D \in \mathbf{D}} \|D\hat{M}D^{-1}\|_\infty > 1,$$

then there exists a function $f \in \ell_2^n$ such that

$$\kappa_i(f) > 0 \quad \text{for } i = 1, 2, \ldots n.$$

Proof. For simplicity, show the result for $n = 2$. Suppose that for every $f \in \ell_2^2$, if $\kappa_1(f) > 0$ then $\kappa_2(f) \leq 0$ or the other way around. We will show that there exists positive numbers d_1, d_2 such that

$$d_1 \kappa_1(f) + d_2 \kappa_2(f) \leq 0.$$

To show this, consider the following set

$$\mathcal{C} = \{(\kappa_1(f), \kappa_2(f)) \mid \text{for all } f \in \ell_2^2\}$$

We first show that $\overline{\mathcal{C}}$ is a convex set in $\mathbb{R}^2$. Let $x_i = \kappa_i(f_1)$, $i = 1, 2$ and $y_i = \kappa_i(f_2)$, $i = 1, 2$. Consider the function $g = \sqrt{\alpha} f_1 + \sqrt{(1 - \alpha)} S_N f_2$ with $0 \leq \alpha \leq 1$ and $N > 0$. It follows that

$$\lim_{N \to \infty} \kappa_i(\sqrt{\alpha} f_1 + \sqrt{(1 - \alpha)} S_N f_2) = \alpha x_i + (1 - \alpha) y_i \in \overline{\mathcal{C}},$$

which establishes convexity. Also, the set $\overline{\mathcal{C}}$ does not intersect the positive cone in $\mathbb{R}^2$, denoted by $\mathcal{P}$. Since both $\overline{\mathcal{C}}$ and $\mathcal{P}$ are convex, they both can be separated by a line (hyperplane in general). This implies that each one of these sets lies on a different side of this line. Let the line (which passes through zero) be given by $d_1 x_1 + d_2 x_2 = 0$, then

$$d_1 x_1 + d_2 x_2 \geq 0 \quad \text{for all } (x_1, x_2) \in \mathcal{P},$$

and

$$d_1 \kappa_1(f) + d_2 \kappa_2(f) \leq 0 \quad \text{for all } f \in \ell_2^2.$$

Also, since $\mathcal{P}$ is a positive cone, it follows that $d_i \geq 0$ (not both are zero). Also, from the hypothesis of the lemma, it follows that $d_i > 0$. (see Exercise 7.11). Then this condition is equivalent to

$$\|DMf\|_2^2 \leq \|Df\|_2^2 \quad \text{with } D = \begin{pmatrix} \sqrt{d_1} & 0 \\ 0 & \sqrt{d_2} \end{pmatrix}, \qquad \text{for all } f \in \ell_2^2$$

implying that

$$\|D\hat{M}D^{-1}\|_\infty \leq 1.$$

which establishes the result.   ∎

Finally we prove Theorem 7.3.4.

Proof of Theorem 7.3.4. If $\inf_{D \in \mathbf{D}} \|D^{-1}\hat{M}D\|_\infty \leq 1$ then the system is stable. Suppose that $\inf_{D \in \mathbf{D}} \|D^{-1}\hat{M}D\|_\infty > 1$. By Lemma 7.3.2 there exists a function $f \in \ell_2^2$ such that $\kappa_i(f) > 0$ for $i = 1, 2$. It follows by Lemma 7.3.1 that there exists a destabilizing perturbation of the system in Figure 7.3.   ∎

Computation.   There is no spectral radius interpretation of the condition in Theorem 7.3.4. For a fixed M, the condition can be computed by optimizing directly over diagonal scales. As we will see shortly (in the time-invariant case), this optimization problem is convex and hence can be computed by various numerical methods. The case where the blocks are MIMO, the condition generalizes exactly as in the ℓ_∞ case.

7.4 STABILITY ROBUSTNESS WITH NONLINEAR PERTURBATIONS

Let

$$\mathbf{\Delta}_{NL} = \{\text{diag}(\Delta_1, \Delta_2, \ldots \Delta_n) \mid \Delta_i \text{ is } p_i \times p_i \text{ causal NLTV operator}\}. \tag{7.9}$$

This represents the class of block diagonal, nonlinear time-varying, causal perturbations. We have assumed that each block is a square matrix of dimension $p_i \times p_i$. Extending the next results for nonsquare matrices is immediate. The various blocks in the perturbations are not related, nor dependent. Finally, the set $\mathbf{B}_{\mathbf{\Delta}_{NL}, p}$ is defined for a given ℓ_p space as in Equation (7.1).

Associated with this class is a set of scalings preserving the norms of elements in the set of perturbations. A reasonable and simple set to pick is:

$$\mathbf{D} = \{\text{diag}(d_1 I_{p_1}, d_2 I_{p_2} \ldots d_n I_{P_n}) \mid d_i \in \mathbb{R}, d_i > 0\}. \tag{7.10}$$

For this class of scales, we have

$$\|D^{-1}\Delta D\|_{\ell_p - \text{ind}} = \|\Delta\|_{\ell_p - \text{ind}}, \qquad \text{for all } \Delta \in \mathbf{\Delta}_{NL}.$$

Notice that the set of scales does not commute with each of the perturbations; however, it preserves the norm. This contrasts the LTV case in which the scales actually commute with each perturbation.

7.4.1 ℓ_∞ Stability Robustness Conditions

It follows from the definition of the structured norm that

$$SN_{\Delta_{NL},\infty}(M) \leq \inf_{D \in \mathbf{D}} \|D^{-1}MD\|_1.$$

However, since this class includes LTV systems, it follows that equality holds. The following theorem shows that equality holds when the perturbations are restricted to be nonlinear time invariant, and thus the stability robustness conditions are identical to the case of LTV perturbations.

Theorem 7.4.1. The structured norm for nonlinear time-invariant perturbations satisfies

$$SN_{\Delta_{NL},\infty}(M) = SN_{\Delta_{LTV},\infty}(M) = \inf_{D \in \mathbf{D}} \|D^{-1}MD\|_1.$$

Proof. The proof of this fact follows exactly as the proof of Theorem 7.3.2, except for the construction of the destabilizing perturbation. Given the signals y and ξ, we show that a nonlinear time-invariant perturbation can be constructed to destabilize the closed-loop system.

Let the signals y_i and ξ_i be given as before ($i = 1, 2$), Δ_i must be such that $\|\Delta_i\|_{\ell_\infty-\mathrm{ind}} < 1$ and $\Delta_i y_i = \xi_i$. Let Δ_i be defined as follows:

$$(\Delta_i f)(k) = \begin{cases} \xi_i(k - j) & \text{if for some integer } j \geq 0, \ P_k f = P_k S_j y_i \\ 0 & \text{otherwise} \end{cases}$$

where S_j is the shift operator by j steps. It can be seen that Δ_i is a nonlinear, time-invariant, and causal system. It has a norm less than one, and maps y_i to ξ_i. ∎

A special case of the above conditions is the standard small gain theorem:

Corollary 7.4.1. The condition $\|M\|_1 \leq 1$ is necessary and sufficient for the stability of $(I - \Delta M)^{-1}$ for all Δ arbitrary with $\|\Delta\|_{\ell_\infty-\mathrm{ind}} < 1$, where Δ is either linear time varying or nonlinear time invariant. ∎

7.4.2 ℓ_2 Stability Robustness Conditions

It follows from the definition of the structured norm that

$$SN_{\Delta_{NL},2}(M) \leq \inf_{D \in \mathbf{D}} \|D^{-1}\hat{M}D\|_\infty.$$

Since this class contains linear time-varying perturbations, equality holds. As in the ℓ_∞-stability case, it follows that equality holds even if the perturbations are restricted to be time invariant.

Theorem 7.4.2. The structured norm for nonlinear time-invariant perturbations satisfies

$$SN_{\Delta_{NL},2}(M) = SN_{\Delta_{LTV},2}(M) = \inf_{D \in \mathbf{D}} \|D^{-1}\hat{M}D\|_\infty.$$

Proof. The proof follows exactly as in the time-varying case if we can show that a nonlinear time-invariant perturbation can be constructed such that

$$\Delta = \begin{pmatrix} \Delta_1 & 0 \\ 0 & \Delta_2 \end{pmatrix}, \quad \Delta_i y_i = (0, 0, f_i, 0, f_i, \ldots) = S_{N_1+N_2}\xi_i, \quad i = 1, 2,$$

with $\|\Delta\|_{\ell_2-\mathrm{ind}} < 1$. Consider the following perturbation

$$(\Delta_i g)(k) = \begin{cases} 0 & \text{if } k < N_1 + N_2 \\ \xi_i(k - j - N_1 - N_2) & \text{if for some integer } j \geq 0, \ P_k g = P_k S_j y_i \\ 0 & \text{otherwise} \end{cases}$$

It can be verified that Δ is a nonlinear, time-invariant, causal perturbation. It satisfies $\|\Delta\|_{\ell_2-\mathrm{ind}} < 1$ with $\Delta_i y_i = S_{N_1+N_2}\xi_i$. This completes the proof. ∎

In summary, the structured norm has the same value for three classes of perturbations:

1. Nonlinear time-varying perturbations.
2. Nonlinear time-invariant perturbations.
3. Linear time-varying perturbations.

7.5 STABILITY ROBUSTNESS WITH LTI PERTURBATIONS

Before we start computing the structured norms for this class of perturbations, it is convenient to define a measure referred to as the *Structured Singular Value* which we will denote by μ. Let Δ_{LTI} denote a class of structured linear time-invariant, stable, finite-dimensional perturbations:

$$\Delta_{LTI} = \left\{ \Delta_1, \ldots, \Delta_n \mid \Delta_i \in \mathcal{RH}_\infty^{p_i \times p_i} \right\}.$$

Again, we have assumed that the blocks are square. The extension to nonsquare blocks is a straightforward exercise (see Exercise 7.2). Notice that we have incorporated stability in the definition of Δ_{LTI}. Although this was unnecessary in defining the $SN_{\Delta_{LTI}, p}(M)$, it will be important in defining the structured singular value function.

Definition 7.5.1. Suppose M and $\Delta \in \Delta_{LTI}$ are connected as in Figure 7.3. The structured singular value function is defined at each frequency as

$$\mu_\Delta[\hat{M}(e^{i\theta})] = \frac{1}{\inf_{\Delta \in \Delta_{LTI}} \left\{ \sigma_{\max}[\hat{\Delta}(e^{i\theta})] \mid det(I - \hat{M}\hat{\Delta})(e^{i\theta}) = 0 \right\}}, \tag{7.11}$$

and if there is no Δ such that $det(I - \hat{M}\hat{\Delta})(e^{i\theta}) = 0$, then $\mu_\Delta[\hat{M}(e^{i\theta})] := 0$. ∎

The Structured Singular Value function can be thought of as the frequency domain parallel to the Structured Norm function. Since $\mu_\Delta[\hat{M}(e^{i\theta})]$ is defined only for LTI perturbations, we will drop the subscript on Δ_{LTI}. The assumption of finite-dimensionality of the perturbations is not crucial and is used for simplicity.

It is more convenient for this class of perturbations to discuss ℓ_2 stability before ℓ_∞ stability. The reason behind this will be clear shortly.

7.5.1 ℓ_2 Stability Robustness Conditions

The next theorem establishes an exact equivalence between $SN_{\Delta_{LTI},2}(M)$ and $\mu_\Delta[\hat{M}(e^{i\theta})]$.

Theorem 7.5.1.   The system in Figure 7.3 is stable for all $\Delta \in \mathbf{B}_{\Delta_{LTI},2}$ if and only if

$$\mu_\Delta[\hat{M}(e^{i\theta})] \leq 1 \quad \text{for all } \theta \in [0, 2\pi].$$

Equivalently, the ℓ_2 structured norm for the class of perturbations Δ_{LTI} is given by:

$$SN_{\Delta_{LTI},2}(M) = \sup_{\theta \in [0,2\pi]} \mu_\Delta[\hat{M}(e^{i\theta})].$$

Proof. To prove the result, it is sufficient to show that the system in Figure 7.3 is stable for every $\Delta \in \mathbf{B}_{\Delta_{LTI},2}$ if and only if $det(I - \hat{M}\hat{\Delta})(e^{i\theta}) \neq 0$ for all $\theta \in [0, 2\pi]$ and all $\Delta \in \mathbf{B}_{\Delta_{LTV},2}$. Necessity is clear. To prove sufficiency, suppose that the system is unstable for some $\hat{\Delta}_0 \in \mathbf{B}_{\Delta_{LTI},2}$ and has a pole at λ_0 in the closed unit disc. If λ_0 lies on the unit circle, then the conclusion follows immediately. Assume that λ_0 lies in the open unit disc. The function f defined as:

$$f(\epsilon, \lambda) = det(I - \hat{M}\epsilon\hat{\Delta}_0)(\lambda)$$

is continuous in $\epsilon \in [0, 1]$ and λ. For ϵ small enough, $f(\epsilon, \lambda)$ has all of its zeros outside the unit disc. Also, for $\epsilon = 1$ we have $f(1, \lambda_0) = 0$. Since the determinant is a continuous function of both λ and ϵ, the zeros of the function f move continuously as a function of ϵ and thus will cross the unit circle for some value of $\epsilon \in [0, 1)$. Since $\epsilon\Delta_0 \in \mathbf{B}_{\Delta_{LTI},2}$ the claim follows.

By noting that the definition of $\mu_\Delta[\hat{M}(e^{i\theta})]$ is consistent with the ℓ_2-induced norm, the theorem follows immediately from the above characterization. ∎

7.5.2 ℓ_∞ Stability Robustness Conditions

In here, we relate the ℓ_∞ stability robustness conditions to the structured singular value function. We give the SISO characterization first.

Theorem 7.5.2.   The system in Figure 7.3 is stable for all $\Delta \in \mathbf{B}_{\Delta_{LTI},\infty}$ if and only if

$$\mu_\Delta[\hat{M}(e^{i\theta})] \leq 1 \quad \text{for all } \theta \in [0, 2\pi].$$

Equivalently, the ℓ_∞ structured norm for the class of perturbations Δ_{LTI} (with SISO blocks) is given by

$$SN_{\Delta_{LTI},\infty}(M) = \sup_{\theta \in [0,2\pi]} \mu_\Delta[\hat{M}(e^{i\theta})].$$

Proof. If $\mu_\Delta[\hat{M}(e^{i\theta})] \leq 1$, then any destabilizing perturbation must have $\|\Delta\|_1 \geq \|\hat{\Delta}\|_\infty \geq 1$ and hence is not admissible. On the other hand, if $\mu_\Delta[\hat{M}(e^{i\theta})] > 1$, then there exists a perturbation Δ such that at θ_0 the $det\,(I - \hat{M}\hat{\Delta})(e^{i\theta_0}) = 0$ and $\sigma_{\max}(\hat{\Delta}(e^{i\theta_0})) < 1$. Let $c_j = \hat{\Delta}_j(e^{i\theta_0})$. To prove the theorem, we need to exhibit a perturbation Δ satisfying:

1. $\hat{\Delta}_j(e^{i\theta_0}) = c_j$ for all $j = 1, \ldots n$.
2. $\|\Delta_j\|_1 < 1$.

This construction is possible and is left as an exercise (see Exercise 10.4). ∎

This is quite a surprising result. The structured singular value function, $\mu_\Delta[\hat{M}(e^{i\theta})]$, provides the necessary and sufficient conditions for stability in both the ℓ_∞ sense, and ℓ_2 sense. The roots of this result go back to Wiener's theorem (Theorem 2.3.3), which indicates that the inverse of an ℓ_1 function is in ℓ_1 if and only if its λ-transform does not vanish in the closed unit disc. This invertibility condition is the same for $\mathcal{H}_\infty$. Similar arguments can be given to show that the same result holds for any ℓ_p. The main point of the proof is constructing a perturbation satisfying:

1. $\hat{\Delta}_j(e^{i\theta_0}) = c_j$ for all $j = 1, \ldots n$.
2. $\|\Delta_j\|_{\ell_p - \mathrm{ind}} < 1$.

We will omit the details of the general ℓ_p problem.

Corollary 7.5.1.   For SISO LTI perturbation blocks, we have for all $1 \leq p \leq \infty$

$$SN_{\Delta_{LTI}, \infty}(M) = SN_{\Delta_{LTI}, 2}(M) = SN_{\Delta_{LTI}, p}(M) = \sup_{\theta \in [0, 2\pi]} \mu_\Delta[\hat{M}(e^{i\theta})]. \qquad ∎$$

MIMO conditions.   The basic result is a straightforward generalization of μ. Since it is no longer true that the $\mathcal{H}_\infty$-norm is bounded by the ℓ_1 norm (it is bounded by a scaled ℓ_1-norm), we cannot get exact necessary and sufficient conditions for stability. We can, however, get upper and lower bounds.

Theorem 7.5.3.   Let $p = \max_i p_i$. Then

$$\frac{1}{\sqrt{p}} \sup_\theta \mu_\Delta[\hat{M}(e^{i\theta})] \leq SN_{\Delta_{LTI}, \infty}(M) \leq \sqrt{p} \sup_\theta \mu_\Delta[\hat{M}(e^{i\theta})].$$

The proof of this theorem is left as an exercise. (See Exercise 7.3.) ∎

7.5.3 Computation of μ

From the previous discussion, it is evident that methods for computing μ are necessary to analyze the ℓ_p stability robustness of a system in the presence of LTI perturbations. In the sequel we present upper and lower bounds of μ that can be computed at each frequency.

At this stage, it is convenient to think of μ as simply operating on complex valued matrices. This is evident from the definition since μ is defined at each frequency. It is also helpful, as will be seen shortly, to consider the augmented set of perturbations defined as follows

$$\mathbf{\Delta}_a = \left\{ \text{diag}(\delta I_r, \Delta_1, \dots \Delta_n) \,\middle|\, \delta \in \mathbf{C}, \, \Delta_i \in \mathbf{C}^{p_i \times p_i} \right\}.$$

The definition of μ is extended in the obvious way for any complex valued matrix $M \in \mathbf{C}^{r+p \times r+p}$, with $p = \sum p_i$:

$$\mu_{\mathbf{\Delta}_a}[M] = \frac{1}{\inf_{\Delta \in \Delta_a} \left\{ \sigma_{\max}[\Delta] \mid det(I - M\Delta) = 0 \right\}}.$$

Let the set $\mathbf{D}_a$ denote a set of matrices that commute with elements in $\mathbf{\Delta}_a$:

$$\mathbf{D}_a = \left\{ \text{diag}(D_r, d_1 I_{p_1}, d_2 I_{p_2} \dots d_n I_{P_n}) \,\middle|\, D_r \in \mathbf{C}^{r \times r}, \, D_r = D_r^* > 0, \, d_i \in \mathbb{R}, \, d_i > 0 \right\}.$$

Also, define the set $\mathbf{U}$ as:

$$\mathbf{U} = \left\{ U \in \mathbf{\Delta}_a \,\middle|\, UU^* = I \right\}. \tag{7.12}$$

A lower bound for μ.   First, we observe that

$$\mu_{\mathbf{\Delta}_a}[M] \geq \mu_{\mathbf{\Delta}}[M] = \rho(M) \quad \mathbf{\Delta} = \left\{ \delta I_{p+r} \mid \delta \in \mathbf{C} \right\} \subset \mathbf{\Delta}_a.$$

If $U \in \mathbf{U}$ and $\Delta \in \mathbf{B}_{\mathbf{\Delta}_a, 2}$, then $U\Delta \in \mathbf{B}_{\mathbf{\Delta}_a, 2}$. It follows directly from the definition that

$$\rho(MU) \leq \mu_{\mathbf{\Delta}_a}[MU] = \mu_{\mathbf{\Delta}_a}[M] \quad \text{for all } U \in \mathbf{U},$$

which implies that

$$\max_{U \in \mathbf{U}} \rho(MU) \leq \mu_{\mathbf{\Delta}_a}[M].$$

In fact, equality always holds. The spectral radius function is not a convex function, and so the optimization problem resulting from this lower bound is a difficult problem. An algorithm similar to the power algorithm discussed earlier can be used.

Upper bound on μ.   From the definition of $\mathbf{D}_a$, it follows directly that

$$\mu_{\mathbf{\Delta}_a}[M] \leq \inf_{D \in \mathbf{D}_a} \sigma_{\max}[D^{-1}MD].$$

While the upper bound is not a convex function in D, another parametrization of the set $\mathbf{D}_a$ using exponentials make the function convex. (See Exercise 7.4.) An easy way to see the quasi convexity of this function is to consider the set

$$\left\{ D^2 \in \mathbf{D} \,\middle|\, \sigma_{\max}[D^{-1}MD] < c \right\},$$

where c is a constant. This is a convex set, since

$$\sigma_{\max}[D^{-1}MD] < c \Leftrightarrow \lambda_{\max}[D^{-1}MDDM^*D^{-1}] < c$$

$$\Leftrightarrow MD^2M^* < cD^2$$

$$\Leftrightarrow MD^2M^* - cD^2 < 0.$$

The later inequality is convex in D^2. The above sets can be thought of as the level sets for the upper bound.

It turns out that this upper bound gives an equality for some special cases. These cases are:

1. $r = 0$ and $n \leq 3$.
2. $r \neq 0$ and $n \leq 1$.

We will not explore these properties any further.

Scalar versus frequency-dependent μ.   If the class of perturbations is given as in $\mathbf{\Delta}_{LTI}$, then there are two approaches to compute estimates of $\mu_{\Delta}[\hat{M}(e^{i\theta})]$ that follow from our previous discussion. The first approach is to get an estimate of $\mu_{\Delta}[\hat{M}(e^{i\theta})]$ at each frequency using the upper and lower bounds:

$$\max_{U \in \mathbf{U}} \rho(\hat{M}(e^{i\theta})U) = \mu_{\Delta}[\hat{M}(e^{i\theta})] \leq \inf_{D \in \mathbf{D}} \sigma_{\max}[D^{-1}\hat{M}(e^{i\theta})D],$$

where the sets $\mathbf{U}$, $\mathbf{D}$ are consistent with the structure

$$\mathbf{\Delta}_{LTI} = \left\{ \mathrm{diag}(\Delta_1, \ldots \Delta_n) \, \big| \, \Delta_i \in \mathbf{C}^{p_i \times p_i} \right\}.$$

The second approach is to consider the frequency dependence of $\hat{M}$ as an extra block of perturbations. An LTI finite-dimensional system can be expressed as an LFT with λI connected in the upper block as shown in Figure 7.7. Let the constant matrix M_c be defined as

$$M_c = \left[\begin{array}{c|c} A & B \\ \hline C & D \end{array} \right]$$

then

$$\hat{M}(\lambda) = F_u(M_c, \lambda I) = D + \lambda C(I - \lambda A)^{-1}B.$$

Stability of $\hat{M}$ is equivalent to $\rho(M_{c_{11}}) = \rho(A) < 1$. This also guarantees that the upper connection is well-posed. The following equivalence can be established:

$$\sup_{\theta \in [0, 2\pi]} \mu_{\Delta}[\hat{M}(e^{i\theta})] \leq 1 \iff \mu_{\Delta_a}[M_c] \leq 1.$$

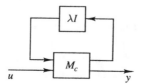

Figure 7.7   A LTI system as a LFT.

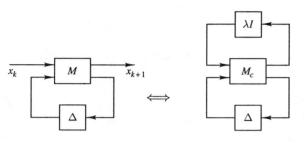

Figure 7.8 Augmenting a problem with λI.

The expense of getting a condition independent of frequency is adding an extra block of perturbations of the form λI. This is depicted in Figure 7.8. Now upper and lower bounds can be computed for $\mu_{\Delta_a}[M_c]$.

Upper bound as a scaled $\mathcal{H}_\infty$ norm.   As discussed earlier

$$\mu_\Delta[\hat{M}(e^{i\theta})] \leq \inf_{D \in \mathbf{D}} \sigma_{\max}[D^{-1}\hat{M}(e^{i\theta})D].$$

To guarantee robustness, the upper bound has to be bounded by one for all frequencies, i.e.,

$$\sup_{\theta \in [0, 2\pi)} \inf_{D \in \mathbf{D}} \sigma_{\max}[D^{-1}\hat{M}(e^{i\theta})D] \leq 1.$$

This condition is equivalent to a scaled $\mathcal{H}_\infty$-norm condition with scales that are frequency dependent, in particular

$$\sup_{\theta \in [0, 2\pi)} \inf_{D \in \mathbf{D}} \sigma_{\max}[D^{-1}\hat{M}(e^{i\theta})D] = \inf_{\hat{D}} \|\hat{D}^{-1}\hat{M}\hat{D}\|_\infty$$

where $\hat{D}$ is an arbitrary diagonal stable system with a stable inverse. This condition should be contrasted with the scaled $\mathcal{H}_\infty$-norm with constant scales that results in the case of time-varying perturbations.

7.6 COMPARISONS BETWEEN STABILITY ROBUSTNESS CONDITIONS

We have already shown the following relations between the structured norms for scalar blocks:

1. $SN_{\Delta_{LTI}, \infty}(M) \leq SN_{\Delta_{LTV}, \infty}(M) = SN_{\Delta_{NL}, \infty}(M).$
2. $SN_{\Delta_{LTI}, 2}(M) \leq SN_{\Delta_{LTV}, 2}(M) = SN_{\Delta_{NL}, 2}(M).$
3. (SISO Blocks) $SN_{\Delta_{LTI}, \infty}(M) = SN_{\Delta_{LTI}, 2}(M) = \sup_\theta \mu_\Delta[\hat{M}(e^{i\theta})].$

As discussed earlier, the first two items are not surprising, while the third is. To obtain a cross comparison between the stability conditions, it is worthwhile comparing the perturbations that have gain less than unity over ℓ_2 with the perturbations that have gain less than unity over ℓ_∞. If the perturbations are restricted to time-invariant ones, the ℓ_∞-stable perturbations with gain less than unity lie inside the unit ball of ℓ_2-stable perturbations (for the multivariable case, the unit ball will be scaled by a constant). This follows directly from the norm inequality between ℓ_1 and $\mathcal{H}_\infty$. If the perturbations are allowed to be time varying, then the two sets are not comparable. Examples illustrating this were given in Chapter 2. These observations are depicted in Figure 7.9.

From (3) above, it follows that the necessary and sufficient condition for stability robustness in the presence of LTI norm-bounded perturbations is given by μ_Δ and is independent of the particular norm considered.

Cross relations between the above conditions.   While the classes of perturbations are different for different conditions, it is possible to obtain a strong relation between all of these conditions. We start with the following lemma.

Lemma 7.6.1.   Assume that A has strictly positive elements. Then

$$\inf_{D \in \mathbf{D}} \sigma_{\max}[D^{-1}AD] = \rho(A).$$

Proof. Since $\rho(D^{-1}AD) = \rho(A)$, it follows that $\sigma_{\max}[D^{-1}AD] \geq \rho(A)$ for every D.

To show the other direction, let $\rho(A) = \beta$. Since A is positive, there exits positive vectors x, y such that

$$Ax = \beta x, \quad y^T A = \beta y^T.$$

Define the diagonal matrix D as follows:

$$d_i = \sqrt{\frac{y_i}{x_i}}.$$

It follows that $y = D^2 x$. Define the positive vector $z = Dx$; then

$$(DA^T D^{-2} AD)z = \beta^2 z.$$

Since the matrix $DA^T D^{-2} AD$ is positive, it follows from Theorem 7.3.1 that $\beta^2 = \rho(DA^T D^{-2} AD)$. This proves the result.   ∎

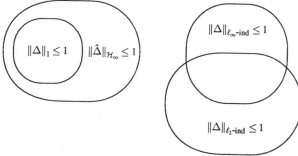

Time-invariant Δ Time-varying Δ **Figure 7.9**   Comparison of unit balls.

The following theorem holds for any stable M, such that $\widehat{M}$ is positive.

Theorem 7.6.1.   Suppose M is stable such that $\widehat{M}$ is positive. Then

$$\mu_{\Delta}[\hat{M}(e^{i\theta})] \leq \inf_{D \in \mathbf{D}} \|D^{-1}\hat{M}D\|_{\infty} \leq \inf_{D \in \mathbf{D}} \sigma_{\max}[D^{-1}\widehat{M}D] = \rho(\widehat{M}). \qquad (7.13)$$

Proof. Notice that if $A \leq B$ componentwise, then $\sigma_{\max}[A] \leq \sigma_{\max}[B]$. The theorem follows directly from Lemma 7.6.1.

This implies that ℓ_{∞}-stability in the presence of ℓ_{∞}-bounded perturbations implies ℓ_2-stability in the presence of ℓ_2-bounded perturbations, and so on. Table 7.2 summarizes the relations of all these conditions.

In terms of computations, the robustness test $\rho(\widehat{M})$ is much easier to compute than the $\mu_{\Delta}[\hat{M}(e^{i\theta})]$ and gives exact answers for any number of perturbation blocks, n. On the other hand, μ is hard to compute especially for $n > 3$ since only upper and lower bounds can be computed.   ∎

TABLE 7.2 CONDITIONS FOR ROBUST STABILITY WITH SISO BLOCKS

Perturbation class	Stability robustness condition		
	$\mu_{\Delta}(M) \leq 1$	$\inf_{D \in \mathbf{D}} \|D^{-1}\| \|D\hat{M}\|_{\mathcal{H}_{\infty}} \leq 1$	$\rho(\widehat{M}) \leq 1$
NLTV, bounded ℓ_2-induced norm	nec	nec and suff	suff
NLTV, bounded ℓ_{∞}-induced norm	nec	nec	nec and suff
NLTI, bounded ℓ_2-induced norm	nec	nec and suff	suff
NLTI, bounded ℓ_{∞}-induced norm	nec	nec	nec and suff
LTV, bounded ℓ_2-induced norm	nec	nec and suff	suff
LTV, bounded ℓ_{∞}-induced norm	nec	nec	nec and suff
LTI, bounded ℓ_2-induced norm	nec and suff	suff	suff
LTI, bounded ℓ_{∞} induced norm	nec and suff	suff	suff

7.7 PERFORMANCE ROBUSTNESS VERSUS STABILITY ROBUSTNESS

The general performance robustness problem is a difficult problem. In certain cases, the problem is equivalent to an augmented stability robustness problem, which has an extra fictitious block of perturbations to accommodate the performance specifications. This equivalence is valid for very particular performance specifications and, in a sense, solves only special problems. For instance, specifications due to direct time domain constraints such as an overshoot constraint cannot be mapped in this fashion and the robust performance problem remains unsolved.

7.7.1 ℓ_∞ Performance Robustness

To motivate the solution of this problem, consider the disturbance rejection example of Figure 3.5. The objective is to find a controller that guarantees good disturbance attenuation for all bounded disturbances, i.e., $\|(I - P_0 K)^{-1} W\|_1 \leq 1$. Can this problem be formulated as a *fictitious* robust stability problem? Consider Figure 7.10. From Corollary 7.4.1, the closed-loop system is stable for all Δ_p of norm less than unity if and only if the transfer function "seen" by the perturbations has ℓ_1 norm less than or equal to unity. Precisely, the system is robustly stable if and only if $\|(I - P_0 K)^{-1} W\|_1 \leq 1$. Thus the fictitious robust stability problem has the same solution as the disturbance rejection problem. This indeed suggests that the robust stability and the robust performance problems may be equivalent.

Before the general result is presented, it is important to note that Corollary 7.4.1 is necessary and sufficient only if the perturbations are allowed to be time varying or nonlinear time invariant. Thus for the above equivalence, the fictitious perturbations have to be either time varying or nonlinear time invariant, but cannot be restricted to be linear time invariant. Guaranteeing the stability for a class of linear time-invariant perturbations may not guarantee that the ℓ_1 norm of the transfer function seen by the perturbations to be bounded by one. In the sequel, the equivalence between the robust stability and robust performance problems will be established for the class Δ_{LTV}. The class Δ_{NL} follows in the same way. For the class Δ_{LTI}, the extra performance block will have to be time varying or nonlinear, thus resulting in a class of mixed perturbations. The analysis of such a class is a difficult problem and remains open.

The next theorem states that performance robustness in one system is equivalent to stability robustness in another system formed by feeding back the regulated outputs to the disturbance inputs through a fictitious perturbation. The usefulness of this theorem stems from the fact that it is enough now to apply the stability robustness results to solve this problem.

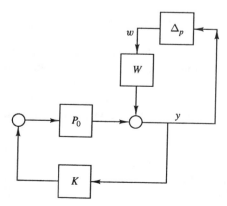

Figure 7.10 Disturbance rejection as a robust stability problem.

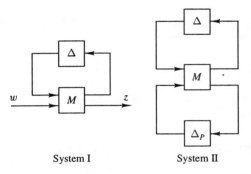

Figure 7.11 Stability robustness vs. performance robustness.

System I   System II

Consider the two systems shown in Figure 7.11, where $M \in \ell_1{}^{p \times q}$ and $\Delta \in \Delta_{LTV}$. In system I, w is an input vector of size $\tilde{p}$ and z is an output vector of size $\tilde{q}$. In system II, Δ_P is linear time-varying of dimension $\tilde{p} \times \tilde{q}$. It follows that $p = \tilde{p} + \sum_i p_i$ and $q = \tilde{q} + \sum_i q_i$. Define the new set of perturbations:

$$\Delta_P = \left\{ \tilde{\Delta} = \mathrm{diag}(\Delta, \Delta_P) \,\Big|\, \Delta \in \Delta_{LTV}, \ \Delta_P \in \mathcal{L}^{\tilde{p} \times \tilde{q}} \right\}.$$

Subdivide M in the following manner:

$$M = \begin{pmatrix} \tilde{M}_{11} & \tilde{M}_{12} \\ \tilde{M}_{21} & \tilde{M}_{22} \end{pmatrix}$$

where $\tilde{M}_{22} \in \ell_1{}^{\tilde{p} \times \tilde{q}}$.

We now state the following theorem establishing the relation between System I and System II.

Theorem 7.7.1.   The following four statements are equivalent:

1. System II achieves ℓ_∞ robust stability for all $\tilde{\Delta} \in \mathbf{B}_{\Delta_P, \infty}$.
2. $(I - M\tilde{\Delta})^{-1}$ is ℓ_∞-stable for all $\tilde{\Delta} \in \mathbf{B}_{\Delta_P, \infty}$.
3. $(I - \tilde{M}_{11}\Delta)^{-1}$ is ℓ_∞-stable and $\|\tilde{M}_{22} + \tilde{M}_{21}\Delta(I - \tilde{M}_{11}\Delta)^{-1}\tilde{M}_{12}\|_{\ell_\infty - \mathrm{ind}} \leq 1$, for all $\Delta \in \mathbf{B}_{\Delta_{LTV}, \infty}$.
4. System I achieves robust performance for all $\Delta \in \mathbf{B}_{\Delta_{LTV}, \infty}$.

Proof. $1 \iff 2$ and $3 \iff 4$ follow by definition. It remains to show that $2 \iff 3$. If 3 holds, then 2 follows since Δ_P sees a system with norm less than or equal to 1. It remains to show that 2 implies 3. To prove this part, we need to develop the stability robustness conditions for time-varying systems in the presence of time-varying perturbations. This can be done similar to the earlier development and will be omitted. ∎

The theorem basically says the following: To achieve robust performance, wrap a fictitious Δ_P mapping z to w, and convert the problem to a robust stability problem. A similar result holds for nonlinear perturbations. The extra block will also be nonlinear.

7.7.2 ℓ_2 Performance Robustness

In the case of ℓ_2, the same result holds.

Theorem 7.7.2.   The conclusions of Theorem 7.7.1 hold with ℓ_2 stability and performance.   ∎

If the set of perturbations are LTI, and the desired performance is in terms of the ℓ_1 norm, then the fictitious robust stability problem requires the extra block to be time varying. The resulting problem is hard to solve and only sufficient conditions can be derived since it contains mixed perturbations. On the other hand, if the performance is given in terms of the $\mathcal{H}_\infty$ norm, then the added block is time invariant, and the robust performance problem will be equivalent to a robust stability problem with time-invariant perturbations., i.e., μ.

Example 7.7.1

Consider the robust disturbance rejection problem in Figure 7.12. The corresponding matrix M is the transfer function from w_1 and w_2 to z_1 and z_2. It is given by

$$M = \begin{pmatrix} W_1(I - P_oK)^{-1} & W_1(I - P_oK)^{-1} \\ W_2P_oK(I - P_oK)^{-1} & W_2P_oK(I - P_oK)^{-1} \end{pmatrix}.$$

Notice that M has two identical rows. If the plant is SISO, then the robust performance condition is given in terms of the spectral radius of $\widehat{M}$; the matrix of norms constructed from M. Performing a nonsingular transformation on $\widehat{M}$ it is easy to see that its spectral radius is equal to the spectral radius of

$$\begin{pmatrix} 0 & \|W_1(I - P_oK)^{-1}\|_1 \\ 0 & \|W_1(I - P_oK)^{-1}\|_1 + \|W_2P_oK(I - P_oK)^{-1}\|_1 \end{pmatrix}.$$

Thus a necessary and sufficient condition for the controller to achieve robust disturbance rejection is

$$\|W_1(I - P_oK)^{-1}\|_1 + \|W_2P_oK(I - P_oK)^{-1}\|_1 \leq 1.$$

Notice that this expression contains the weighted nominal sensitivity function and the nominal weighted closed-loop function (complementary sensitivity). In many places, such an objective function is termed the mixed sensitivity problem.

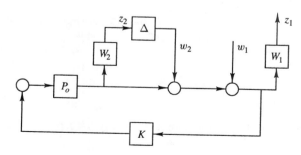

Figure 7.12   Robust disturbance rejection problem.

For the MIMO case, manipulations with the spectral radius condition shows that the robust performance condition is still the sum of the two nominal norms, as shown above.

Suppose now that the perturbations were known to be SISO linear time invariant. Then the stability robustness condition is given by (since the number of blocks equals 2):

$$\mu_\Delta[\hat{M}(e^{i\theta})] = \inf_{D \in \mathbf{D}} \rho^{\frac{1}{2}}(D^{-1}\hat{M}(e^{i\theta})D^2\hat{M}^*(e^{i\theta})D^{-1}) \leq 1.$$

Expanding out the expression for the spectral radius, it is easy to show that

$$\rho(\alpha) = |\hat{W}_1(I - \hat{P}_o\hat{K})^{-1}(e^{i\theta})|^2(1 + \alpha^2) + |\hat{W}_2\hat{P}_o\hat{K}(I - \hat{P}_o\hat{K})^{-1}(e^{i\theta})|^2 \left(1 + \frac{1}{\alpha^2}\right),$$

where $\alpha := d_2/d_1$. Then, finding the minimum ρ results in the following condition

$$|\hat{W}_1(I - \hat{P}_o\hat{K})^{-1}(e^{i\theta})| + |\hat{W}_2\hat{P}_o\hat{K}(I - \hat{P}_o\hat{K})^{-1}(e^{i\theta})| \leq 1, \qquad (7.14)$$

which should hold for all frequency points. It is evident that the first condition resulting from the spectral radius condition is harder to satisfy than the second. This is consistent with the fact that the first guarantees stability for a larger class of perturbations.

There is a nice graphical interpretation of condition 7.14, which exploits the Nyquist stability criterion for SISO systems. Consider the Nyquist plot for the loop dynamics, $\hat{P}_o\hat{K}$. Condition 7.14 can be rewritten as follows:

$$|\hat{W}_1(e^{i\theta})| + |(\hat{W}_2\hat{P}_o\hat{K})(e^{i\theta})| \leq |1 - (\hat{P}_o\hat{K})(e^{i\theta})|.$$

In other words, the distance between the minus one point and $(\hat{P}_0\hat{K})(e^{i\theta})$ should be greater than $|\hat{W}_1(e^{i\theta})| + |(\hat{W}_2\hat{P}_o\hat{K})(e^{i\theta})|$, for all $\theta \in [-\pi, \pi]$. This condition corresponds to nonoverlapping circles in Figure 7.13.

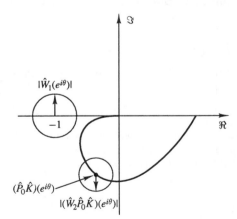

Figure 7.13 Graphical interpretation of performance robustness.

7.8 RANK-ONE PROBLEMS

Rank-one problems define an interesting class of robust stability problems in which the value of $SN_{\Delta, p}(M)$ is quite simple. We only discuss SISO blocks.

Definition 7.8.1.   A robust stability problem that reduces to the problem in Figure 7.3 with $\Delta = (\Delta_1 \quad \ldots \quad \Delta_n)$ and M of dimension $n \times 1$ is known as a rank-one problem. Similarly, if $\Delta = (\Delta_1 \quad \ldots \quad \Delta_n)^T$ and M is of dimension $1 \times n$. ∎

Example 7.8.1

The robust performance problem in Example 7.7.1 is a rank-one problem. It is evident that M is given by

$$M = \begin{pmatrix} W_1(I - P_oK)^{-1} \\ W_2 P_o K (I - P_oK)^{-1} \end{pmatrix} (1 \quad 1).$$

It follows that $(I - M\Delta)$ has a stable inverse if and only if

$$1 - (\Delta_1 \quad \Delta_2) \begin{pmatrix} W_1(I - P_oK)^{-1} \\ W_2 P_o K (I - P_oK)^{-1} \end{pmatrix}$$

has a stable inverse. Thus it is a rank-one problem.

Example 7.8.2

The problem of stability robustness in the presence of coprime factor perturbations is a rank-one problem (see Figure 3.11). From Equation 3.16, the stability of the system is equivalent to the stability of the operator

$$\left(1 - (X \quad Y) \begin{pmatrix} \Delta_1 \\ \Delta_2 \end{pmatrix} \right)^{-1}.$$

7.8.1 Linear Time-Varying Perturbations

ℓ_∞ stability robustness.   If Δ_i is ℓ_∞-stable with $\|\Delta_i\|_{\ell_\infty-\text{ind}} < 1$ then

$$SN_{\Delta_{LTV}, \infty}(M) = \sum_{i=1}^{n} \|m_i\|_1$$

where $M = (m_1 \quad \ldots m_n)^T$. It is interesting that the stability robustness condition is a norm condition on the system M.

ℓ_2 stability robustness.   If Δ_i is ℓ_2-stable with $\|\Delta_i\|_{\ell_2-\text{ind}} < 1$ then

$$SN_{\Delta_{LTV}, 2}(M) = \inf_{d_1, \ldots d_n > 0} \sum_{j=1}^{n} d_j^{-2} \left(\sup_{\theta \in [0, ,2\pi)} \sum_{k=1}^{n} |\hat{m}_k(e^{i\theta})|^2 d_k^2 \right)$$

where $M = (m_1 \quad \ldots m_n)^T$. This can be derived by direct substitution.

7.8.2 Time-Invariant Perturbations

For time-invariant perturbations, we need to compute $\mu_\Delta[\hat{M}(e^{i\theta})]$ when M is a column matrix. This is trivially given by

$$\mu_\Delta[\hat{M}(e^{i\theta})] = \sum_{k=1}^{n} |\hat{m}_k(e^{i\theta})| = |\hat{M}(e^{i\theta})|_1.$$

This condition defines a norm condition on the vector $\hat{M}$ similar to the result in Example 7.8.1.

7.8.3 Real Parameter Perturbations

Throughout this chapter we have not discussed real parametric uncertainty. Such uncertainty is quite common in control problems and results from the lack of precise information about certain parameters in the system. The analysis of the general stability robustness problem in the presence of real parameter uncertainty is quite involved. However, the class of rank-one problems can be easily analyzed and is presented next.

Let the perturbation set Δ be given by

$$\Delta_{\text{real}} = \{\delta | \delta = (\delta_1 \quad \ldots \quad \delta_n), \ \delta_i \in \mathbb{R}\}$$

The norm on this set is given by $|\delta|_\infty$.

The structured norm with respect to this class of perturbations is given by $\mu_{\Delta_{\text{real}}}[M]$ at each frequency point. We will drop the frequency argument and treat M as a complex vector.

Theorem 7.8.1.   The structured singular value for the class Δ_{real} is given by

$$\mu_{\Delta_{\text{real}}}[M] = \min_{\alpha \in \mathbb{R}} |\Re M + \alpha \Im M|_1.$$

Proof. The simplest proof of this result follows from a duality argument. It is a special case of the example in Section 9.4.2.   ∎

Example 7.8.3

A good application of this result is the stability robustness of a class of interval plants. Consider a class of systems with closed loop polynomials given by

$$a(\lambda, \delta) = \lambda^n - (\delta A + b) \begin{pmatrix} \lambda^{n-1} \\ \vdots \\ 1 \end{pmatrix}$$

where $\delta \in \Delta_{\text{real}}$, $A \in \mathbb{R}^{n \times n}$ and $b \in \mathbb{R}^{1 \times n}$. The problem is to find the maximum norm of δ for which the system is stable. Define

$$\hat{M}_1(\lambda) = A \begin{pmatrix} \lambda^{n-1} \\ \vdots \\ 1 \end{pmatrix}, \quad \hat{r}(\lambda) = \lambda^n - b \begin{pmatrix} \lambda^{n-1} \\ \vdots \\ 1 \end{pmatrix}.$$

Since Δ_{real} includes the point 0, it follows that $\hat{r}(\lambda)$ has zeros outside the unit disc. Define

$$\hat{M} = \frac{1}{\hat{r}(\lambda)} \hat{M}_1(\lambda).$$

It is immediate now that the solution is given by Theorem 7.8.1.

7.8.4 Mixed Real and Complex Perturbations

Now we consider a rank-one robust stability problem with combined real and complex perturbations. Complex perturbations arise from the evaluation of a time-invariant perturbation at a frequency point. Define

$$\mathbf{\Delta}_{\text{mixed}} = \{\Delta_m = (\begin{array}{cccccc} \delta_1 & \cdots & \delta_{n_1} & \Delta_{n_1+1} & \cdots & \Delta_n \end{array}) = (\begin{array}{cc} \delta & \Delta \end{array}),\ \delta_i \in \mathbb{R} \text{ and } \Delta_j \in \mathbb{C}\}.$$

The norm on this set is given by

$$\|\Delta\| = \max\{|\delta|_\infty, |\Delta|_\infty\}.$$

Notice that for the complex part of the perturbations, the norm is given by the maximum magnitude of all components. Decompose M comformally with Δ_m:

$$M = \begin{pmatrix} R \\ S \end{pmatrix}.$$

Theorem 7.8.2.   The structured singular value for the class $\mathbf{\Delta}_{\text{mixed}}$ is given by

$$\mu_{\mathbf{\Delta}_{\text{mixed}}}[M] = \min_{\alpha \in \mathbb{R}} |\Re R + \alpha \Im R|_1 + \sqrt{1 + \alpha^2}|S|_1.$$

Proof. The proof of this result follows directly from the example in Section 9.4.2.
∎

Example 7.8.4

Let a set of SISO plants be characterized by the following sets of numerators and denominators:

$$N = N_o + \delta N_\delta + \Delta_N, \quad D = D_o - \delta D_\delta - \Delta_D$$

with N_o, D_o coprime. Define

$$\Omega = \{P = ND^{-1}|\delta \in \mathbb{R}^{1 \times n},\ \Delta_N,\ \Delta_D \in \mathcal{RH}_\infty\}.$$

The set of plants is described in Figure 7.14. It is clear that this is a rank-one problem. To find M, let K be any stabilizing controller with coprime factorization Y/X such that

$$D_o X - N_o Y = 1.$$

Then it follows that

$$M = \begin{pmatrix} D_\delta X + N_\delta Y \\ X \\ Y \end{pmatrix}$$

and

$$\mathbf{\Delta}_{\text{mixed}} = \{(\begin{array}{ccc} \delta & \Delta_D & \Delta_N \end{array})\}.$$

It follows from Theorem 7.8.2 that the robust stability analysis for this class of problems can be computed quite efficiently. Uncertain descriptions as in this example are quite natural, especially when the nominal plant is obtained from standard identification methods that estimate the coefficients of the numerator and denomnator directly. The parametric uncertainty will then represent the deviation of the nominal model from the actual system if it has the same degree, and the complex perturbations represent the undermodeling of the McMillan degree of the system.

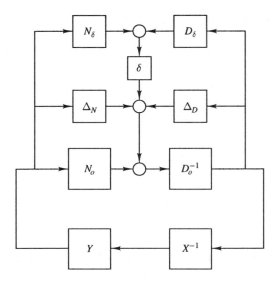

Figure 7.14 Mixed coprime factor perturbations.

7.9 SUMMARY

In this chapter we have analyzed the stability robustness of LTI systems in the presence of various classes of structured perturbations. Results are summarized in Table 7.3.

The performance robustness problems that have known solutions are problems that can be converted to a robust stability problem by adding a fictitious perturbation to represent the performance. Table 7.4 shows how such a perturbation can be selected to make the equivalence between stability robustness and performance robustness exact. Some of the resulting problems have mixed perturbations.

The synthesis problem now becomes an optimization problem involving the structured norm as an objective function. In the case of time-varying or nonlinear perturbations, the optimization problem is given by

$$\inf_{K \text{ stabilizing}} \quad \inf_{D \in \mathbf{D}} \| D^{-1} F_\ell(G, K) D \|_{\ell_p - \text{ind}}.$$

For linear time-invariant perturbations, the problem is given by

$$\inf_{K \text{ stabilizing}} \sup_{\theta \in [0, 2\pi]} \mu_\Delta [F_\ell(\hat{G}, \hat{K})(e^{i\theta})].$$

TABLE 7.3   ROBUST STABILITY CONDITIONS

Stability notion	Class of perturbations			
	LTI	LTV	NL	Parametric
ℓ_2-Stability	$\mu_\Delta[\hat{M}(e^{i\theta})]$	$\min_{D \in \mathbf{D}} \| D^{-1} \hat{M} D \|_\infty$	$\min_{D \in \mathbf{D}} \| D^{-1} \hat{M} D \|_\infty$	$\mu_{\Delta_{\text{real}}}[M]$
ℓ_∞-Stability	$\mu_\Delta[\hat{M}(e^{i\theta})]$	$\rho(\widehat{M})$	$\rho(\widehat{M})$	$\mu_{\Delta_{\text{real}}}[M]$

TABLE 7.4   PERTURBATION BLOCKS
FOR PERFORMANCE

	Fictitious Perturbation block Δ_P	
Plant Perturbation	Performance ℓ_∞-induced	Performance ℓ_2-induced
LTV	LTV	LTV
NL	NL	NL
LTI	LTV	LTI
Real Parametric	LTV	LTI

In general, μ_Δ is replaced by the upper bound

$$\inf_{K \text{ stabilizing}} \quad \inf_{\hat{D} \text{ frequency dependent}} \quad \|\hat{D}^{-1}[F_\ell(\hat{G}, \hat{K})\hat{D}\|_\infty.$$

These computations will be discussed in Chapter 14.1. It is interesting to note that the robustness condition for LTI perturbations involves scales that are frequency dependent. This indicates that the trade-offs are given at each frequency point. This contrasts LTV perturbations where the scales are constant for all frequencies.

EXERCISES

7.1. Show that the destabilizing perturbation in Theorem 7.3.2 can be constructed so that it is strictly causal.

7.2. Generalize the stability robustness results for nonsquare blocks. (Hint: Pick the diagonal scalings to have compatible dimensions with the perturbations.)

7.3. Prove the result in Theorem 7.5.3.

7.4. Consider the following parametrization of positive diagonal elements:

$$D = \exp(X), \quad X \text{ any diagonal matrix.}$$

Is the problem

$$\inf_{X \text{ diagonal}} \| \exp(X)\widehat{M} \exp(-X)\|_1$$

a convex minimization problem?

7.5. Suppose that M has a row with all elements equal to zero, except for the diagonal element. Clearly, this matrix does not satisfy the conditions required for the proof of Theorem 7.3.2. Show that the theorem is still valid for such a case.

7.6. Complete the proof of Theorem 7.3.3.

7.7. Consider the class of plants described as

$$\Omega = \left\{ P \,|\, P = (1 + \Delta_1 W_1)P_o + \Delta_2 W_2, \|\Delta_i\|_{\ell_\infty - \text{ind}} < 1 \right\}.$$

Show that a necessary and sufficient condition for a controller to robustly stabilize this set and robustly reject disturbances at the output of the plant (i.e., satisfy $\|W_3(I - PK)^{-1}\| \leq 1$ for all $P \in \Omega$) is given by

$$\|W_3(I - P_oK)^{-1}\|_1 + \|W_1P_oK(I - P_oK)^{-1}\|_1 + \|W_2K(I - P_oK)^{-1}\|_1 \leq 1.$$

7.8. Consider the class of plants

$$\Omega = \left\{ P|P = P_o(1 + W_1\Delta), \ \|\Delta\|_{\ell_\infty-\text{ind}} < 1 \right\}.$$

Assume the reference inputs are modeled as W_2w_2, with $w_2 \in \ell_\infty$. Show that the robust tracking condition is given by

$$\|[(I - KP_o)^{-1}W_2 \quad (I - KP_o)^{-1}KP_oW_1]\|_1 \leq 1.$$

Compare to Example 7.7.1 in both SISO and MIMO cases.

7.9. Given the finite set of systems:

$$\Omega = \{P_1, \ldots, P_n \mid P_i \text{ is FDLTI system for } i = 1, \ldots, n\}$$

Assume that there exists one FDLTI controller that stabilizes every plant in Ω. Show that there exists a 2-input 2-output system, P, such that
(a) $\Omega \subset \{F_\ell(P, \Delta) \mid \Delta \in \mathcal{RH}_\infty, \|\hat{\Delta}\|_\infty < 1\}$
(b) K stabilizes all plants in $\{F_\ell(P, \Delta) \mid \Delta \in \mathcal{RH}_\infty, \|\hat{\Delta}\|_\infty < 1\}$.

7.10. Suppose A is a positive matrix. Show that

$$\mu_\Delta(A) = \rho(A)$$

where

$$\Delta = \{\text{diag}(\Delta_1, \ldots, \Delta_n)|\Delta_i \in \mathbf{C}\}.$$

Does the answer change if the perturbations are real?

7.11. Consider the proof of Lemma 7.3.2. Show that if d_2 is equal to zero in the separating hyperplane, then $M_{12} = 0$. Show how this contradicts the hypothesis of the Lemma. Prove that Theorem 7.3.4 still holds.

7.12. Let M be an $n \times n$ system such that $\hat{M}(e^{i\theta})$ has rank equal to one at each θ. Compute $\mu_{\Delta_{\text{real}}}[\hat{M}(e^{i\theta})]$.

7.13. In this problem, we are interested in studying the stability of linear time-invariant systems when the A matrix is perturbed by a *real matrix*.
(a) Show that for any $n \times n$ complex matrix F with rank greater than or equal to r,

$$\min_{B}\{\sigma_{\max}[B]| \text{ rank of } (F + B) = r < n\} = \sigma_{r+1}[F]$$

where $\sigma_{r+1}[.]$ denotes the $r + 1$ singular value. Construct a minimizing matrix B from the singular value decomposition of F. Will the matrix B be a real matrix, or a complex matrix?
Now suppose we have the following unforced dynamic system:

$$x(k + 1) = (A + \Delta)x(k)$$

where A is a fixed stable matrix, and Δ is an unknown real perturbation matrix. The *stability margin* of this system is defined as

$$\mu[A] = \frac{1}{\min_{\Delta \in \mathbb{R}^{n \times n}}\{\sigma_{\max}[\Delta]|A + \Delta \text{ is unstable}\}}.$$

(b) Argue that the minimizing solution Δ of the above problem results in $A + \Delta$ having eigenvalues on the unit circle.

(c) Computing μ may be a hard task. We can get an upper bound on μ by dropping the condition that Δ is a real matrix, and allow complex matrices. Show that

$$\mu[A] \leq \frac{1}{\min_{\Delta \in \mathbb{C}^{n \times n}} \{\sigma_{\max}[\Delta] | A + \Delta \text{ is unstable }\}} = \frac{1}{\min_{\theta \in [0, 2\pi)} \sigma_{\min}(A - e^{i\theta}I)}$$

where $\sigma_{\min}[.]$ denotes the smallest singular value.

(d) To improve the above upper bound, we use the information that if Δ is real, then poles appear in complex-conjugate pairs. Define

$$A_\theta = \begin{pmatrix} A - \cos\theta I & \sin\theta I \\ -\sin\theta I & A - \cos\theta I \end{pmatrix}.$$

Show that

$$\mu[A] \leq \frac{1}{\min_{\theta \in [0, 2\pi)} \sigma_{\min}[A_\theta]}.$$

(e) The above bound can be improved. In fact, it follows that

$$\mu[A] \leq \frac{1}{\min_{\theta \in [0, 2\pi)} \sigma_{2n-1}[A_\theta]}$$

where σ_{2n-1} is the next to last singular value. Show this result.

(f) For every $\alpha \in \mathbb{R}$, define

$$A_\theta(\alpha) = \begin{pmatrix} A - \cos\theta I & \alpha \sin\theta I \\ \frac{-1}{\alpha} \sin\theta I & A - \cos\theta I \end{pmatrix}.$$

Show that

$$\mu[A] \leq \frac{1}{\min_{\theta \in [0, 2\pi)} \inf_{\alpha \neq 0} \sigma_{2n-1}[A_\theta(\alpha)]}.$$

NOTES AND REFERENCES

The general setup for robust stability and performance using LFT and the formulation of the structured uncertainty problem (μ) can be found in [DS81, Doy82, Doy85, DWS82]. The treatment was constrained to ℓ_2 stability in the presence of time-invariant perturbations. Simultaneously, the idea of minimizing the conservatism of the small-gain theorem by introducing diagonal scalings was suggested in [Saf82]. The structured singular value function, μ, as described in this chapter, is generally known as the complex μ [PD93]. For textbooks discussing μ see [MZ89, Mac89, DFT92].

The use of singular values to analyze the stability robustness of systems was proposed in [DS81], which is a specialization of the general small-gain approach proposed in [Zam66]. In fact, the argument that the maximum singular value provides the least conservative condition for stability in the presence of perturbations of bounded singular values was made in [DS81] and in more detail in [Che84].

In the case of ℓ_∞ stability, it was evident that the small-gain theorem is conservative if the perturbations are time invariant. It was shown in [DO88] that if the perturbations are enlarged to contain time-varying norm-bounded perturbations, then the small-gain condition is both necessary and sufficient. The ℓ_∞ stability robustness in the presence of structured uncertainty was first analyzed in [KP90] for a special case. A general theory was then developed in [KP91, KP93] in which it was proved that the diagonally scaled norm with constant scales is both necessary and sufficient for stability robustness in the presence of time-varying norm-bounded perturbations. The proofs presented in this chapter follow the ones in [DK93]. The equivalence between robust stability and robust performance was also shown in [KP91]. The relations of this condition to the spectral radius of certain matrices are analyzed in [DK93, KP93] (see also [Saf82, Vid81]). Generalizations to time-varying systems can be found in [KD92, Kha93, SK91, SD91].

The role that time-varying perturbations played in the ℓ_∞ theory motivated the results in [Meg, Sha94], in which it was shown that the diagonally scaled norm with constant scales is also a necessary condition for robustness in ℓ_2 if the perturbations are time varying. Also, it was shown in [BD92, DK93] that μ gives the necessary and sufficient condition for stability in the presence of LTI perturbations regardless of the signal norm.

Cases of rank-one problems can be found in [QD89, CFN91]. For more general treatment of mixed real and complex uncertainty, see [FTD91, You93]. Kharitonov-like results for continuous-time plants are found in [Kha78, Bar89, Bha87, CB89, DCB90, DTV93]. In these references, it is shown that the problem of evaluating the stability of an interval set of plants with independent real parameters is equivalent to checking the stability of only four polynomials. Generalization to the linearly dependent parameters have also been given. Some results that generalize to discrete-time systems are reported in [Ran92]. Synthesis with rank-one mixed perturbations is shown to be equivalent to a quasi-convex problem in [RM]. Exact synthesis for the Example 7.7.1 using duality theory can be found in [OZ93].

CHAPTER 8 ———————————

Linear Programming

We have seen from previous chapters that controller synthesis involves solving constrained optimization problems. In Chapter 3 it was demonstrated that many of these problems can be formulated in terms of infinite-dimensional linear programming problems. In the next two chapters we will develop a rich theory for infinite-dimensional optimization problems that will play an important role in either giving exact solutions to certain classes of problems, or providing good methods for obtaining approximate solutions. The main tool presented in these chapters is *duality theory*.

We start by presenting some fundamental concepts and results from finite-dimensional linear programming. This is quite a rich topic, and a detailed exposition is beyond the scope of this book. The presentation in this chapter is intended to motivate the developments in the next chapter and highlight the main results in linear programming that will be useful later on. As we shall see in later chapters, linear programming theory will be instrumental to the computational procedures involved in actual controller designs.

8.1 THE LINEAR PROGRAMMING PROBLEM

A linear programming (LP) problem is an optimization problem in $\mathbb{R}^n$, where the objective function is linear in the unknowns, and the unknowns have to satisfy a set of linear equality and/or inequality constraints. There are several equivalent ways of stating such problems. Here we adopt the following *standard form*:

$$\min_{x} \ c^T x$$

subject to

$$Ax = b,$$
$$x_i \geq 0, \quad i = 1, \ldots, n,$$

(8.1)

where $x, c \in \mathbb{R}^n$, $b \in \mathbb{R}^m$ and $A \in \mathbb{R}^{m \times n}$. It should be noted that any LP problem can be transformed into the above form.

The set of inequalities $x_i \geq 0$ for $i = 1, \ldots, n$ specify a convex set in $\mathbb{R}^n$ known as a *convex cone*. In general, a set C is a cone if for every x in C, αx is in C for any real $\alpha \geq 0$. To simplify the notation, the convex cone in Equation (8.1) will be denoted by $x \geq 0$.

The set of constraints in Equation (8.1) define a convex region in $\mathbb{R}^n$. Indeed, if $x, y \in \mathbb{R}^n$ satisfy such constraints then a convex combination will also satisfy them:

$$A(\alpha x + (1 - \alpha)y) = \alpha Ax + Ay - \alpha Ay = b$$

and $\alpha x + (1 - \alpha)y \geq 0$ for $0 \leq \alpha \leq 1$ since $x, y \geq 0$.

8.2 BASIC SOLUTIONS AND RELATED PROPERTIES

Next, we state a number of standard definitions in linear programming. To simplify the exposition we will assume that $n > m$ and that A is full rank (i.e., rank of A is m).

Definition 8.2.1. An element $x \in \mathbb{R}^n$ is a feasible point (or a feasible solution) if it satisfies the set of constraints in Equation (8.1). Then, the collection of all feasible points defines the feasible set. ∎

Definition 8.2.2. Given the set of m equality constraints $Ax = b$, let B be any nonsingular submatrix made up with m columns of A. Then, $x_B = B^{-1}b$ defines a basic solution of $Ax = b$. Such a solution will have $n - m$ components equal to zero, corresponding to those columns of A not in B. The m components that correspond to the columns of B are called basic variables and are given in x_B. ∎

Consequently, if the columns of A are ordered such that the first m columns define a basis B, then the corresponding basic solution takes the form $x = (x_B^T \ 0 \ \cdots \ 0)^T$.

Definition 8.2.3. A basic feasible solution is a basic solution where $x \geq 0$. ∎

Definition 8.2.4. If in a basic solution one or more basic variables are zeros, then such a solution is said to be a degenerate basic solution. ∎

Note that in general a nondegenerate basic solution x, defines a unique basis B simply by collecting those columns of A corresponding to the strictly positive components of x. If the solution is degenerate, however, there is ambiguity in the choice of the basis.

Let $\mathcal{F}$ denote the (convex) feasible set. Then three situations are possible: either $\mathcal{F}$ is empty, bounded, or unbounded. Clearly, if the feasible set is empty, then the LP problem has no solution. If it is bounded then there is at least one optimal solution, and all optimal solutions are finite. Finally, if $\mathcal{F}$ is unbounded then the optimal solution may be unbounded.

In general, a convex set defined by the intersection of a finite number of linear equalities and/or inequalities is known as a *convex polytope*. Such a set enjoys the property of having a finite number of *extreme points* (a point in a convex set is extreme if it cannot be expressed as the convex combination of any other two different points in the set). In fact, it can be shown that x is an extreme point of $\mathcal{F}$ if and only if x is a basic feasible solution.

Theorem 8.2.1.   If $\mathcal{F}$ is not empty then it has at least a basic feasible solution (i.e., an extreme point).

Proof. Let a_i denote the ith column of A. Suppose x is a feasible solution. Assume, for simplicity, that the first p components are strictly positive and the rest are zero, i.e., $x = (x_1 \cdots x_p \, 0 \cdots 0)$. Then,

$$x_1 a_1 + x_2 a_2 + \cdots + x_p a_p = b.$$

First we treat the trivial case. If $a_1, \ldots, a_p$ are linearly independent then $p \leq m$ since A is a full rank $m \times n$ matrix. Then, x is clearly a basic solution (degenerate if $p < m$) and the proof is complete. If $a_1, \ldots, a_p$ are not linearly independent then there exists a nonzero $y \in \mathbb{R}^n$, with at least one strictly positive component, such that

$$y_1 a_1 + y_2 a_2 + \cdots + y_p a_p = 0$$

and with the rest $n - p$ components equal to zero. Then $A(x - \epsilon y) = b$ for any scalar ϵ, since $Ay = 0$. Furthermore, note that the last $n - p$ components of $(x - \epsilon y)$ remain equal to zero for all ϵ. Now, set

$$\epsilon = \min_{\{i \mid y_i > 0\}} \frac{x_i}{y_i}.$$

Clearly, this choice of ϵ will make $x_i - \epsilon y_i = 0$ for some $i \in \{1, \ldots, p\}$ without violating feasibility, that is, maintaining $(x - \epsilon y) \geq 0$. Then, the new feasible point will have $p - 1$ strictly positive components. This process can be repeated until the number of strictly positive components is less than or equal to m and the corresponding columns of A are linearly independent (i.e., until the feasible solution is also basic). ∎

The following corollary to the above theorem states the fundamental principle behind linear programming:

Corollary 8.2.1. Given a linear program in standard form, if an optimal solution exists, then the optimal value is attained at an extreme point of the feasible set (i.e., at a basic feasible solution).

Proof. Consider the proof of Theorem 8.2.1 but starting with an optimal feasible solution, x. The trivial case is exactly as before so we will only proof the case where $a_1, \ldots, a_p$ are linearly dependent.

The objective function of the perturbed optimal solution is given by $c^T x - \epsilon c^T y$. Assume $c^T y \neq 0$, then there exists an $\epsilon \neq 0$ such that $(x - \epsilon y)$ remains feasible and $\epsilon c^T y > 0$. But this implies that the cost can be reduced without violating feasibility, which contradicts the fact that $c^T x$ attained the optimal value. Therefore, we conclude that $c^T y = 0$. Because this procedure does not affect the optimality condition, the rest of the proof is exactly as in Theorem 8.2.1.   ∎

This result suggests a straightforward algorithm for finding an optimal solution to Equation (8.1). We know that the optimal value is attained at an extreme point, and that there is a finite number of them. In fact, the number of extreme points is bounded from above by the number of m-dimensional bases that can be formed with the columns of A, a number not greater than $\frac{n!}{m!(n-m)!}$. Then, the algorithm goes as follows: (1) compute all possible basic solutions, (2) discard those that are infeasible (i.e., those with negative components), (3) compute the objective function on the remaining set, and (4) pick the one that has the minimum value.

However simple, such algorithm can be highly inefficient. For instance, a relatively small linear program may have $n = 50$ and $m = 20$, and the number of basic solutions is in the order of 10^{13}. Moreover, the computation of each basic solution involves (in principle) the inversion of the corresponding basis, an $m \times m$ matrix. It turns out, by expanding the ideas in the above proofs we will be able to define a very efficient algorithm known as the *simplex method*.

Example 8.2.1

Consider the following LP problem in standard form:

$$\min_{x} 2x_1 - x_2 + 3x_3$$

subject to

$$x_1 + x_2 + x_3 = 1,$$
$$x \geq 0.$$

$$(8.2)$$

The equality constraint defines a plane in $\mathbb{R}^3$ whose intersection with the convex cone $x \geq 0$ determines the feasible set, see Figure 8.1. Clearly, $\mathcal{F}$ has three extreme points corresponding to the three vertices of a triangle. Indeed, the three basic feasible solutions are given by the points $(1, 0, 0)$, $(0, 1, 0)$ and $(0, 0, 1)$. By evaluating the objective function at each point we find that the optimal is attained at $(0, 1, 0)$ with a value of -1. Note that none of the basic feasible solutions are degenerate.

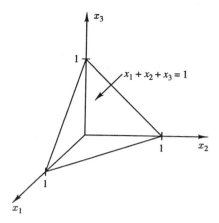

Figure 8.1   Feasible set in $\mathbb{R}^3$.

Example 8.2.2

Consider the problem of finding a solution of minimum $|\cdot|_1$-norm to an underdetermined system of linear equations. That is, for $x \in \mathbb{R}^n$ and $A \in \mathbb{R}^{m \times n}$ of full rank:

$$\min_x |x|_1$$

$$\text{subject to} \tag{8.3}$$

$$Ax = b.$$

Next, we will show that this problem can be transformed into a standard LP.

There are two reasons why Problem (8.3) is not in standard form: First, the objective function is not linear in x, and second, the x_i's are not constrained to be nonnegative. As in the case concerning infinite sequences, consider expressing each x_i as the difference of two nonnegative variables, that is: $x_i = x_i^+ - x_i^-$ where x_i^+ and x_i^- are restricted to be nonnegative. With this notation, the feasible set is defined by:

$$A(x^+ - x^-) = b,$$
$$x^+, x^- \geq 0.$$

Note that this transformation doubles the number of variables (i.e., the problem is now in $\mathbb{R}^{2n}$) without changing the number of equality constraints. This implies that the transformed feasible set will have more extreme points in general. However, an appropriate transformation of the objective function will make such extra basic feasible solutions not optimal. Indeed, consider the following (standard) LP problem:

$$\min_{x^+, x^-} x_1^+ + x_1^- + \cdots + x_n^+ + x_n^-$$

$$\text{subject to}$$

$$A(x^+ - x^-) = b, \tag{8.4}$$

$$x^+, x^- \geq 0.$$

We claim that for any A and b, all optimal solutions of (8.4) are such that either x_i^+ or x_i^- is equal to zero for all i. For, if an optimal solution is such that x_i^+ and x_i^- are strictly positive, then subtracting $\min(x_i^+, x_i^-)$ from each of these variables does not violate feasibility and reduces the cost, contradicting the optimality assumption. This implies that any optimal solution is such that

$$x_1^+ + x_1^- + x_2^+ + x_2^- + \cdots + x_n^+ + x_n^- = |x|_1.$$

Therefore, Problem (8.4) is the standard form equivalent of Problem (8.3).

Example 8.2.3

This example illustrates another common transformation in linear programming. Consider the following problem in $\mathbb{R}^2$ where the linear constraints are all inequalities:

$$\min_{x} \ -x_1 - 2x_2$$

subject to

$$
\begin{aligned}
x_1 + x_2 &\leq 4, \\
x_1 &\leq 3, \\
x_2 &\leq 3, \\
x &\geq 0.
\end{aligned}
\tag{8.5}
$$

To be able to bring the above problem into standard form we need to introduce artificial variables known as *slack variables*. In general, given a set of m inequalities of the form $Ax \leq b$, then $x \in \mathbb{R}^n$ satisfies the set if and only if there exists a nonnegative vector of slack variables, $y \in \mathbb{R}^m$, such that $Ax + y = b$.

Thus, problem (8.5) transforms into the following standard form:

$$\min_{x, y} \ -x_1 - 2x_2$$

subject to

$$
\begin{aligned}
x_1 + x_2 + y_1 &= 4, \\
x_1 + y_2 &= 3, \\
x_2 + y_3 &= 3, \\
x, y &\geq 0.
\end{aligned}
\tag{8.6}
$$

The feasible set for problem (8.5) is depicted in Figure 8.2. Clearly, a basic solution of the transformed problem requires that at least two of the five variables be zero (as there are three equality constraints). It is evident from Figure 8.2 that there are five extreme points:

x	y	$c^T x$
$(0, 0)$	$(4, 3, 3)$	0
$(0, 3)$	$(1, 3, 0)$	-6
$(1, 3)$	$(0, 2, 0)$	-7
$(3, 1)$	$(0, 0, 2)$	-5
$(3, 0)$	$(1, 0, 3)$	-3

and that the optimal value of -7 is attained at $x = (1, 3)$.

The vector in Figure 8.2 represents the gradient of the objective function with the sign changed, i.e., $-c$. In such context, there is an interesting physical interpretation of the optimal solution. If one views $-c$ as a gravity force, then a free particle inside the feasible set would come to rest at an optimal extreme point (point of minimum potential energy).

To illustrate the occurrence of multiple optimal solutions, consider the above problem but with $c^T = (-1 \ \ -1)$. Then, the gradient is normal to the edge of the feasible space corresponding to $y_1 = 0$, and the extreme points $(1, 3)$ and $(3, 1)$ are both optimal. It should

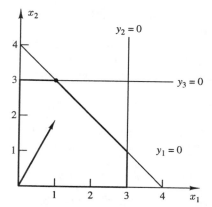

Figure 8.2   Feasible set in $\mathbb{R}^2$.

be noted that if there are two optimal solutions, then there is an infinite number of them, since any convex combination of optimal solutions is also optimal. This results directly from the fact that convex combinations of optimal solutions are feasible (due to the convexity of the feasible set) and attain the same value of the objective function. In the above example, the optimal solutions are given by points on the edge of the feasible set corresponding to $y_1 = 0$.

8.3 PIVOTING

To develop the simplex algorithm we need to have a better understanding of *pivoting*: the mechanism by which a sequence of basic solutions can be constructed efficiently by simple row and column operations. Such procedure has its roots in the solution of linear equations by Gaussian elimination. Instead of inverting a full $m \times m$ matrix every time a new basic solution is computed, pivoting can be used to exchange a single basic variable at a time with a minimum amount of computations.

A solution to the system of equations $Ax = b$ can be viewed as a representation of the vector $b \in \mathbb{R}^m$ via a linear combination of the columns of A, with coefficients x_i. A basic solution is then obtained when at most m coefficients x_i are not zero. Let us assume as before that the basic variables correspond to the first m components of x. Partition the matrix A accordingly, i.e.,

$$A = (B \; \bar{B}),$$

where B is an $m \times m$ full rank matrix and $\bar{B}$ is $m \times (n - m)$. Then, the infinite set of solutions to $Ax = b$ can be expressed as follows:

$$x_B + B^{-1} \bar{B} \bar{x}_B = B^{-1} b \tag{8.7}$$

with $x = (x_B \; \bar{x}_B)$. Clearly, a basic solution is obtained by setting the nonbasic variables, $\bar{x}_B$, to zero.

Equivalently, Equation (8.7) can be viewed as the original set of equations, but expressed in terms of a different basis, that is, the basis corresponding to the columns of B. With this interpretation, pivoting is the process of replacing a vector of the basis by another not already in the basis.

At this point it is convenient to introduce a common way of arranging the information contained in Equation (8.7) in the form of a tableau. Note that every column of A can be represented as a linear combination of the columns of B:

$$a_j = \varphi_{1j} a_1 + \varphi_{2j} a_2 + \cdots + \varphi_{mj} a_m, \tag{8.8}$$

where $j \in \{1, \ldots, n\}$. The coefficient φ_{ij} corresponds to the jth component of the ith column of A expressed in terms of the basis B. Similarly,

$$b = \varphi_{10} a_1 + \varphi_{20} a_2 + \cdots + \varphi_{m0} a_m. \tag{8.9}$$

Let us arrange the above coefficients in the following tableau:

$$
\begin{array}{cccccccc}
a_1 & a_2 & \cdots & a_m & a_{m+1} & a_{m+2} & \cdots & a_n & b \\
\\
1 & 0 & \cdots & 0 & \varphi_{1,m+1} & \varphi_{1,m+2} & \cdots & \varphi_{1,n} & \varphi_{10} \\
0 & 1 & \cdots & 0 & \varphi_{2,m+1} & \varphi_{2,m+2} & \cdots & \varphi_{2,n} & \varphi_{20} \\
\vdots & & \ddots & \vdots & \vdots & \vdots & & \vdots & \vdots \\
0 & 0 & \cdots & 1 & \varphi_{m,m+1} & \varphi_{m,m+2} & \cdots & \varphi_{m,n} & \varphi_{m0}
\end{array} \tag{8.10}
$$

Now, suppose that we want to replace vector a_p of the basis, where $0 \leq p \leq m$, by the vector a_q, where $(m+1) \leq q \leq n$. Assume that the new set of vectors define a basis in $\mathbb{R}^m$ (note that this is so if and only if the pth component of the qth vector, φ_{pq}, is not zero). Then, a_q can be expressed in terms of the old basis as (see Equation (8.8)):

$$a_q = \varphi_{1q} a_1 + \cdots + \varphi_{pq} a_p + \cdots + \varphi_{mq} a_m.$$

By solving for a_p we find an expression of such vector in terms of the new basis:

$$a_p = \frac{1}{\varphi_{pq}} a_q - \frac{1}{\varphi_{pq}} \sum_{\substack{i=1 \\ i \neq p}}^{m} \varphi_{iq} a_i.$$

Then, replacing all occurrences of a_p in Equations (8.8) and (8.9) we construct the updated tableau. It is easy to show that, in terms of the new basis, the (i, j) coefficient is given by

$$\varphi_{ij} \longleftarrow \begin{cases} \varphi_{ij} - \dfrac{\varphi_{iq}}{\varphi_{pq}} \varphi_{pj} & \text{for } i \neq p \\ \dfrac{\varphi_{pj}}{\varphi_{pq}} & \text{for } i = p \end{cases}. \tag{8.11}$$

Clearly, the new representation of a_q will have all zero coefficients but φ_{pq}, which will be equal to one. Also, the last column of the tableau will give the value of the new set of basic variables.

8.4 MAINTAINING FEASIBILITY

In the previous section we have described a simple procedure by which a sequence of basic solutions can be computed, but without regard to the feasibility of such sequence. Here we impose the extra condition of maintaining the nonnegativity of the representation of b in the new basis (i.e., the feasibility of the new basic solution).

We will see that this requirement dictates which vector has to leave the basis, given that a new vector has been chosen to enter the basis. Indeed, assume that the current tableau is as in (8.10), and further, that the current basic solution is feasible (i.e., $\varphi_{i0} \geq 0$ for $i = 1, \ldots, m$). That is,

$$\varphi_{10}a_1 + \varphi_{20}a_2 + \cdots + \varphi_{m0}a_m = b. \tag{8.12}$$

Suppose that we want to bring a_q (with $q > m$) to the basis, then a_q can be represented in terms of the old basis as

$$a_q = \varphi_{1q}a_1 + \varphi_{2q}a_2 + \cdots + \varphi_{mq}a_m. \tag{8.13}$$

Now we use an argument similar to that of Theorem 8.2.1: Multiply Equation (8.13) by $\epsilon > 0$ and subtract it from Equation (8.12) to get

$$(\varphi_{10} - \epsilon\varphi_{1q})a_1 + (\varphi_{20} - \epsilon\varphi_{2q})a_2 + \cdots + (\varphi_{m0} - \epsilon\varphi_{mq})a_m + \epsilon a_q = b. \tag{8.14}$$

As ϵ increases, the coefficients of the old basis vectors will increase, decrease, or remain the same depending on the sign of the corresponding φ_{iq}. Assuming that at least one φ_{iq} is strictly positive, set ϵ such that it makes one of the coefficients zero while keeping the rest nonnegative. That is, set

$$\epsilon = \min_{\{i \,|\, \varphi_{iq} > 0\}} \frac{\varphi_{i0}}{\varphi_{iq}}. \tag{8.15}$$

Note that the minimizing index in the above equation determines which vector has to leave the basis in order to preserve feasibility. Say that p is such index; then the tableau is updated by pivoting on the element φ_{pq}.

It is interesting to point out that two special cases may arise when the above procedure is applied: First, when no φ_{iq} is strictly positive and therefore ϵ is not defined, and second, when more than one coefficient $(\varphi_{i0} - \epsilon\varphi_{iq})$ becomes zero for the ϵ given by Equation (8.15). The first situation implies that ϵ can be made arbitrarily high without violating feasibility. Therefore, such situation corresponds to the case where the feasible set is unbounded (note that at this point it is not clear whether the optimal solution will be unbounded or not). The second case implies that the new basic feasible solution is degenerate (i.e., b is represented by combining less than m vectors of the new basis). In such case any of the vectors associated with the zero coefficients can be removed from the basis.

8.5 OPTIMALITY CONDITION

In the previous section we have shown that, in order to preserve feasibility, the vector that leaves the basis cannot be arbitrary. Indeed, once the new basis vector is selected, it has to be interchanged with a specific vector of the old basis (assuming nondegeneracy).

Note, however, that there is a certain degree of freedom in the choice of the new basis vector. Clearly, one could use that freedom to generate a basic feasible solution with a lower value of the objective function, if possible.

Consider Equation (8.7), then

$$x_B = B^{-1}b - B^{-1}\bar{B}\bar{x}_B \qquad (8.16)$$

gives an expression relating the basic and nonbasic variables so that $Ax = b$. The current basic feasible solution is obtained by letting $\bar{x}_B = 0$. Let us partition c according to x_B and $\bar{x}_B$:

$$\mu := c^T x = c_B^T x_B + \bar{c}_B^T \bar{x}_B. \qquad (8.17)$$

By substituting Equation (8.16) in Equation (8.17) we get an expression for the cost as a function of the nonbasic variables, with the condition $Ax = b$ built-in:

$$\mu = \mu_B + r_B^T \bar{x}_B, \qquad (8.18)$$

where $\mu_B := c_B^T B^{-1} b$ corresponds to the current value of the objective function, i.e., $c_B^T x_B = c_B^T B^{-1} b$, and $r_B := (\bar{c}_B^T - c_B^T B^{-1} \bar{B})^T \in \mathbb{R}^{n-m}$ is known as the *relative cost vector*. The term $r_B^T \bar{x}_B$ reflects the way the nonbasic variables affect the cost. Note that Equation (8.18) gives all possible values of the objective function, such that $Ax = b$, as a function of the nonbasic variables, $\bar{x}_B$. The only constraint not included in that expression is the nonnegativity of $\bar{x}_B$.

Say that the jth nonbasic variable is brought into the basis; then the sign of the jth component of the vector r_B determines whether the cost will increase, decrease, or stay the same with such change. Since we seek the minimum of the objective function, the nonbasic variable with the most negative coefficient will constitute a good choice for a new basic variable.

Theorem 8.5.1. Given a basic feasible solution with objective value $\mu_B = c_B^T B^{-1} b$, if the jth component of r_B is strictly negative, then there exists a feasible solution with objective value less than μ_B.

Proof. It follows directly from the previous discussion.   ∎

Corollary 8.5.1. If for some basic feasible solution $r_B \geq 0$, then such solution is optimal.

Proof. From Equation (8.18), increasing the value of any nonbasic variable would only increase the cost. Thus, the current basic feasible solution must be optimal.   ∎

Corollary 8.5.2. Given the conditions of Theorem 8.5.1, if x_q is a nonbasic variable corresponding to a strictly negative component of r_B, and if $\varphi_{iq} \leq 0$ for $i = 1, \ldots, m$, then the objective function can be made unboundedly negative.

Proof. If $\varphi_{iq} \leq 0$ for $i = 1, \ldots, m$, then, by Equation (8.14), ϵ can be made arbitrarily high, implying that the feasible set is unbounded. Furthermore, for any positive

ϵ, the component of the feasible solution corresponding to x_q is precisely given by ϵ. Therefore, by Equation (8.18), the cost can be arbitrarily reduced.   ∎

8.6 THE SIMPLEX METHOD

In the previous sections we have shown some of the fundamental properties of LP problems by using constructive arguments. Here we piece those arguments together and define a well-known method for solving LP problems known as the *simplex method*.

Consider augmenting the original set of equations, $Ax = b$, with the equation $c^T x - \mu = 0$, that is,

$$\begin{pmatrix} A & 0 \\ c^T & -1 \end{pmatrix} \begin{pmatrix} x \\ \mu \end{pmatrix} = \begin{pmatrix} b \\ 0 \end{pmatrix}. \tag{8.19}$$

The number of equations is $m + 1$, and we are interested in finding basic solutions to such array where the variable μ is always taken as basic. If the first m variables and μ are assumed to be basic, then, after performing the corresponding change of basis, we get the following transformed set of equations:

$$\begin{pmatrix} I & B^{-1}\bar{B} & 0 \\ 0 & \bar{c}_B^T - c_B^T B^{-1}\bar{B} & -1 \end{pmatrix} \begin{pmatrix} x_B \\ \bar{x}_B \\ \mu \end{pmatrix} = \begin{pmatrix} B^{-1}b \\ -\mu_B \end{pmatrix}. \tag{8.20}$$

In particular, note that the last row of the above equation is obtained by applying the standard rules of pivoting presented before. This observation motivates the introduction of the *simplex tableau*. To this end, let us extend the definition of relative cost on basic variables. Clearly, if a currently basic variable is "brought into the basis," the cost does not change (since the basic solution is obviously the same). Thus, we assign a value of zero to the relative cost of basic variables. In other words, define $r \in \mathbb{R}^n$ to be the (extended) relative cost vector, then

$$r^T := \begin{pmatrix} 0 & \cdots & 0 & r_B^T \end{pmatrix}$$

and Equation (8.20) can be written as follows:

$$\begin{pmatrix} I & B^{-1}\bar{B} & 0 \\ & r^T & -1 \end{pmatrix} \begin{pmatrix} x_B \\ \bar{x}_B \\ \mu \end{pmatrix} = \begin{pmatrix} B^{-1}b \\ -\mu_B \end{pmatrix}. \tag{8.21}$$

The simplex tableau is a direct representation of the above equation, where the column corresponding to the basic variable μ is left out:

$$\begin{array}{cc|c} I & B^{-1}\bar{B} & B^{-1}b \\ \hline & r^T & -\mu_B \end{array} \tag{8.22}$$

The pivoting procedure can be applied to the tableau directly, so that a sequence of basic feasible solutions are generated. As a by-product, the tableau keeps track of the current value of the objective function, μ_B, the basic solution that achieves it, $B^{-1}b$, and the current value of the reduced cost vector, r^T. The tableau need not carry the column corresponding to the variable μ since such variable is always taken as basic.

Note that the array in Tableau 8.22 depicts the special case where the basic variables are ordered as the first m components of x (in order to simplify the notation). In general, however, the basic variables will be shuffled among the n components of x and have to be identified through the equally shuffled $m \times m$ identity matrix present in the first m rows of the tableau.

Having defined the simplex tableau we are ready to describe the simplex algorithm. Typically, the simplex method is divided into two phases: *phase one*, in which a single basic feasible solution is computed; and *phase two*, in which a sequence of basic feasible solutions is constructed such that the cost is reduced monotonically until the optimal solution is found or the unboundedness of the solution is established.

It is convenient to describe phase two of the algorithm first. Assume that a basic feasible solution with its corresponding tableau is given. Then, phase two of the simplex algorithm is given by the following procedure:

Step 1:   If $r \geq 0$, stop (the current basic feasible solution is optimal).

Step 2:   Pick q such that $r_q < 0$. (This index points to a nonbasic variable that, if brought into the basis, reduces the value of the objective function.)

Step 3:   If $\varphi_{iq} \leq 0$ for $i = 1, \ldots, m$, stop (the solution is unbounded).

Step 4:   Select index $p \leq m$ such that

$$\frac{\varphi_{p0}}{\varphi_{pq}} = \min_{\{i | \varphi_{iq} > 0\}} \frac{\varphi_{i0}}{\varphi_{iq}}.$$

This index points to the variable that can leave the basis without violating feasibility.

Step 5:   Pivot the full tableau on φ_{pq} and return to Step 1.

The convergence characteristics of this procedure are presented in the next theorem.

Theorem 8.6.1.   Given a LP problem in standard form, if every basic feasible solution is nondegenerate, then phase two of the simplex algorithm reaches a terminating condition (i.e., optimality or unboundedness) after a finite number of iterations.

Proof. If an iteration is such that neither terminating conditions are satisfied, then the value of the objective function is strictly reduced since the corresponding reduced cost, r_q, is strictly negative and all basic feasible solutions are assumed to be nondegenerate (i.e., each new basic variable is strictly positive). This implies that the same basic feasible solution cannot be repeated in the iterative process. Therefore, a terminating condition must be met in a finite number of iterations since the feasible set has only a finite number of extreme points.   ∎

To illustrate why the nondegeneracy assumption is required in the above theorem, consider the following. Assume that at some point in the iterative process a degenerate solution is obtained. In particular, say that $\varphi_{k0} = 0$ (recall that the basic variables can be read directly from the last column of the tableau, in this case the kth entry is zero). Assume that in Step 2 we pick q with $r_q < 0$, such that $\varphi_{kq} > 0$. Clearly, Step 4 will result in φ_{kq} as the pivot element with $\epsilon = \varphi_{k0}/\varphi_{kq} = 0$. Thus, the degenerate basic variable leaves the basis and is replaced by x_q with a value of zero (since $\epsilon = 0$). Thus, degeneracy is maintained and, most importantly, the value of the objective function does not change. It is possible for this process to continue until the original (degenerate) basic solution is obtained again, completing what is known as a *cycle*. In such cases, the simplex algorithm does not terminate and, in theory, runs indefinitely. In practice, however, small perturbations due to round-off numerical errors make this phenomenon very rare. In any case, the simplex algorithm can be modified to avoid such situations entirely. Finally, let us point out that, in most cases, degenerate solutions can be handled as nondegenerate ones without resulting in cycling.

The proof of Theorem 8.6.1 was based on the fact that the feasible set has a finite number of extreme points. In this sense, the simplex algorithm seems to be no different from the algorithm proposed initially, where all extreme points were listed. However, practice has shown that moving from one extreme point to another, such that the associated cost is decreased, is very advantageous. Indeed, the number of iterations required for convergence is typically proportional to the number of equality constraints (and much smaller than the total number of extreme points).

Example 8.6.1

To illustrate the workings of phase two, we apply the procedure to the problem in Example 8.2.3. Recall that such a problem has only inequality constraints that required the addition of slack variables. This characteristic will enable us to read a starting basic feasible solution directly from the initial tableau. We rewrite the statement of the problem (8.6) for convenience:

$$\min_{x,y} -x_1 - 2x_2$$

subject to
$$\begin{aligned}
x_1 + x_2 + y_1 &= 4, \\
x_1 + y_2 &= 3, \\
x_2 + y_3 &= 3, \\
x, y &\geq 0.
\end{aligned} \qquad (8.23)$$

Clearly, $(y, x) = (4, 3, 3, 0, 0)$ is a valid basic feasible solution of the augmented problem. Then, the initial tableau is given by:

y_1	y_2	y_3	x_1	x_2	b	
1	0	0	1	1	4	
0	1	0	1	0	3	(8.24)
0	0	1	0	1	3	
r^T 0	0	0	−1	−2	0	$-\mu_B$

and we can proceed with phase two of the simplex algorithm. It is standard practice to choose the nonbasic variable with the most negative reduced cost as the new basic variable. Consequently, we take x_2 into the basis. Step 4 dictates that y_3 should leave the basis, which is accomplished by pivoting on φ_{35} (shown bolded in the tableau). Updating the tableau results in:

	y_1	y_2	y_3	x_1	x_2	b	
	1	0	-1	1	0	1	
	0	1	0	1	0	3	(8.25)
	0	0	1	0	1	3	
r^T	0	0	2	-1	0	6	$-\mu_B$

At this point, the basic feasible solution is $(y, x) = (1, 3, 0, 0, 3)$, and it is clearly not optimal since there is still a strictly negative component of r^T corresponding to x_1. Again, we bring x_1 into the basis and, consequently, remove y_1 from the basis (in order to preserve feasibility) by pivoting on φ_{14}. This iteration results in the final tableau:

	y_1	y_2	y_3	x_1	x_2	b	
	1	0	-1	1	0	1	
	-1	1	-1	0	0	2	(8.26)
	0	0	0	0	1	3	
r^T	1	0	1	0	0	7	$-\mu_B$

Clearly, the current basic feasible solution, $(y, x) = (0, 2, 0, 1, 3)$, is optimal since $r^T > 0$. Further, the corresponding value of the objective function is $\mu_B = -7$. Note that this is exactly the same result that was obtained before (in Example 8.2.3), but required the construction of only two basic feasible solutions from a total of five.

Now, let us go back to phase one of the simplex algorithm. Given an LP problem in standard form,

$$\min_x c^T x$$
$$\text{subject to}$$
$$Ax = b,$$
$$x \geq 0,$$

(8.27)

phase one generates one basic feasible solution, which is then used as the starting point of phase two. This can be done by solving an *artificial* LP problem. Assume, without loss of generality, that $b \geq 0$, and consider the following LP problem:

$$\min_{x, y} y_1 + y_2 + \cdots + y_m$$
$$\text{subject to}$$
$$y + Ax = b,$$
$$x \geq 0,$$
$$y \geq 0,$$

(8.28)

where $y \in \mathbb{R}^m$ is a nonnegative vector of *artificial* variables. Clearly, phase two of the simplex algorithm can be readily applied to the artificial LP problem since $y = b$ and $x = 0$ constitute a legitimate basic feasible solution. If the optimal solution of the artificial problem yields an objective value greater than zero (i.e., $y > 0$), then it is clear that there is no solution to $Ax = b$ with $x \geq 0$ and the feasible set of the original problem is empty. Similarly, if the optimal solution of the artificial problem results in an objective value of zero (i.e., $y = 0$), then the corresponding value of x constitutes a basic feasible solution of the original problem. (Note: if the optimal solution to the artificial problem is degenerate and such that some y_i's are basic, they can be replaced by nonbasic x_i's with zero value, so that the resulting basic solution is in terms of x only.)

Example 8.6.2

Consider the following LP problem in standard form:

$$\min_{x} \; x_1 - 2x_2 + x_3$$

subject to

$$-x_1 + 3x_3 = 1, \tag{8.29}$$
$$2x_1 + x_2 + x_3 = 2,$$
$$x \geq 0.$$

We will solve it by applying both phases of the simplex algorithm. In phase one, we add two auxiliary variables and solve the following problem:

$$\min_{x,y} \; y_1 + y_2$$

subject to

$$y_1 - x_1 + 3x_3 = 1, \tag{8.30}$$
$$y_2 + 2x_1 + x_2 + x_3 = 2,$$
$$x \geq 0,$$
$$y \geq 0.$$

Then, the corresponding initial tableau is

y_1	y_2	x_1	x_2	x_3	b	
1	0	−1	0	3	1	
0	1	2	1	1	2	(8.31)

	c^T	1	1	0	0	0	0	$-\mu_B$

Although a basic feasible solution is available for the auxiliary problem, note that the last row in the tableau is not in canonical form (i.e., the reduced cost of the basic variables are not zero). In fact, the last row must be updated by completing the pivoting operation. After doing so, the following tableau is obtained:

y_1	y_2	x_1	x_2	x_3	b	
1	0	−1	0	**3**	1	
0	1	2	1	1	2	(8.32)

	r^T	0	0	−1	−1	−4	−3	$-\mu_B$

We choose x_3 as the new basic variable (since it has the most negative reduced cost) and remove y_1, then

$$
\begin{array}{cccccc|c}
y_1 & y_2 & x_1 & x_2 & x_3 & & b \\
\end{array}
$$

y_1	y_2	x_1	x_2	x_3	b
1/3	0	−1/3	0	1	1/3
−1/3	1	**7/3**	1	0	5/3

r^T 4/3   0   −7/3   −1   0   −5/3 $-\mu_B$ (8.33)

Next, we bring x_1 into the basis and remove y_2:

y_1	y_2	x_1	x_2	x_3	b
2/7	1/7	0	1/7	1	4/7
−1/7	3/7	1	3/7	0	5/7

r^T 1   1   0   0   0   0 $-\mu_B$ (8.34)

The positivity of r^T indicates the end of phase one. Clearly, the optimal cost of the auxiliary problem is zero, so we conclude that the feasible set is not empty. Further, a basic feasible solution for the original problem is given by the above tableau, i.e., $x = (5/7, 0, 4/7)$. Then, the initial tableau of phase two reads as follows:

x_1	x_2	x_3	b
0	1/7	1	4/7
1	3/7	0	5/7

c^T 1   −2   1   0 $-\mu_B$ (8.35)

Again, as in phase one, the last row needs to be updated:

x_1	x_2	x_3	b
0	1/7	1	4/7
1	**3/7**	0	5/7

r^T 0   −18/7   0   −9/7 $-\mu_B$ (8.36)

The current solution is clearly not optimal since x_2 has a strictly negative reduced cost. Then, we bring x_2 into the basis and remove x_1 (by pivoting on φ_{22}), to arrive to the final tableau:

x_1	x_2	x_3	b
1/3	0	1	1/3
7/3	1	0	5/3

r^T 6   0   0   3 $-\mu_B$ (8.37)

Therefore, the optimal solution is $x = (0, 5/3, 1/3)$, which yields an objective value $\mu_B = -3$.

8.7 LINEAR PROGRAMMING AND DUALITY

Given a general linear minimization problem, known as the *primal problem*, it is always possible to define an associated linear maximization problem, known as the *dual problem*, that is based on the same problem data and whose solution bears strong connections to the original one. Such problems are known as *primal-dual pairs*, and can take different equivalent forms. Here we will specify the primal problem as one given in standard form; then the corresponding primal-dual pair is given by

$$
\begin{array}{llll}
(Primal) & \min_{x} \ c^T x & (Dual) & \max_{\gamma} \ \gamma^T b \\
\text{subject to} & & \text{subject to} & \\
& Ax = b, & & \gamma^T A \leq c^T, \\
& x \geq 0, & &
\end{array}
\tag{8.38}
$$

where γ is the vector of *dual variables* in $\mathbb{R}^m$ (i.e., in the dual space). There is a strong symmetry between the primal and dual problem in the following sense: If the dual problem is rewritten in standard form and then dualized, the corresponding primal problem is obtained. In other words, the role of primal or dual is quite arbitrary and can be interchanged.

Here we derive the intimate connection between an LP problem and its dual by exploiting the properties of the simplex method of solution. First, consider the objective values attained by a primal-dual pair at arbitrary feasible points.

Lemma 8.7.1.   If x and γ are feasible for the primal-dual pair of Equation (8.38), respectively, then $c^T x \geq \gamma^T b$.

Proof. Since $Ax = b$ we have that $\gamma^T b = \gamma^T Ax$. But $\gamma^T A \leq c^T$, then $\gamma^T b \leq c^T x$ since x is nonnegative. ∎

This result, known as *weak duality*, implies that each feasible solution of the primal problem yields an upper bound on the dual problem, and, analogously, each feasible solution of the dual problem yields a lower bound on the primal. This has an important consequence: If x and γ are feasible such that $c^T x = \gamma^T b$, then they must be the optimal solutions for their respective problems. Otherwise, the optimal solutions would violate Lemma 8.7.1 owing to the fact that $c^T x$ is minimized and $\gamma^T b$ is maximized. That the converse is true is not obvious, and its proof constitutes the main duality theorem of linear programming.

Theorem 8.7.1.   The primal problem has an optimal solution if and only if the dual problem has an optimal solution, and further, both objectives achieve the same optimal value. Additionally, if either problem has an unbounded optimal solution, the other problem has an empty feasible set.

Proof. Without loss of generality, assume that the optimal primal solution, x, is basic, and that the columns of A and c^T are reordered such that $x = (x_B, 0)$. Then, using the same notation as before, we have that $r_B^T = \bar{c}_B^T - c_B^T B^{-1} \bar{B} \geq 0$ since x is optimal. Therefore,

$$c_B^T B^{-1} \bar{B} \leq \bar{c}_B^T.$$

Now, let $\gamma^T = c_B^T B^{-1}$, then

$$\gamma^T A = (\gamma^T B \quad \gamma^T \bar{B}) = (c_B^T \quad c_B^T B^{-1} \bar{B}) \leq (c_B^T \quad \bar{c}_B^T) = c^T,$$

which implies that γ is a feasible point of the dual problem. Furthermore,

$$\gamma^T b = c_B^T B^{-1} b = c_B^T x_B = c^T x,$$

thus, by Lemma 8.7.1, γ is optimal.

Given an optimal γ, the existence of an optimal x follows directly from the symmetry of the primal-dual pair (i.e., by viewing the dual problem as a primal one).

The last claim of the theorem is proved by contradiction. Consider the case where the primal problem has an unbounded objective, and assume that there exists a feasible point in the dual problem, say γ. Then, let M be the value of the dual objective function at γ, that is, $M = \gamma^T b$. Now, pick a feasible x such that $c^T x < M$ (this is always possible owing to the unboundedness of the solution). Clearly, this contradicts Lemma 8.7.1 since $c^T x < \gamma^T b$, so we conclude that the dual feasible set is empty. As before, the other case follows from the symmetry of the primal-dual pair. ∎

In other words, the above theorem shows that there is no gap (i.e., interval on the real line) between the set of achievable values of the primal and dual problems. In fact, both sets share only one common point, which corresponds to the optimal value of their respective objectives.

Interestingly, the solution of the dual problem can be obtained directly from the solution of the primal problem, which stresses the notion that these two problems are equivalent. Indeed, the columns of A corresponding to the basic variables of the (primal) optimal solution define the full rank matrix B and the vector c_B. Then, the solution of the dual problem is simply given by $\gamma^T = c_B^T B^{-1}$.

Example 8.7.1

Consider the solution to the problem in Example 8.6.2. The optimal basic variables turned out to be x_2 and x_3; consequently, the matrix B is given by

$$B = \begin{pmatrix} 0 & 3 \\ 1 & 1 \end{pmatrix}$$

and $c_B^T = (-2 \quad 1)$. The solution to the dual problem is

$$\gamma^T = c_B^T B^{-1} = (1 \quad -2).$$

Note that $\gamma^T b = c^T x = -3$ as indicated in Theorem 8.7.1.

We end this chapter by presenting another important relationship between the optimal solutions of the primal-dual pair known as *complementary slackness*.

Theorem 8.7.2. Given feasible solutions x and γ, they both are optimal solutions of their respective problems if and only if, for all $i = 1, \ldots, m$,

1. $x_i > 0 \Rightarrow \gamma^T a_i = c_i$,
2. $x_i = 0 \Leftarrow \gamma^T a_i < c_i$,

where a_i is the ith column of A.

Proof. x and γ are optimal if and only if, by Theorem 8.7.1, $\gamma^T b = c^T x$ or equivalently, $(\gamma^T A - c^T)x = 0$. Consider the ith equation:

$$(\gamma^T a_i - c_i)x_i = 0.$$

This equation holds if and only if both conditions of the theorem are satisfied. ∎

The conditions in the previous theorem are sometimes referred to as the *alignment conditions*.

8.8 SUMMARY

We have summarized the basic ideas arising in finite-dimensional linear programming problems. Duality of linear programs is a special case of a more general duality theory for optimization problems. Note that if the primal problem has n variables and m constraints, then the dual problem has m variables and n constraints. In the next chapter we will discuss classes of infinite-dimensional linear programs that have a similar duality representation. For such problems, it is possible to convert a problem with infinitely many variables but finitely many constraints to a finite-dimensional linear program, to which an exact solution can be found.

EXERCISES

8.1. Consider the problem of finding a solution of minimum $|\cdot|_\infty$-norm to an underdetermined system of linear equations. That is, for $x \in \mathbb{R}^n$ and $A \in \mathbb{R}^{m \times n}$ of full row rank:

$$\min_x |x|_\infty$$
$$\text{subject to}$$
$$Ax = b.$$

Show that this problem is equivalent to a linear program.

8.2. (Farkas Lemma) Let $A \in \mathbb{R}^{m \times n}$, $b \in \mathbb{R}^m$ be given. Show that there exists a solution x such that

$$Ax = b, \quad x \geq 0$$

if and only if every $y \in \mathbb{R}^m$ with the property $A^T y \geq 0$ (pointwise inequality) satisfies $y^T b \geq 0$.

8.3. In a nondegenerate situation, show that the basis does not change if b is perturbed with a small enough vector Δb. Show that the change in the cost is given by $\gamma^T \Delta b$, where γ is the solution of the dual problem.

NOTES AND REFERENCES

Material in this chapter is standard and can be found in many linear programming books [Chv83, Lue84].

CHAPTER 9 ———————————

Infinite-Dimensional Optimization

In this chapter we introduce results from optimization theory for infinite-dimensional spaces. A fair number of optimization problems that arise in control applications are distance problems, i.e., problems of the form

$$\inf_{x \in M} \|x_0 - x\|, \tag{9.1}$$

where X is a normed linear space, $x_0 \in X$ is fixed and $M \subset X$ is a convex set, possibly a subspace (see Figure 9.1). One of the obstacles for solving such a problem is that M is usually a subset of an infinite-dimensional vector space, making the optimization problem infinite dimensional in general. On the other hand, these problems have more structure than do arbitrary function minimization problems, which in turn enables us to give general results about existence and computation of optimal solutions. The main tool that provides such information is known as *duality theory*, which is discussed in detail in this chapter.

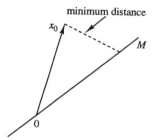

Figure 9.1 Minimum distance problem.

In the previous chapter we discussed the duality theory for finite-dimensional linear programs. Extension of these results to arbitrary infinite-dimensional linear programs is not at all straightforward. In this chapter we set forth a special class of infinite-dimensional linear programs, formulated inside the space ℓ_1, and present their dual representations. For more general classes of problems, we will be content with weak duality results, which can be easily shown.

Finally, this chapter contains a fair amount of mathematical results. These results are quite standard, and we prove only the key ones needed in the development of future chapters. The rest are mentioned without proof.

9.1 DUAL SPACES

Definition 9.1.1. Let X be a normed linear space; f is said to be a bounded linear functional on X if f is a continuous linear operator from X to $\mathbb{R}$. The induced norm of f is given by

$$\|f\| = \sup_{x \in X, x \neq 0} \frac{|f(x)|}{\|x\|}.$$

The dual space of X, denoted by X^*, is the collection of all bounded linear functionals on X, equipped with the natural induced norm. ∎

The space of all bounded linear functionals on X^* is given by X^{**}. It is easy to show that $X \subseteq X^{**}$. In the case where $X^{**} = X$, we say X is reflexive. Whenever f is represented by some element $x^* \in X^*$, we use the notation $\langle x, x^* \rangle$ to represent the value $f(x)$.

Examples of dual spaces

Example 9.1.1

Dual of $(\mathbb{R}^n, |\cdot|_p)$: The dual space of $(\mathbb{R}^n, |\cdot|_p)$, $1 \leq p \leq \infty$, is given by $(\mathbb{R}^n, |\cdot|_q)$ where $\frac{1}{p} + \frac{1}{q} = 1$. In particular, every bounded linear functional, f, is representable uniquely as $f(x) = y^T x$ for some $y \in \mathbb{R}^n$.

Example 9.1.2

Dual of ℓ_p, $1 \leq p < \infty$: The dual of the space ℓ_p is given by the space ℓ_q, where $\frac{1}{p} + \frac{1}{q} = 1$. The characterization is given by the following theorem.

Theorem 9.1.1. Every bounded linear functional on ℓ_p, $1 \leq p < \infty$, is representable uniquely in the form

$$f(x) = \sum_{k=0}^{\infty} x(k) y(k),$$

where $y = \{y(k)\}$ is an element in ℓ_q. Furthermore, every element in ℓ_q defines a member of ℓ_p^* in this way and

$$\|f\| = \|y\|_q.$$ ∎

The spaces ℓ_p, $1 < p < \infty$ are all reflexive. The space ℓ_1, is not reflexive, i.e., $\ell_1^* = \ell_\infty$ and $\ell_\infty^* \neq \ell_1$. It is interesting to note that the subspace of ℓ_∞ denoted by c_0 defined as

$$c_0 = \{x \in \ell_\infty | \lim_{k \to \infty} x(k) = 0\}$$

satisfies $c_0^* = \ell_1$ (see Exercises 9.1–9.4). This will prove to be of significant importance.

Example 9.1.3

Dual of $\mathcal{L}_p(B)$, $1 \leq p < \infty$: The dual of the space $\mathcal{L}_p(B)$, where B denotes any closed interval of $\mathbb{R}$ (possibly all of $\mathbb{R}$), is given by the space $\mathcal{L}_q(B)$, where $\frac{1}{p} + \frac{1}{q} = 1$. The characterization is given by the following theorem.

Theorem 9.1.2.   Every bounded linear functional on $\mathcal{L}_p(B)$, $1 \leq p < \infty$, is representable uniquely in the form

$$f(x) = \int_B x(t)y(t)dt,$$

where y is an element in $\mathcal{L}_q(B)$. Furthermore, every element of $\mathcal{L}_q(B)$ defines a member of $\mathcal{L}_p(B)^*$ in this way and

$$\|f\| = \|y\|_q.$$ ∎

Example 9.1.4

Dual of $\ell_1^{m \times n}$: The dual space of $\ell_1^{m \times n}$ is given by the space $\ell_\infty^{m \times n}$. Every bounded linear functional can be represented by an element in $\ell_\infty^{m \times n}$ with the operation

$$\langle x, y \rangle = \sum_{i,j} \langle x_{ij}, y_{ij} \rangle.$$

The induced norm is given by

$$\|y\| = \sum_i \max_j \|y_{ij}\|_\infty.$$

The result is a natural extension of Theorem 9.1.1. Let $c_0^{m \times n}$ denote all matrices of sequences that converge to zero. Then the dual of $c_0^{m \times n}$ is given by $\ell_1^{m \times n}$.

It is interesting to apply the above result to ℓ_1^m. The dual space is given by ℓ_∞^m, with the induced norm

$$\|y\| = \sum_i \|y_i\|_\infty.$$

This induced norm is not consistent with our standard definition of the ℓ_∞ norm of a vector of signals. We will not, however, make a distinction in the notation between the two norms. It will be clear which one is used from the context.

Example 9.1.5

Dual of a Hilbert space: The dual of any Hilbert space is the space itself. Every element in the space defines a bounded linear functional on the whole space through the inner product operation. The induced norm of each of these functionals is equal to the norm of the element. This is known as the Frechet-Reisz representation theorem. (See Exercise 9.5.)

9.2 LINEAR OPERATORS AND ADJOINTS

Let T be a bounded linear operator from X to Y. We can associate with T an operator T^* from Y^* to X^*. This operator is linear, bounded, and has the same norm as T.

Definition 9.2.1.   To each bounded linear operator $T : X \to Y$ corresponds a unique linear bounded operator $T^* : Y^* \to X^*$ known as the adjoint operator, such that for all $x \in X$ and all $y^* \in Y^*$,

$$\langle Tx, y^* \rangle = \langle x, T^*y^* \rangle. \qquad \blacksquare$$

Example 9.2.1

Let A be a $n \times m$ matrix of real numbers. Then

$$A^* = A^T.$$

This result is independent of the particular norm taken on $\mathbb{R}^n$.

Example 9.2.2

Let $H \in \ell_1$ be viewed as a convolution operator on c_0. The adjoint of H, H^*, is an operator on ℓ_1 and is defined from the relation

$$\langle Hx, y^* \rangle = \langle x, H^*y^* \rangle.$$

By manipulating the definition, we get

$$H^*y^*(t) = \sum_{s=0}^{\infty} H^T(s)y^*(t+s).$$

Notice that H is a causal operator on c_0; however, H^* is not a causal operator on ℓ_1. Also, one can verify by direct computation that $\|H\| = \|H^*\|$. This is not particular to this example, and it can be shown to be a general property of adjoint operators.

Theorem 9.2.1.   Let $T : X \to Y$ be a bounded linear operator, and $T^* : Y^* \to X^*$ be its adjoint. Then

$$\|T\| = \|T^*\|. \qquad \blacksquare$$

9.3 MINIMUM DISTANCE PROBLEMS

9.3.1 Hahn-Banach Theorem

Given a functional f defined on a subspace M, we would like to know if it is possible to find another function F, defined on the whole space X, such that it agrees with f on M and does not increase its norm. Thus, F is said to be an extension of f that preserves the norm. The solution of this problem is provided by the Hahn-Banach Theorem.

Theorem 9.3.1.   Let f be a bounded linear functional defined on a subspace M of a normed vector space X. Then there is a bounded linear functional F defined on X, which is an extension of f; i.e.,

1. $F(x) = f(x)$ for all $x \in M$.
2. $\|F\| = \sup_{x \in M, \, x \neq 0} \frac{|f(x)|}{\|x\|}$.   ∎

9.3.2 Geometric Interpretation

To obtain a geometric picture, it is necessary that bounded linear functionals on X are viewed as subsets of the space X. This requires a geometric interpretation of Hahn-Banach theorem.

Hyperplanes

Definition 9.3.1.   A linear variety in a vector space X is a translated subspace (i.e., of the form $x_0 + M$ and M is a subspace). A hyperplane in X is a maximal linear variety; i.e., cannot be properly contained in any linear variety other than the whole space.   ∎

In finite-dimensional vector spaces, it is evident that a hyperplane H is associated with one vector y, such that $H = \{x \mid y^T x = c\}$ for some constant c. Of course, y can be thought of as a linear functional on $X = \mathbb{R}^n$, and the hyperplane H is simply some translation of the null space of y.

The situation is quite similar for infinite-dimensional spaces. Let H be a hyperplane in X, expressed as $H = x_0 + M$, where M is a subspace. Assume $x_0 \neq 0$ and denote by $[x_0 + M]$ the subspace generated by $x_0 + M$. Since H is a hyperplane, $[x_0 + M] = X$. Define the linear functional f on X by the relation

$$f(\alpha x_0 + m) = \alpha.$$

Then f is a linear functional and $H = \{x \mid f(x) = 1\}$. If $x_0 = 0$, then pick any $x \in X$ but not in H and repeat the above argument.

On the other hand, let f be a linear functional. Then the null space of f is a hyperplane that goes through zero. The set $H = \{x \mid f(x) = c\}$ is simply a translation of the null space of f. Thus, the equivalence can be summarized as

All Hyperplanes $\Longleftrightarrow$ All Linear Functionals.

Notice that such functionals may not be bounded. If f is a bounded linear functional, then any associated hyperplane is closed. The equivalence becomes

All Closed Hyperplanes $\Longleftrightarrow$ All Bounded Linear Functionals.

The advantage of this interpretation is evident for infinite-dimensional spaces. The dual space of an infinite-dimensional space is in general a different space; however, with this interpretation, a bounded linear functional is viewed as a hyperplane in the space itself.

Separating hyperplanes. A hyperplane H, associated with a bounded linear functional x^*, defines two half spaces:

$$\{x | \langle x, x^* \rangle \geq c\}, \quad \{x | \langle x, x^* \rangle \leq c\}, \quad c \in \mathbb{R}.$$

In $\mathbb{R}^n$, if C_1 and C_2 are two disjoint convex sets, then there exists a hyperplane H that separates C_1 and C_2, i.e., each set lies inside one half space. This result is true in general spaces, and in a sense, is equivalent to Hahn-Banach theorem. We will present this theorem without proof.

Theorem 9.3.2. Let C_1 and C_2 be two convex sets with C_1 having interior points, and C_2 contains no interior points of C_1. Then there exits a closed hyperplane separating C_1 and C_2. Equivalently, there exists a bounded linear functional x^* such that

$$\sup_{x \in C_1} \langle x, x^* \rangle \leq \inf_{x \in C_2} \langle x, x^* \rangle. \qquad \blacksquare$$

Notice that this theorem differs from the standard result in $\mathbb{R}^n$ in that at least one of the convex sets has to have an interior point.

Example 9.3.1

If C is closed and convex, and a point x lies outside C, then x and C can be separated by a hyperplane. To show this, notice that there exists a ball around x that does not intersect C. Since the ball contains interior points, the above theorem can be used to separate it from C.

9.3.3 Duality Theorems

Using Hahn-Banach theorem, we can derive two important theorems that characterize solutions to minimum distance problems. We will start with the following definitions.

Definition 9.3.2. A vector $x^* \in X^*$ is said to be aligned with a vector $x \in X$ if $\langle x, x^* \rangle = \|x\| \, \|x^*\|$. $\qquad \blacksquare$

A consequence of this definition is the following theorem concerning alignment between the spaces ℓ_1 and ℓ_∞.

Theorem 9.3.3. If $G \in \ell_\infty^{m \times n}$ is aligned with $H \in \ell_1^{m \times n}$, then G and H satisfy the following conditions:

1. if $|g_{ij}(t)| < \max_{1 \le j \le n} \|g_{ij}\|_\infty$, then $h_{ij}(t) = 0$.
2. $h_{ij}(t)g_{ij}(t) \ge 0$.
3. let $I = \{i \in [1, 2, \dots, m] \mid (G)_i \equiv 0\}$, then $\|(H)_i\|_1 = \nu = $ constant for all i not in I, $(G)_i$ denotes the ith row of a matrix.
4. for all $i \in I$, $(H)_i$ can be anything such that $\|(H)_i\|_1 \le \nu$.   ∎

Definition 9.3.3.   Let S be a subset of X. The annihilator of S, denoted $S^\perp$, consists of all elements $x^* \in X^*$ such that $\langle x, x^* \rangle = 0$ for all $x \in S$.   ∎

Notice that $S^\perp$ is a subspace of X^* even though S may not be a subspace. Finally, we introduce the notion of left annihilators for sets that lie in X^*.

Definition 9.3.4.   Let $U \subset X^*$. The left annihilator of U, denoted $^\perp U$, consists of all elements $x \in X$ such that $\langle x, x^* \rangle = 0$ for all $x^* \in U$.   ∎

Example 9.3.2

If X is a Hilbert Space and $S \subset X$, then $S^\perp = {}^\perp S$, which is the orthogonal complement of S.

The following two theorems characterize optimal solutions for minimum distance problems. It is shown that a minimum distance problem in the primal space X is equivalent to a maximization problem in the dual space X^*, in particular, inside the annihilator subspace. A similar statement holds for minimum distance problems in the dual space. In both cases, the minimum distance problem is referred to as the primal problem, and the maximization problem as the dual problem. The optimal solution for the dual (if it exists) is referred to as the extremal functional. As will be seen from the next theorems, a solution to the minimum distance problem is guaranteed if it is posed inside a dual space.

Theorem 9.3.4.   Let x be an element in a normed linear space X and let d denote its distance from the subspace M $(\overline{M} \ne X)$. Then

$$d := \inf_{m \in M} \|x - m\| = \max_{x^* \in \overline{BM}^\perp} \langle x, x^* \rangle,$$

where the maximum on the right is achieved for some $x_0^* \in M^\perp$ with $\|x_0^*\| = 1$. If the infimum on the left is achieved for some $m_0 \in M$, then $x - m_0$ is aligned with x_0^*.

Proof. The dual problem is a lower bound of the primal problem since

$$\max_{x^* \in \overline{BM}^\perp} \langle x, x^* \rangle = \max_{x^* \in \overline{BM}^\perp} \langle x - m, x^* \rangle \le \|x - m\| \quad \text{for all } m \in M.$$

To show equality, we have to exhibit a functional $x^* \in \overline{BM}^\perp$ that achieves the value d. Let $[x + M]$ be the subspace generated by x and M, and consider the functional $f : [x + M] \to \mathbb{R}$ defined as follows:

$$f(\alpha x + m) = \alpha d \quad \text{for all} \ \alpha \in \mathbb{R}.$$

Then f is a bounded linear functional on $[x + M]$ with induced norm ($\|f\|_M$)

$$\|f\|_M := \sup_{\alpha \in \mathbb{R}, m \in M} \frac{|f(x)|}{\|\alpha x + m\|}$$

$$= \sup_{\alpha \in \mathbb{R}, m \in M} \frac{|\alpha| d}{|\alpha| \|x + \frac{1}{\alpha} m\|}$$

$$= \sup_{m \in M} \frac{d}{\|x + m\|} \quad \text{Since } M \text{ is a subspace}$$

$$= \frac{d}{\inf_{m \in M} \|x + m\|} = 1.$$

Define x_0^* to be the Hahn-Banach extension of f. It follows immediately that $\|x_0^*\| = 1$ and $\langle x, x_0^* \rangle = d$. The alignment follows from the fact that

$$d = \|x - m_0\| \|x_0^*\| = \langle x, x_0^* \rangle = \langle x - m, x_0^* \rangle$$

for any minimizing solution m_0. $\blacksquare$

Notice that an extremal functional of this problem always exists regardless whether the primal has a solution or not. In general, we would like to know when the primal problem has a solution. The next theorem shows that this is possible if the primal is formulated in a dual space.

Theorem 9.3.5. Let M be a subspace in a real normed linear space X. Let $x^* \in X^*$ and let d denote its distance from $M^\perp$. Then

$$d := \min_{m^* \in M^\perp} \|x^* - m^*\| = \sup_{x \in \overline{BM}} \langle x, x^* \rangle,$$

where the minimum on the left is achieved for some $m_0^* \in M^\perp$. If the supremum on the right is achieved for some $x_0 \in M$, then $x^* - m_0^*$ is aligned with x_0.

Proof. The dual problem is a lower bound to the primal problem; i.e.,

$$d \geq \sup_{x \in \overline{BM}} \langle x, x^* \rangle =: d_1.$$

We need to construct $m_0^* \in M^\perp$ such that $\|x^* - m_0^*\| = d_1$. Let y^* be the restriction of x^* to the subspace M. Then its norm over the smaller subspace is precisely

$$\|y^*\|_M := \sup_{x \in \overline{BM}} \langle x, x^* \rangle = d_1.$$

Let y_0^* be the Hahn-Banach extension of y^*. Then $\|y_0^*\| = d_1$. More so, $m_0^* := x^* - y_0^* \in M^\perp$. Hence, this is the desired m_0^*. Alignment follows as in the previous theorem. ∎

Example 9.3.3

If X is a Hilbert space, then both of the duality theorems give the standard Projection theorem. Let $\Pi_{M^\perp}$ denote the projection on the subspace $M^\perp$. The above theorems imply that

$$\inf_{m \in M} \|x - m\| = \|\Pi_{M^\perp} x\|,$$

which says that the minimum distance from a point to a subspace M is equal to the length of the orthogonal projection of x on $M^\perp$. This is precisely the projection theorem.

9.4 EXAMPLES

In this section we present a few examples that demonstrate the utility of the above theorems.

9.4.1 Minimum Norm Solution for $Ax = b$

Let A be $m \times n$ matrix with rank equal to m. The objective is to compute a solution to the problem:

$$v = \min_x |x|_1$$
$$\text{subject to}$$
$$Ax = b.$$

We have shown in the previous chapter that the solution of this problem is equivalent to solving a linear program. This can be demonstrated by using the above duality theorems.

Let x_0 be one solution satisfying $Ax_0 = b$. All solutions to the equation are given by $x_0 - \tilde{x}$ with $A\tilde{x} = 0$. The problem becomes

$$v = \min_{\tilde{x}} |x_0 - \tilde{x}|_1$$
$$\text{subject to}$$
$$A\tilde{x} = 0.$$

The equation $A\tilde{x} = 0$ can be interpreted as follows: Let a_i^T be the ith row of the matrix A. $A\tilde{x} = 0$ if and only if $a_i^T \tilde{x} = 0$. Define the set M as

$$M = \text{span}\{a_i, \; i = 1, ..., m\}.$$

The problem is equivalent to

$$v = \min_{\tilde{x} \in M^\perp} |x_0 - \tilde{x}|_1.$$

By Theorem 9.3.5, the problem is equivalent to a maximization problem inside M; i.e.,

$$v = \max_{r \in \overline{BM}} \langle x_0, r \rangle.$$

Given any element $r \in M$, r has the representation

$$r = \sum_{i=1}^{m} \alpha_i a_i.$$

Substituting in the above equation and noticing that $a_i^T x_0 = b_i$, we get the following:

$$v = \max_{\alpha = (\alpha_1 \ldots \alpha_m)} \alpha^T b$$

subject to

$$|r|_\infty \leq 1 \iff -1 \leq \sum_{i=1}^{m} \alpha_i a_i \leq 1,$$

where the above inequalities are pointwise. This is evidently an LP. Existence of a solution to the dual problem is easy to prove. All optimal solutions are characterized through the alignment condition.

9.4.2 Maximum Radius of a Stabilizing Perturbation

In Chapter 7 we discussed the solution to the robust stability problem in the presence of both real parametric and complex perturbations for a class of rank-one problems. The computation of the maximum stability margin was given by the constrained problem

$$v = \min \max\{|\delta|_\infty, |\Delta|_\infty\}$$

subject to

$$1 - \delta R - \Delta S = 0,$$

$$\delta \in \mathbb{R}^{1 \times n_1}, \ \Delta \in \mathbb{C}^{1 \times n_2}.$$

To obtain a solution to this problem, consider the space $\mathbb{R}^{n_1 + 2n_2}$ equipped with the norm

$$\|x\| = \max\{\max_{i \leq n_1} |x_i|, \ \max_{i \leq n_2} \sqrt{(x_{n_1 + i})^2 + (x_{n_1 + n_2 + i})^2}\}.$$

The dual space is given by $\mathbb{R}^{n_1 + 2n_2}$ equipped with the norm

$$\|y\|_{\text{dual}} = \sum_{i=1}^{n_1} |y_i| + \sum_{i=1}^{n_2} \sqrt{(y_{n_1 + i})^2 + (y_{n_1 + n_2 + i})^2}.$$

The constraints can be written as follows:

$$(\Re R \quad \Re S \quad -\Im S) x = 1, \quad (\Im R \quad \Im S \quad \Re S) x = 0.$$

The dual problem is then given by:

$$v = \max \alpha_1$$

subject to

$$\left\| \alpha_1 \begin{pmatrix} \Re R \\ \Re S \\ -\Im S \end{pmatrix} + \alpha_2 \begin{pmatrix} \Im R \\ \Im S \\ \Re S \end{pmatrix} \right\|_{\text{dual}} \leq 1$$

By dividing the constraints in the dual problem by α_1 and defining $\alpha = \alpha_2/\alpha_1$, and then computing the dual norm, we get:

$$|\Re R + \alpha \Im R|_1 + \sqrt{1+\alpha^2}|S|_1 \leq \frac{1}{\alpha_1},$$

where the $|S|_1$ denotes the sum of magnitudes of the components of S. From the last inequality, it follows that

$$v = \frac{1}{\min_\alpha\{|\Re R + \alpha \Im R|_1 + \sqrt{1+\alpha^2}|S|_1\}}.$$

9.4.3 A Minimum Effort Control Problem

This example illustrates the use of duality theory for solving minimum effort control problems for continuous-time systems.

Suppose we wish to bring a rocket car of unit mass, and subject only to the force of the rocket thrust, to rest at $x = 0$ in time T with a minimum effort control input. Assume that initially $x(0) = 0$ and $\dot{x}(0) = 1$. The equation of motion is given by

$$\ddot{x} = -u.$$

The solution for this equation is given by

$$x(t) = t - \int_0^t (t-\tau)u(\tau)d\tau,$$

and

$$\dot{x}(t) = 1 - \int_0^t u(\tau)d\tau.$$

Three cases are discussed below: minimum Peak Control, minimum Energy Control, and minimum Absolute-Sum Control.

Formulation of the minimum norm problem.   The problem will be solved in three different spaces $\mathcal{L}_\infty$, $\mathcal{L}_2$, and $\mathcal{L}_1$ on the interval $[0, T]$. The conditions on the final states are equivalent to two linear constraints on the minimization problem. The problem is given by:

$$v = \min \|u\|_p$$

subject to

$$T = \int_0^T (T-\tau)u(\tau)d\tau, \qquad (9.2)$$

and

$$1 = \int_0^T u(\tau)d\tau, \qquad (9.3)$$

for $p = \infty, 2, 1$. Let u_0 be one function that satisfies the linear constraints. This is possible for any T since the system is controllable. The problem becomes

$$v = \min_{\tilde{u}} \|u_0 - \tilde{u}\|_p,$$

subject to

$$\int_0^T (T - \tau)\tilde{u}(\tau)d\tau = 0,$$

and

$$\int_0^T \tilde{u}(\tau)d\tau = 0.$$

Define the set M as the set of all possible linear combinations of the functions $T - \tau$ and 1 on the interval $[0, T]$; i.e.,

$$M = \text{Span}\{T - \tau, 1\}.$$

Minimum peak control. In this case, the problem is formulated inside $\mathcal{L}_\infty[0, T]$. M can be viewed as a subset of $\mathcal{L}_1[0, T]$. All functions inside $\mathcal{L}_\infty[0, T]$ satisfying the constraints are elements inside $M^\perp$. The problem becomes

$$v = \min_{\tilde{u} \in M^\perp} \|u_0 - \tilde{u}\|_\infty.$$

Using Theorem 9.3.5, v is given by

$$v = \max_{\alpha_1, \alpha_2} \alpha_1 T + \alpha_2,$$

subject to

$$\int_0^T |\alpha_1(T - \tau) + \alpha_2|d\tau \le 1.$$

This is a maximization problem with two parameters, and its solution is (conceptually) straightforward. Without solving this problem, we know from the alignment conditions that the optimal $u_0 - \tilde{u}$ is aligned with the extremal functional. This implies that $u_0 - \tilde{u}$ is a constant and switches signs at most once, say at t_0. The general form of the optimal $u_0 - \tilde{u}$ is

$$u_0 - \tilde{u} = \begin{cases} U & 0 \le t \le t_0 \\ -U & t_0 < t \le T \end{cases}.$$

This function has to satisfy the integral constraints in Equations (9.2, 9.3). Substituting in the constraints, we get

$$t_0 = \frac{T}{\sqrt{2}} \quad \text{and} \quad U = \frac{1}{(\sqrt{2} - 1)T}.$$

Minimum energy control. In this case, the problem is formulated inside $\mathcal{L}_2[0, T]$. M can be viewed as a subset of $\mathcal{L}_2[0, T]$. The problem is stated as follows:

$$v = \min_{\tilde{u} \in M^\perp} \|u_0 - \tilde{u}\|_2.$$

From the Projection theorem stated in Example 9.3.3, the optimal solution $u_0 - \tilde{u}$ is orthogonal to $M^\perp$, i.e., $u_0 - \tilde{u} \in M$ and has the form

$$(u_0 - \tilde{u})(\tau) = r_1(T - \tau) + r_2, \quad r_1, r_2 \in \mathbb{R}.$$

To evaluate r_1 and r_2 note that $u_0 - \tilde{u}$ satisfies Equations (9.2, 9.3). Solving, we get $r_1 = 6T^{-2}$ and $r_2 = -2T^{-1}$.

Minimum absolute-sum control. In both cases discussed above, the minimization problem was formulated in a dual space, which immediately guarantees the existence of a minimizing solution. The space $\mathcal{L}_1[0, T]$ is not the dual of any space. Instead, the problem can be formulated inside the space of functions of bounded variations denoted by $BV[0, T]$, which is the dual of the space of continuous functions on $[0, 1]$, denoted by $C[0, 1]$. However, we will not do that.

Initially, the problem is formulated in $\mathcal{L}_1[0, 1]$. The annihilator subspace of all $\tilde{u}$ is given by $M \in \mathcal{L}_\infty$. The dual problem is

$$\nu = \max_{\alpha_1, \alpha_2} \alpha_1 T + \alpha_2,$$

subject to

$$|\alpha_1(T - \tau) + \alpha_2| \leq 1 \quad \text{for all } \tau \in [0, T].$$

This problem is also a finite-dimensional optimization problem in the sense that it has a finite number of variables. In fact, it is a semi-infinite linear program since it has a continuum of linear constraints. From the alignment conditions, the optimal solution $u = u_0 - \tilde{u}$ is nonzero only when the extremal functional is maximized. This can occur at only two points $\tau = 0$ and $\tau = T$ (assuming α_1 is not zero). Clearly, there does not exist a function in $\mathcal{L}_1[0, T]$ with nonzero norm that is supported at exactly two points. Hence, a solution in $\mathcal{L}_1[0, T]$ does not exist; however, it exists in the larger space $BV[0, T]$. The solution has the form

$$(u_0 - \tilde{u})(\tau) = r_1\delta(\tau) + r_2\delta(\tau - T),$$

where δ is the standard delta function; r_1 and r_2, which satisfy Equations (9.2, 9.3), are given by $r_1 = 1$ and $r_2 = 1 - T$. The solutions of these three cases are compared in Figures 9.2 and 9.3.

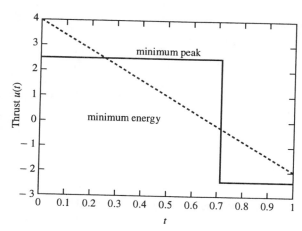

Figure 9.2 Minimum effort control problem: Optimal thrust for $T = 1$.

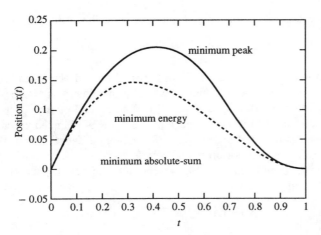

Figure 9.3 Minimum effort control problem: Optimal position for $T = 1$.

9.5 EXISTENCE OF OPTIMAL SOLUTIONS

It is clear from the two duality theorems for minimum distance problems that it is always an advantage to formulate an optimization problem inside a dual space of some space. It is worthwhile noting that existence of optimal solutions is not necessarily a topological property; however, most of the well-known conditions that guarantee existence are topological. One of the basic results on existence of a minimizer is based on continuity over compact sets. This is given in the following well-known theorem.

Theorem 9.5.1. Let X be a topological vector space (i.e., a vector space with a collection of open sets that defines a topology), and f be a lower semi-continuous function from X to $\mathbb{R}$. Then, the following optimization problem

$$\inf_{x \in S} f(x) \quad S \subset X, \quad S \text{ is compact,}$$

has a solution $x_0 \in S$.

Proof. Let $x_n \in S$ be a sequence such that

$$\lim_{n \to \infty} f(x_n) = \inf_{x \in S} f(x).$$

Since S is compact, x_n has a convergent subsequence x_{n_k} converging to x_0. By lower semi-continuity of f, it follows that

$$f(x_0) \leq \lim_{k \to \infty} f(x_{n_k}) = \inf_{x \in S} f(x),$$

implying that x_0 is a minimizer.   ■

To use this result, it is necessary that S be a bounded and closed set. This, however, is not sufficient if X is an infinite-dimensional normed linear space, with the topology inherited from the norm. On the other hand, if the topology is replaced by a weaker one, such that the function f remains continuous, and S is compact, then a solution can be guaranteed to exist. In other words, to guarantee existence, it is not necessary to stay with the topology inherited from the norm—any other weaker topology will suffice as long as f remains continuous.

The notion of a weaker topology is connected with the notion of weak convergence. Next, different notions for convergence in normed linear spaces are introduced.

9.5.1 Weak Convergence

The standard notion of norm convergence is defined below.

Definition 9.5.1.   A sequence $\{x_n\}$ in a normed linear space X is said to converge strongly (or in the norm) to $x \in X$ if $\|x_n - x\| \to 0$. This is denoted by $x_n \to x$.   ∎

Two notions of weak convergence are defined below.

Definition 9.5.2.   A sequence $\{x_n\}$ in a normed linear space X is said to converge weakly to $x \in X$ if for every $x^* \in X^*$, $< x_n, x^* > \to < x, x^* >$. This is denoted by $x_n \overset{w}{\to} x$.   ∎

Definition 9.5.3.   A sequence $\{x_n^*\}$ in a normed linear space X^* is said to converge weak-star (or $weak^*$) to $x^* \in X^*$ if for every $x \in X$, $< x, x_n^* > \to < x, x^* >$. This is denoted by $x_n^* \overset{w^*}{\to} x^*$.   ∎

It follows from the above definitions that in X^*:

$$\text{Strong Convergence} \implies \text{Weak Convergence} \implies \text{Weak}^* \text{ Convergence.}$$

Example 9.5.1

For the space ℓ_1, the notions of strong convergence and weak convergence coincide. On the other hand, a sequence $h_n \in \ell_1$ converges $weak^*$ to $h \in \ell_1$ if and only if h_n converges to h on compact intervals, i.e., for any integer K,

$$\lim_{n \to \infty} \| P_K (h_n - h) \|_1 = 0.$$

If in addition $\|h_n\|_1$ converges to $\|h\|_1$, then h_n is strongly convergent to h.

9.5.2 Compactness and Closure

The above definition gives rise to the following, less severe, notion of compactness as well as a compatible notion of closure.

Definition 9.5.4. A set $S \subset X^*$ is *weak**-compact if every sequence in S has a *weak** convergent subsequence. ∎

Definition 9.5.5. A set $S \subset X^*$ is *weak**-closed if it contains all of its *weak** limit points. ∎

Notice that a compact set is immediately *weak** compact while a closed set may not be *weak** closed. To illustrate the latter, consider the following example.

Example 9.5.2

Consider the set

$$S = \{x \in \ell_1 | \langle x, \underline{1} \rangle = 0\}$$

where $\underline{1} = (1, 1, \ldots)$. Since $\underline{1} \in \ell_\infty = \ell_1^*$, then S is closed. However, S is not *weak** closed. To see this let $x_n = \{(1 - 1/n)^k\}$ and $y_n(k) = x_n(k) - x_n(k-1)$. The sequence $y_n \in S$ and its *weak** limit is equal to the sequence $(1, 0 \ldots)$ which is not in S.

Example 9.5.3

Let $x_1, \ldots, x_N$ be N-sequences inside c_0. The set

$$S = \{y \in \ell_1 | \langle x_i, y \rangle = 0, \quad \text{for all } i = 1, \ldots N\}$$

is *weak** closed.

The central result needed for existence arguments is given in the next theorem, known as Alaoglu's theorem.

Theorem 9.5.2. The closed unit ball in X^* is *weak**-compact. ∎

This theorem, together with Theorem 9.5.1, provides a way for showing the existence of minimizing solutions. The set S needs to be compact and the function f needs to be lower semi-continuous, both with respect to the *weak** topology.

9.5.3 Distance Problems

When dealing with distance problems, the function f is simply the norm. To prove existence of solutions to Problem (9.1), it is sufficient to show that M is *weak** closed. The intuitive justification for this is:

1. $\| \cdot \|$ is lower semi-continuous with respect to the *weak** topology.
2. M is *weak**-closed.
3. Even though M is only *weak**-closed, effectively, only a bounded subset of M is necessary. In fact, it is easily argued that all minimizing solutions are bounded.

From Theorem 9.3.5, a minimizing solution exists if the problem is posed in a dual space, and the constraint set is expressed as some annihilator subspace. To relate this to the above discussion, the following theorem is introduced.

Theorem 9.5.3.   Given $M \subset X$ and $N \subset X^*$, then

1. $M^\perp$ is *weak**-closed in X^*,
2. $^\perp N$ is norm-closed in X,
3. $(^\perp N)^\perp$ is equal to the *weak**-closure of N.   ∎

Theorems 9.3.4, 9.3.5, and 9.5.3 can be combined in the following way:

Theorem 9.5.4.   Let $S \subset X^*$ be a subspace, $x^* \in X^*$ is fixed, and define the (primal) distance problem in X^*:

$$d = \inf_{y^* \in S} \|x^* - y^*\|.$$

Then the following are true:

1. If S is *weak**-closed, then there exists a minimizing solution y_o^*.
2. If S is *weak**-closed, then

$$d = \sup_{x \in \overline{B^\perp S}} \langle x, x^* \rangle.$$

A maximizing solution inside $^\perp S$ may not exist. This problem is known as the pre-dual problem.
3. For any S

$$d = \max_{x^{**} \in \overline{BS^\perp}} \langle x^*, x^{**} \rangle.$$

A maximizing solution will always exist inside $S^\perp$. This problem is known as the dual problem.

Proof. Define $M = {}^\perp S$. Then the *weak**-closure of $S = M^\perp$. The above results follow directly from Theorems 9.3.4 and 9.3.5.   ∎

Theorem 9.5.4 will be repeatedly used in the context of ℓ_1 model matching problems. In such cases, the distance problem is defined inside $X^* = \ell_1$. Two dual problems are defined: one inside $X = c_0$, which may not have a solution, and the other is inside $X^{**} = \ell_\infty$, which will have a solution. The existence of a minimizing solution depends on the *weak** closure of the constraint set.

In the case when the subspace S is defined as the null-space of an operator T, it is possible to show that $\mathcal{N}(T)$ is *weak**-closed if T can be written as the adjoint of some other operator.

Theorem 9.5.5.   Given Banach spaces X and Y, and bounded linear operator $T : X \rightarrow Y$. Then

$$\mathcal{N}(T^*) = \mathcal{R}(T)^{\perp},$$

which implies that $\mathcal{N}(T^*)$ is *weak**-closed in Y^*.   ∎

9.6 DUALITY IN LINEAR PROGRAMMING

In Chapter 8 we discussed duality theory for finite-dimensional linear programming problems. The dual representation helps in reducing the complexity of the optimization problem, as well as characterizing optimal solutions. Because computations can only be done for finite problems, Chapter 8 presented a complete account of the simplex algorithm.

We have shown in Chapter 3 that a large class of design problems can be posed as an ℓ_1 minimization problem subject to both equality and inequality constraints. These constraints include the feasibility conditions as well as added time-domain and frequency-domain constraints. Our aim here is to develop a duality theory for such problems that parallels the finite-dimensional theory.

9.6.1 General Primal-Dual Pair

Definition 9.6.1.   A set $\mathcal{P}$ is a convex cone (or simply a cone) in a vector space X if for every $x \in \mathcal{P}$ $\alpha x \in \mathcal{P}$,   for all $\alpha \geq 0$. The positive conjugate cone of $\mathcal{P}$, denoted by $\mathcal{P}^{\oplus}$, is defined as

$$\mathcal{P}^{\oplus} = \{x^* \in X^* | \langle x, x^* \rangle \geq 0, \text{ for all } x \in \mathcal{P}\}.$$

The negative conjugate cone of $\mathcal{P}$, denoted by $\mathcal{P}^{\ominus}$ is equal to $-\mathcal{P}^{\oplus}$.   ∎

Let X and Z be normed linear spaces, $c \in X^*$; $\mathcal{A}$ is an operator from X to Z and $b \in Z$. Let X, Z be equipped with positive cones $\mathcal{P}_X$, $\mathcal{P}_Z$, respectively. A general linear programming problem is given by

$$(PLP) \qquad\qquad v^p = \inf \langle x, c \rangle$$
$$\text{subject to}$$
$$\mathcal{A}x \leq b,$$
$$x \geq 0.$$

Associated with this problem is a dual linear program, with variables $\gamma \in Z^*$, defined as

$$(DLP) \qquad\qquad v^d = \sup \langle b, \gamma \rangle$$
$$\text{subject to}$$
$$\mathcal{A}^* \gamma \leq c,$$
$$\gamma \leq 0,$$

where the inequalities in the last linear program are with respect to the positive conjugate cone in Z^*.

In the finite-dimensional case, PLP and DLP have the same value, and the solutions are aligned. The situation is not as simple for infinite-dimensional problems.

Example 9.6.1

This example is not motivated by a specific control problem. It is discussed simply to highlight the fact that infinite-dimensional linear programs behave differently from finite-dimensional ones. Consider the following optimization problem:

$$\min x_0$$

subject to $\qquad\qquad\qquad\qquad\qquad\qquad\qquad\qquad\qquad\qquad$ (9.4)

$$x_0 + \sum_{i=1}^{\infty} i x_i = 1,$$

$$\sum_{i=1}^{\infty} x_i = 0,$$

$$x_i \geq 0, \qquad i = 0, 1, 2, \ldots$$

The primal space X is the set of sequences with only a finite number of nonzero terms. The only feasible solution for Problem (9.4) is $x_0 = 1, x_1 = x_2 = \ldots = 0$, and the value of the program is 1. The dual of Problem (9.4) is

$$\max z_1^*$$

subject to $\qquad\qquad\qquad\qquad\qquad\qquad\qquad\qquad\qquad\qquad$ (9.5)

$$z_1^* \leq 1,$$

$$n z_1^* + z_2^* \leq 0, \qquad n = 1, 2, \ldots$$

Since the last constraint must hold for all n, it follows that $z_1^* \leq 0$ and the solution for problem (9.5) is 0.

A general theory that will address the duality of PLP is beyond the scope of this book. However, we will be content with establishing a weak duality result for the general problem.

Theorem 9.6.1. The value of DLP is a lower bound for PLP, i.e.,

$$v^p \geq v^d.$$

Proof. For any feasible x and γ, it follows that

$$\langle x, c \rangle \geq \langle x, A^* \gamma \rangle \geq \langle Ax, \gamma \rangle \geq \langle b, \gamma \rangle.$$

The inequality remains valid if we minimize the left-hand side and maximize the right-hand side. ∎

The previously derived duality theorem for minimum distance problems to subspaces gives a dual characterization of a class of linear programming problems.

9.6.2 A Class of Infinite-Dimensional Linear Programs

We have seen in Chapter 3 that the general ℓ_1 minimization problem gives rise to an infinite-dimensional linear programming problem. In here, we show that the dual formulation is also a linear program. Consider the following problem:

$$
\inf \| \Phi \|_1
$$
$$
\text{subject to} \tag{9.6}
$$
$$
\mathcal{A}\Phi = b,
$$

where $\mathcal{A}$ is a linear operator on $\ell_1^{n_z \times n_w}$, and $b \in \ell_1$. This problem resembles the finite dimensional problem solved in Section 9.4.1. Since the range of the operator $\mathcal{A}$ is in ℓ_1, we cannot follow the same procedure to characterize the annihilator subspace. The following theorem gives such a characterization.

     Theorem 9.6.2.   Given Banach spaces X and Y, and bounded linear operator $T : X \to Y$. If the $\mathcal{R}(T)$ is closed, then

$$
\mathcal{N}(T)^\perp = \mathcal{R}(T^*). \qquad\blacksquare
$$

     We first note that $\mathcal{A}$ is an operator from $\ell_1^{n_w \times n_z}$ to ℓ_1. Assume that $\mathcal{R}(\mathcal{A})$ is closed. The operator $\mathcal{A}^*$ maps ℓ_∞ to $\ell_\infty^{n_w \times n_z}$. The dual problem of Problem (9.6) is

$$
\max \langle b, \gamma \rangle
$$
$$
\text{subject to}
$$
$$
\| \mathcal{A}^* \gamma \|_\infty \le 1, \tag{9.7}
$$
$$
\gamma \in \ell_\infty.
$$

The dual Problem (9.7) has a linear objective function, with linear constraints, and has the same value as Problem (9.6). However, since $\gamma \in \ell_\infty$, the problem is infinite dimensional. This is not a surprising result as we have shown in Chapter 3 that the primal problem can be written as

$$
\inf \nu
$$
$$
\text{subject to}
$$
$$
\mathcal{A}_{\ell_1}(\Phi^+ + \Phi^-) \le \nu \mathbf{1},
$$
$$
\mathcal{A}(\Phi^+ - \Phi^-) = b, \tag{9.8}
$$
$$
\Phi^+, \ \Phi^- \in \ell_1^{n_w \times n_z},
$$
$$
\Phi^+, \ \Phi^- \ge 0,
$$

which is readily a linear program, also infinite dimensional. It is straightforward to verify that

$$
\| \mathcal{A}^* \gamma \|_\infty \le 1 \iff
\begin{cases}
-\mathcal{A}_{\ell_1}^* \gamma_0 \le \mathcal{A}^* \gamma \le \mathcal{A}_{\ell_1}^* \gamma_0 \\
\mathbf{1}^T \gamma_0 \le 1 \\
\gamma_0 \ge 0
\end{cases} .
$$

This implies that Problem (9.7) is equivalent to

$$\max \langle b, \gamma \rangle$$
$$\text{subject to}$$
$$-\mathcal{A}_{\ell_1}^* \gamma_0 \le \mathcal{A}^* \gamma \le \mathcal{A}_{\ell_1}^* \gamma_0,$$
$$\mathbf{1}^T \gamma_0 \le 1,$$
$$\gamma \in \ell_\infty,$$
$$\gamma_0 \ge 0.$$

(9.9)

The discussion above shows that Problem (9.8) and Problem (9.9) are a primal-dual pair of linear programs without a duality gap.

9.7 SUMMARY

In this chapter we have highlighted the basic theory behind duality in minimum distance problems. Duality provides an alternate formulation of the optimization problem that can give information which is not transparent from the primal problem. Duality in linear programming is a more elaborate topic in the sense that there could exist a gap between the primal and the dual problems.

Duality theory is exploited in the future chapters in two different ways. First, to show that certain classes of control synthesis problems can be solved exactly (i.e., with a finite number of operations) even though the problem is infinite dimensional. Second, to show that when exact solutions cannot be obtained, approximate ones can be computed with an estimate of optimality. Without employing duality, it is difficult to know how far from optimal a given solution is. For more general problems involving multiple objectives and constraints, weak duality is used.

EXERCISES

9.1. A Schauder basis of a normed linear space X is a countable set

$$E = \{e_0, e_1, \ldots\}$$

such that every element in X can be expressed as a linear combination of elements in E, i.e.,

$$x = \sum_{i=0}^{\infty} \alpha_i e_i,$$

and

$$\lim_{n \to \infty} \|x - \sum_{i=0}^{n} \alpha_i e_i\| = 0.$$

Show that the standard basis with $e_i = (0, 0, \ldots 1, 0 \ldots)$ form a Schauder basis for ℓ_p with $p < \infty$. Show the same for c_0. Is this result true for ℓ_∞? Argue that ℓ_∞ does not have a Schauder basis.

9.2. Show that $\ell_1^* = \ell_\infty$. (Hint: For a given functional, find its representation by expressing elements in ℓ_1 in terms of a Schauder basis). Also compute the dual space of $\ell_1^{m \times n}$.

9.3. Consider the following function on ℓ_∞:

$$f(x) = \limsup_{n \to \infty} \frac{1}{n} \sum_{k=0}^{n-1} x(k).$$

Is this a bounded linear functional on ℓ_∞? If so, can it be represented by an element in ℓ_1?

9.4. Show that $c_0^* = \ell_1$.

9.5. Let X be a Hilbert space. We want to show that $X^* = X$.
   (a) Show that every $x \in X$ defines a bounded linear functional on X through the inner product. Compute its induced norm.
   (b) Given $f \in X^*$, show that the null space of f, $\mathcal{N}(f)$, is closed.
   (c) Show that

$$X = \mathcal{N}(f) \oplus Z$$

   with Z being a one-dimensional subspace in X.
   (d) Show that there exists an element $z \in Z$ such that

$$f(x) = \langle x, z \rangle.$$

9.6. Verify the results in Examples 9.5.1 and 9.5.3.

9.7. (A steering control problem) The open loop system in Figure 9.4 consists of a discrete-time plant P, and a saturation nonlinearity at the input of the plant described as

$$\text{Sat}(u) = \begin{cases} u & |u| \le U_{\max} \\ U_{\max} \text{sgn}(u) & |u| > U_{\max} \end{cases}.$$

The plant P is governed by the difference equation

$$x(k+1) = Ax(k) + bu(k), \quad x(0) = x_0,$$

where A has no eigenvalues on the unit circle, and (A, b) is reachable. We would like to know whether we can find an input $v \in \ell_\infty$ such that

$$\lim_{k \to \infty} x(k) = 0.$$

This is called a steering control problem.
   (a) Ignoring saturations, find one possible u_0 that steers the initial condition, x_0, to zero. Show that all such inputs can be parametrized as $u_0 + \tilde{u}$, with $\tilde{u}$ satisfying a finite number of interpolation conditions (use λ-transform).
   (b) Show that every u in the above parametrization lies in c_0.
   (c) Show that the existence of v can be posed as an optimization problem over the u's.
   (d) Find the smallest saturation level, $U_{\max}$, as a function of the initial condition x_0, such that there exists an input that steers the initial condition to zero.

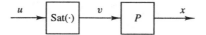

Figure 9.4 Steering problem.

9.8. Consider the reachability operator associated with a stable reachable system. The operator can be viewed as a multiplication by an infinite matrix, i.e.,

$$\mathcal{C} : \ell_p \to \mathbb{R}^n$$
$$u \to (\, B \quad AB \quad A^2 B \quad \ldots \,)\, u.$$

Compute the minimum energy input such that $\mathcal{C}u = x_f$, where x_f is a fixed final state.

9.9. In the previous problem, consider the minimization problem

$$\min_u \|u\|_1$$
subject to
$$\mathcal{C}u = x_f.$$

Show that:

(a) The minimum value of $\|u\|_1$ can be computed by solving a finite-dimensional LP, with infinitely many constraints. (Hint: Consider the dual problem.)

(b) Show that only a finite number of such constraints can be active.

(c) Show how to compute the minimizing solution of the above problem.

(d) Does a minimizing solution to this problem exist?

9.10. In the previous problem, consider the minimization problem

$$\inf_u \|u\|_\infty$$
subject to
$$\mathcal{C}u = x_f.$$

(a) Find the dual of the above minimization problem.

(b) Show how the dual problem can be computed to any desired degree of accuracy.

(c) Show that suboptimal solutions, with norms arbitrarily close to the minimum norm, can be computed.

(d) Does a minimizing solution to this problem exist?

9.11. (Farkas Lemma) Let A be an operator from ℓ_1 to $\mathbb{R}^n$. Let $\mathcal{P}$ be a positive cone in ℓ_1, and $A(\mathcal{P})$ is closed. Show that there exists a solution to the equation $Ax = b$ with $x \in \mathcal{P}$ if and only if every $y^* \in \mathbb{R}^n$ with the property $A^* y^* \geq 0$ (with respect to the conjugate cone) satisfies $b^T y^* \geq 0$.

9.12. Let $f_1, f_2, \ldots f_n$ be linearly independent functionals on a space X. Suppose that there exists a functional f with the property that for every $x \in X$ satisfying $f_i(x) = 0$ for all $i = 1, \ldots n$, we have $f(x) = 0$. Show that there exists constants $\alpha_1, \ldots \alpha_n$ such that

$$f = \sum_{i=1}^n \alpha_i f_i$$

9.13. Let $\mathcal{L}_{TV}$ denote the space of all linear, causal, ℓ_∞-stable systems. Show that $\mathcal{L}_{TV}$ is the dual space of some space $\mathcal{L}_0$. Characterize $\mathcal{L}_0$.

9.14. Let G be any linear, causal, ℓ_∞-stable system. Define

$$\overline{G}_N = \frac{1}{N} \sum_{k=0}^{N-1} S_{-k} G S_k.$$

Let $\tilde{G}$ be a *weak** limit of $\overline{G}_N$. Show that $\tilde{G}$ is time invariant.

9.15. In this exercise we would like to verify that time-varying controllers do not improve the performance when measured by some induced norm, even though they enlarge the set of all achievable closed-loop maps. This problem is the extension of Exercise 6.6 for periodic controllers. We will only consider the ℓ_∞ case. Consider the following two problems

$$\nu_{TI} = \inf_{Q \text{ stable, time invariant}} \|H - UQV\|_{\ell_\infty - \text{ind}},$$

$$\nu_{TV} = \inf_{Q \text{ stable, time varying}} \|H - UQV\|_{\ell_\infty - \text{ind}}.$$

Show that

$$\nu_{TI} = \nu_{TV}$$

Hint: Show that for any time-varying Q there exists a time-invariant Q such that its performance is at least as good as the time-varying one. Use the average-delayed operator in Exercise 9.14

NOTES AND REFERENCES

A complete and lucid presentation of optimization in infinite-dimensional spaces can be found in [Lue69], which this chapter follows closely. Duality theory extends to more general function minimization problems through Lagrange multiplier theory. The above reference is an excellent source for such extensions. Much of the theory can be found in standard functional analysis books; see [Con85, Kre89, Rud73b].

The use of duality in minimum effort control problems appeared in [Neu62] in which problems similar to the examples shown in this chapter were solved.

We have not presented a complete theory for duality in linear programming. If the linear program arises from a distance problem to a subspace, then the results can be obtained from the duality of minimum distance problems. Otherwise, we showed only weak duality. For a more detailed study of duality in infinite-dimensional linear programs see [AP84, AN87] and references therein.

The steering control problem in Exercise 9.7 is solved in [Mey90a]. The result in Exercise 9.15 is derived in [SD91], which extends to all ℓ_p-induced norms.

CHAPTER 10 —————————————

SISO Model Matching Problems

In the following chapters we present a complete solution of the ℓ_1 model matching problem. We intend to achieve two goals from this study. The first is to obtain a deeper understanding of the ℓ_1 problem and the different methods for computing controllers, and the second is to show that many of these methods extend to solve general performance problems involving mixed constraints.

In this chapter, however, we present a complete solution of the SISO ℓ_1 model matching problem. The solution is a very nice example of the power of duality theory for solving minimum distance problems. We then present the solution of the $\mathcal{H}_2$ model matching problem using the projection theorem. Finally, the $\mathcal{H}_\infty$ model matching problem is solved, using Nehari's theorem. The three solutions are compared in terms of their structure, controller order, and the location of closed-loop poles. Extreme examples showing the gap between optimal $\mathcal{H}_\infty$ and ℓ_1 solutions are presented. Such examples motivate the general mixed optimization problems discussed in later chapters.

10.1 PROBLEM DEFINITION

The general model matching problem is shown in Figure 10.1. The optimal model matching problem is defined as

$$\inf_{Q \in \ell_1} \| H - U Q \|,$$

where H, U are fixed elements in ℓ_1. The norm $\| \cdot \|$ will denote

$$\| \cdot \| = \begin{cases} \ell_1 \text{ norm} \\ \mathcal{H}_2 \text{ norm} \\ \mathcal{H}_\infty \text{ norm} \end{cases}.$$

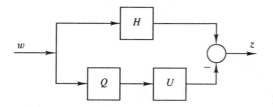

Figure 10.1 Model matching problem.

Let S denote the subspace in ℓ_1 defined as

$$S = \{R \in \ell_1 | R = UQ \text{ for some } Q \in \ell_1\}.$$

The SISO model matching problem can then be stated as a minimum distance problem given by

$$\inf_{R \in S} \|H - R\|. \tag{10.1}$$

10.2 THE SUBSPACE S

To characterize the subspace S, we consider two cases. In the first case, we assume that $\hat{U}$ has no zeros on the unit circle. By Wiener's theorem (Theorem 6.2.1), the subspace S is characterized by zero interpolation conditions. Let $\hat{U}(\lambda)$ have N zeros inside the open unit disc, denoted by a_i, each of multiplicity r_i. Then $R \in S$ if and only if $R \in \ell_1$ and

$$\hat{R}^{(k)}(a_i) = 0 \quad \text{for} \quad k = 0, \ldots r_i - 1, \quad i = 1, \ldots, N.$$

The second case is the general case where $\hat{U}$ has zeros on the unit circle. We have shown in Chapter 6 that S cannot be characterized by interpolation conditions. However, if S is restricted to rational elements, then the above characterization is still valid.

In the sequel, the solutions of the three model matching problems are derived when all the zeros are in the disc, but not on the boundary. We will discuss the boundary zeros in the ℓ_1 problem only.

Assumption. $\hat{U}$ has no zeros on the unit circle. Also, for simplicity, it is assumed that the a_i's are simple.

The evaluation of the λ-transform of a function at a point inside the unit disc can be interpreted as a complex-valued linear functional on the space ℓ_1. Explicitly, if $a \in \mathcal{D}$, define $\underline{a} = (1, a, a^2, \ldots)$, $\Re\underline{a} = (1, \Re(a), \Re(a^2), \ldots)$ and $\Im\underline{a} = (0, \Im(a), \Im(a^2), \ldots)$, then both $\Re\underline{a}$ and $\Im\underline{a}$ are elements in c_0. Define

$$M = \text{Span}\{\Re(\underline{a}_i), \Im(\underline{a}_i)\}.$$

M is a finite dimensional subspace in c_0. The next theorem gives a characterization of the subspace S.

Theorem 10.2.1

$$S = M^\perp.$$

Proof. Let $R \in S$. Then $\hat{R}(a_i) = 0$ for all $i = 1, \ldots, N$. Recall that

then

$$\hat{R}(\lambda) = \sum_{j=0}^{\infty} R(j)\lambda^j,$$

$$\hat{R}(a_i) = \sum_{j=0}^{\infty} R(j)a_i^j = \langle R, \underline{a} \rangle = 0.$$

Hence, $\langle R, \Re(\underline{a}) \rangle = 0$ and $\langle R, \Im(\underline{a}) \rangle = 0$. Thus $R \in M^\perp$. The converse of this theorem follows by reversing the above arguments. ∎

The above theorem establishes an important fact: The feasible subspace S is the annihilator subspace of some finite-dimensional subspace inside c_0. This suggests that Problem 10.1, even though posed inside an infinite-dimensional space, is in fact finite dimensional. This will be apparent from the duality theory.

10.3 THE ℓ_1 MODEL MATCHING PROBLEM

In this section the characterization of the subspace S is used to calculate the minimum value of Problem 10.1 and to compute the minimizing solution in the ℓ_1 case. The central ideas lie in Theorem 9.3.5.

10.3.1 Invoking Duality

The problem defined is equivalent to

$$v^o = \inf_{R \in M^\perp} \|H - R\|_1,$$

where $H \in \ell_1$ and $R \in M^\perp \subset \ell_1$. Theorem 9.3.5 states that

$$v^o = \min_{R \in M^\perp} \|H - R\| = \sup_{x \in \overline{BM}} \langle x, H \rangle.$$

Notice that the "inf" is replaced by a "min" since Theorem 9.3.5 guarantees the existence of solutions to the primal problem. To compute v^o, we expand $\langle x, H \rangle$ explicitly for any $x \in M$. Since x has the representation

$$x = \sum_{i=1}^{N} \alpha_i \Re(\underline{a}_i) + \alpha_{i+N} \Im(\underline{a}_i), \qquad \alpha_i \in \mathbb{R},$$

it follows that

$$\langle x, H \rangle = \sum_{i=1}^{N} \alpha_i \Re[\hat{H}(a_i)] + \alpha_{i+N} \Im[\hat{H}(a_i)].$$

Hence

$$v^o = \sup_{\alpha_1, \ldots, \alpha_{2N}} \sum_{i=1}^{N} \alpha_i \Re[\hat{H}(a_i)] + \alpha_{i+N} \Im[\hat{H}(a_i)],$$

subject to

$$\|x\|_\infty = \sup_k \left[\left| \sum_{i=1}^{N} \alpha_i \Re(a_i^k) + \alpha_{i+N} \Im(a_i^k) \right| \right] \le 1.$$

The above calculations are summarized in the following theorem.

Theorem 10.3.1.   The minimum value of the distance problem

$$v^o = \min_{Q \in \ell_1} \|H - UQ\|_1$$

is given by

$$v^o = \max_{\alpha_1, \ldots, \alpha_{2N}} \sum_{i=1}^{N} \alpha_i \Re[\hat{H}(a_i)] + \alpha_{i+N} \Im[\hat{H}(a_i)],$$

subject to

$$\left| \sum_{i=1}^{N} \alpha_i \Re(a_i^k) + \alpha_{i+N} \Im(a_i^k) \right| \le 1 \quad \text{for all } k = 0, 1, 2, \ldots.$$

Proof. The calculations follow from the discussion preceding the theorem. It is possible to replace the supremum by a maximization in the dual problem since the dual objective is a continuous function on a compact set, and thus has a maximizer (Theorem 9.5.1). ∎

Theorem 10.3.1 shows that the computation of v^o is converted to a linear programming problem with at most $2N$ variables. Hence, by using duality, an infinite-dimensional optimization problem has been converted to a finite-dimensional one. The only troubling factor is that the LP has infinitely many constraints. In the next section it is shown that only a finite number of these constraints are necessary. A bound on the number of these constraints can be derived as a function of the problem data, and hence is computed a priori.

10.3.2 A Finite-Dimensional LP

Because $|a_i| < 1$, it is evident that for large enough k, a_i^k will be small enough that the remaining constraints are no longer active. More precisely, we can compute an integer, L^*, such that any α that satisfies the first L^* constraints also satisfies the remaining ones. Define

$$V_L = \begin{pmatrix} \Re(a_1^0) & \cdots & \Re(a_N^0) & | & \Im(a_1^0) & \cdots & \Im(a_N^0) \\ \Re(a_1^1) & \cdots & \Re(a_N^1) & | & \Im(a_1^1) & \cdots & \Im(a_N^1) \\ \vdots & & \vdots & | & \vdots & & \vdots \\ \Re(a_1^L) & \cdots & \Re(a_N^L) & | & \Im(a_1^L) & \cdots & \Im(a_N^L) \end{pmatrix}.$$

Notice that to construct this matrix, if a_i is real, then the column of zeros corresponding to the imaginary part is dropped. Also, whenever a_i is complex, either a_i or $\bar{a}_i$ is considered. Since the a_i's are assumed to be distinct, V_L has full column rank for all $L \geq 2N$ (V_L is a real Vandermonde matrix.). The complete set of constraints ($L = \infty$) is given by

$$\|V_\infty \underline{\alpha}\|_\infty \leq 1,$$

where $\underline{\alpha} = (\alpha_1, ..., \alpha_{2N})^T$. If the first L constraints are satisfied, i.e.,

$$|V_L \underline{\alpha}|_\infty \leq 1$$

and since $\underline{\alpha} = V_L^{-\ell} V_L \underline{\alpha}$ where $V_L^{-\ell}$ is the left inverse of V_L, then a bound on the $|\underline{\alpha}|_1$ is given by

$$|\underline{\alpha}|_1 \leq |V_L^{-\ell}|_{\infty,1} |V_L \underline{\alpha}|_\infty \leq |V_L^{-\ell}|_{\infty,1},$$

where $|\cdot|_{\infty,1}$ is the induced norm of a matrix mapping $(\mathbb{R}^{L+1}, |\cdot|_\infty)$ to $(\mathbb{R}^N, |\cdot|_1)$. The above indicates that satisfying the first L constraints (for any $L \geq 2N$) guarantees an upper bound on $|\underline{\alpha}|_1$. However, for any k,

$$\left| \sum_{i=1}^{N} \alpha_i \Re(a_i^k) + \alpha_{i+N} \Im(a_i^k) \right| \leq \max_i |a_i|^k |\underline{\alpha}|_1.$$

Hence, if L^* is picked such that

$$\max_i |a_i|^{L^*} |V_L^{-\ell}|_{\infty,1} < 1$$

then it follows that

$$|V_{L^*} \underline{\alpha}|_\infty \leq 1 \Longleftrightarrow \|V_\infty \underline{\alpha}\|_\infty \leq 1.$$

Notice that when V_L has a large dimension, $|V_L^{-\ell}|_{\infty,1}$ may be hard to compute. However, it can always be upper-bounded by, say, $L|V_L^{-\ell}|_1$.

10.3.3 Construction of the Minimizer

Let $\tilde{x}$ be the extremal functional from the dual problem. Its pointwise values are

$$\tilde{x}(k) = \sum_{i=1}^{N} \tilde{\alpha}_i \Re(a_i^k) + \tilde{\alpha}_{i+N} \Im(a_i^k),$$

where the $\tilde{\alpha}_i$'s are the solution of the dual LP problem. From the bound derived in the previous section, it follows that $|\tilde{x}(k)| < 1$ for all $k > L^*$. The construction of the optimal solution follows from the second part of Theorem 9.3.5, which asserts that any optimal solution is aligned with the extremal element $\tilde{x}$. If $H \in \ell_1$ is aligned with $y \in c_0$, then the following two conditions are satisfied (Theorem 9.3.3):

1. $H(k)y(k) \geq 0$.
2. $H(k) = 0$ if $|y(i)| < \|y\|_\infty$.

This characterization can be used to find all possible optimal solutions.

Theorem 10.3.2. The optimal solution $\Phi^o = H - R^o$ satisfies

1. $\Phi^o(k)\tilde{x}(k) \geq 0$.
2. $\Phi^o(k) = 0$ if $|\tilde{x}(k)| < 1$.
3. $\sum_{k=0}^{L^*} |\Phi^o(k)| = v^o$.   ∎

The above conditions define a set of linear constraints to which a solution is guaranteed to exist. The existence is established from the duality theorem.

It is worthwhile noting that if we consider the dual problem as a linear program and take its dual, we get back the primal problem. Because many of the algorithms used to solve LP problems compute both the primal and dual solutions, the construction step is not necessary in general.

10.4 DIRECT SOLUTION

As a consequence of the above characterization, the ℓ_1 optimal solution is a finite impulse response (FIR) of length less than or equal to L^*. Hence we can pose the optimization problem directly in the primal space in the following way: Let $\Phi \in \mathbb{R}^{L^*+1}$, V_{L^*} be defined as before and $b_i = \Re[\hat{H}(a_i)]$, $b_{i+N} = \Im[\hat{H}(a_i)]$ for $i = 1, \ldots, N$. Then

$$v^o = \min_{V_{L^*}^T \Phi = b} |\Phi|_1.$$

Let $\Phi = \Phi^+ - \Phi^-$, $\Phi^+(k), \Phi^-(k) \geq 0$. The above problem is equivalent to the LP

$$v^o = \min_{V_{L^*}^T (\Phi^+ - \Phi^-) = b} \sum_{k=0}^{L^*} \Phi^+(k) + \Phi^-(k).$$

The equivalence of the two problems follows from standard LP arguments and is discussed in detail in Example 8.2.2 and in Section 9.4.1.

10.5 PROPERTIES OF ℓ_1 OPTIMAL SOLUTIONS

In this section general properties of ℓ_1 optimal solutions are discussed.

Property 1: Optimal solutions are FIR. Optimal solutions are of finite length. Controllers that exhibit such closed-loop maps are know as deadbeat controllers. A bound on the length of any solution was derived in Section 10.3.2 and was shown to be a function of the location of the interpolation conditions, not the number of interpolation points. In fact, for a problem with two interpolation conditions, the solution can be of arbitrary high order. Consider the following example.

Example 10.5.1

Let $\hat{H} = -\epsilon^{-1}(\lambda - (1 - \epsilon))$ and $U = (\lambda - (1 - \epsilon))(\lambda - (1 - 2\epsilon))$ with $0 < \epsilon < .5$. The dual problem is given by

$$\nu^o = \max_{\alpha_1, \alpha_2} \alpha_1,$$

subject to

$$|\alpha_1(1 - 2\epsilon)^k + \alpha_2(1 - \epsilon)^k| \le 1, \qquad \text{for all } k \ge 0.$$

The solution of this problem results in at most two active constraints, one of which occurs at $k = 0$ and the other at some k. The value of α_1 can be computed to be

$$\alpha_1 = \frac{(1 - \epsilon)^k + 1}{(1 - \epsilon)^k - (1 - 2\epsilon)^k} \approx \frac{e^{-\epsilon k} + 1}{e^{-\epsilon k} - e^{-2\epsilon k}}$$

where the last approximation holds for small ϵ. Maximizing over k yields an optimal $k^* \approx \epsilon^{-1}\sinh(1)$ as ϵ approaches zero. Clearly, k^* can be made arbitrarily high with an appropriate choice of ϵ.

Property 2: Existence.   Optimal solutions are guaranteed to exist if all the interpolation conditions are in the unit disc. This follows directly from the application of Theorem 9.3.5. The lack of boundary interpolation conditions enables us to formulate the problem in a dual space, namely the dual of c_0. The case where there are boundary interpolations is discussed shortly.

Property 3: Uniqueness.   Optimal solutions are not necessarily unique. Consider the following example.

Example 10.5.2

Let $\hat{H}$ and $\hat{U}$ be as follows

$$\hat{H} = \frac{49(3 + 2\lambda)^2}{(4 - \lambda)^2(31 - 30\lambda)^2}, \qquad \hat{U} = \left(\lambda - \frac{1}{2}\right)\left(\lambda - \frac{1}{3}\right).$$

The dual problem is given by

$$\nu_0 = \max_{\alpha_1, \alpha_2} \frac{1}{4}\alpha_1 + \frac{1}{9}\alpha_2,$$

subject to

$$\left|\left(\frac{1}{2}\right)^k \alpha_1 + \left(\frac{1}{3}\right)^k \alpha_2\right| \le 1, \qquad \text{for all } k \ge 0.$$

By solving it we find that only the first three constraints are active and the optimal solution has the form $\hat{\Phi}(\lambda) = \Phi(0) + \Phi(1)\lambda + \Phi(2)\lambda^2$ with $\nu^o = 1$. The extremal functional has the form $\tilde{x} = (-1, 1, 1, *, *...)$ (* indicates a number with magnitude less than 1). Using the alignment conditions, all optimal solutions are characterized as follows:

1. $-\Phi(0) + \Phi(1) + \Phi(2) = 1$
2. $\Phi(0) \le 0, \Phi(1), \Phi(2) \ge 0$

3. $\Phi(0) + 0.5\Phi(1) + 0.25\Phi(2) = .25$

4. $\Phi(0) + \frac{1}{3}\Phi(1) + \frac{1}{9}\Phi(2) = \frac{1}{9}$

There are infinitely many solutions for the above system of inequalities. One solution is $\hat{\Phi} = \lambda^2$; another solution is $\hat{\Phi} = -\frac{1}{6} + \frac{5}{6}\lambda$.

In the general case, given any solution to the dual problem, all optimal solutions to the primal problem are characterized by the alignment conditions. This means that any such characterization is simply given by a finite number of linear constraints, with finitely many variables. This makes it possible to superimpose another criterion that can be minimized over all ℓ_1 optimal solutions. The following is an example.

Example 10.5.3

Consider the minimization problem:

$$\min \sum_{k=0}^{\infty} |\Phi(k)|^2,$$

subject to

$$\Phi = H - R, \ R \in \mathcal{S}, \ \|\Phi\|_1 = \min_{R \in \mathcal{S}} \|H - R\|_1.$$

This problem is a finite-dimensional quadratic programming problem. The constraints result in a finite-dimensional linear characterization of all optimal ℓ_1 solutions. The objective function is a quadratic function. Such problems can be easily solved.

10.6 EXTENSIONS

In this section we address two important extensions. The first is the formulation of the dual problem when $\hat{\Phi}$ has to interpolate zeros with higher multiplicity than one. The second is the formulation of the dual problem when there are interpolation points on the unit circle.

10.6.1 Higher Zero Multiplicity

Earlier, we assumed that the zeros of $\hat{U}$ in the disc are simple. If a zero has multiplicity r, then the interpolation conditions due to this zero (Theorem 6.2.1) are given by $\hat{R}^{(i)}(a) = 0$, $i = 0, 1, ..., r - 1$. Each one of these interpolation conditions defines two bounded linear functionals on R, which will have representations in ℓ_∞. The set of linear functionals can be taken to be the real and imaginary parts of

$$f_i(k) = \begin{cases} 0 & k < i \\ k^{(i)} a^{k-i} & k \geq i \end{cases}. \tag{10.2}$$

In fact, f_i lies inside c_0. The subspace M will have to be modified to include such functionals. The general solution follows exactly as in Theorem 10.3.1.

10.6.2 Zeros on the Unit Circle

The problem of boundary interpolation is an important problem that arises quite often since many control systems have built-in integrators. In the previous discussion we assumed that $\hat{U}$ has no zeros on the unit circle. When this assumption does not hold, the previous analysis breaks down at two important steps. The first step is Theorem 6.2.1. We have already seen that if $\hat{U}$ has a zero on the unit circle, then not every stable system $\hat{R}$ interpolating this zero can be written as $\hat{U}\hat{f}$, with $\hat{f}$ stable. This is possible, however, if $\hat{R} \in \mathcal{RH}_\infty$. The second step is the characterization of the annihilator subspace. We will deal with these problems separately.

Theorem 10.6.1.   For any $\hat{H}$, $\hat{U} \in \ell_1$, it follows that

$$\nu^o = \inf_{Q \in \ell_1} \|H - UQ\|_1 = \inf_{Q \in \mathcal{RH}_\infty} \|H - UQ\|_1.$$

Also, if $\hat{U} \in \mathcal{RH}_\infty$ then for any $\epsilon > 0$, Q can be chosen such that $H - UQ$ is FIR and $\|H - UQ\|_1 \leq \nu^o + \epsilon$.

Proof. Let $Q_\epsilon \in \ell_1$ be any suboptimal solution of the minimization problem, with $\|H - UQ_\epsilon\|_1 = \nu^o + \epsilon/2$. Let Q_a be a rational approximation of Q_ϵ, such that $\|Q_a - Q_\epsilon\|_1 \leq \epsilon/2\|U\|_1$. This is always possible, and Q_a can be chosen to be FIR. It follows that

$$\|H - UQ_a\|_1 \leq \nu^o + \|U(Q_\epsilon - Q_a)\|_1 + \frac{\epsilon}{2} \leq \nu^o + \frac{\epsilon}{2\|U\|_1}\|U\|_1 + \frac{\epsilon}{2} = \nu^o + \epsilon.$$

Since ϵ is arbitrarily small, the first result follows. To verify the second assertion, notice that if $\hat{U}$ is rational, then UQ_a can be approximated arbitrarily closely by an FIR system, which has zeros at the same locations as the zeros of $\hat{U}$ (this may not be possible if U is irrational). Because H can also be approximated by a FIR system, the result follows. ∎

Next we show the dual problem.

Theorem 10.6.2.   Let $\tilde{U}$ denote the part of U such that

1. If "a" is a zero of $\hat{U}$ in the disc with multiplicity l, then "a" is a zero of $\hat{\tilde{U}}$ with the same multiplicity.

2. If "a" is a zero of $\hat{U}$ on the unit circle with multiplicity l, then "a" is a zero of $\hat{\tilde{U}}$ with multiplicity equal to one.

Then it follows that

$$\nu^o = \inf_{Q \in \mathcal{RH}_\infty} \|H - UQ\|_1 = \inf_{Q \in \mathcal{RH}_\infty} \|H - \tilde{U}Q\|_1.$$

Proof. We will prove this theorem by means of a simplified problem. This captures the most general case. Let $\hat{U}$ have n_1 real and distinct zeros, $a_1, \ldots a_{n_1}$, in the disc and one zero, b, on the unit circle with multiplicity equal to 2. Consider the following minimization problem

$$\nu_N = \min \|\Phi\|_1$$

subject to

$$\hat{\Phi}(a_i) = \hat{H}(a_i), \quad i = 1, \ldots n_1,$$
$$\hat{\Phi}(b) = \hat{H}(b),$$
$$\hat{\Phi}'(b) = \hat{H}'(b),$$
$$\Phi(k) = 0, \quad \text{for all } k \geq N.$$

The dual problem is given by

$$\nu_N = \max \sum_{i=1}^{n_1} \alpha_i \hat{H}(a_i) + \beta_1 \hat{H}(b) + \beta_2 \hat{H}'(b)$$

subject to

$$| \sum_{i=1}^{n_1} \alpha_i a_i^k + \beta_1 b^k + \beta_2 k b^{k-1} | \leq 1, \quad k = 0, 1, \ldots N - 1.$$

Since the term kb^{k-1} is unbounded as k increases, it follows that β_2 will go to zero as N goes to infinity. This implies that the dual problem for the infinite length FIR (which is the original problem) does not involve the functionals resulting from the higher multiplicities of zeros on the unit circle. But this is the dual to the minimization problem with U replaced by $\tilde{U}$. Thus the result follows. ∎

It follows from the above theorem that the dual problem lies in ℓ_∞ (not c_0), and that the subspace $\mathcal{S}$ containing only rational elements is characterized through the annihilator subspace, i.e.,

$$\mathcal{S}^\perp = \tilde{M}$$

where $\tilde{M}$ is the subspace of functionals resulting from the zeros of $\tilde{U}$. With boundary interpolations, optimal solutions to the primal problem are not guaranteed to exist; however, such solutions can be approximated arbitrarily closely (see Exercise 10.6). The following is an example showing that existence may fail.

Example 10.6.1

Consider the minimization problem:

$$\inf_{Q \in \ell_1} \|H - UQ\|_1,$$

where $\hat{H} = 2(1 - \lambda)$ and $\hat{U} = (0.5 - \lambda)(1 - \lambda)$. The dual problem in ℓ_∞ is given by:

$$\nu^o = \max_{\alpha_1, \alpha_2} \alpha_1,$$

subject to

$$|\alpha_1 (.5)^k + \alpha_2| \leq 1, \quad k = 0, 1, \ldots.$$

It is clear that the maximum is attained at $\tilde{\alpha}_1 = 2$ and $\tilde{\alpha}_2 = -1$ and $\nu^o = 2$. The only active constraint is the first constraint; the rest do not touch the boundary. From the alignment condition, the optimal solution is aligned with the extremal functional, and thus its only nonzero component is $\Phi(0)$. But the optimal solution has to satisfy $\hat{\Phi}(1) = 0$ and $\hat{\Phi}(.5) = 1$, which is impossible to satisfy with a constant. Hence a solution does not exist.

10.7 A PRIMAL-DUAL PAIR OF LINEAR PROGRAMS

Denote by V_{∞}^T, the operator representing the interpolation conditions, and view elements in ℓ_1 as infinite column vectors. Then, $b = V_{\infty}^T H$. Following the development of Section 9.6.2, Problem 10.1 is equivalent to

$$\min \nu$$

subject to

$$\sum_0^{\infty} \Phi^+(k) + \Phi^-(k) - \nu \leq 0,$$
$$V_{\infty}^T(\Phi^+ - \Phi^-) = b,$$
$$\Phi^+(k), \ \Phi^-(k) \geq 0.$$

Let $\mathbf{1} = (1, 1, 1, \ldots)^T$. This functional is precisely the operator $\mathcal{A}_{\ell_1}$ introduced in Chapter 3. The matrix V_{∞}^T has at most $2N$ rows and an infinite number of columns, and so it has a finite-dimensional range that is immediately closed. Using the dual linear program from Section 9.6.2, we get

$$\max \alpha^T b$$

subject to

$$-\mathbf{1} \leq V_{\infty}\alpha \leq \mathbf{1},$$

where the inequality is taken pointwise. This is exactly the dual problem obtained from the duality theory for distance problems.

10.8 THE $\mathcal{H}_2$ MODEL MATCHING PROBLEM

In this problem, the "energy" contained in a transfer function's pulse response is minimized. This can also be viewed as minimizing the variance of the output due to a white noise input. The problem is stated as follows:

$$\inf_{Q \in \ell_1} \|H - UQ\|_2 = \inf_{R \in \mathcal{S}} \|H - R\|_2. \tag{10.3}$$

We assume that $\hat{U}$ has no zeros on the unit circle. This problem is a distance problem inside the space ℓ_2 (complex-valued space), and Theorem 9.3.5 can be used to provide a solution for this problem. To do that, it is convenient to enlarge the subspace $\mathcal{S} \subset \ell_1$ to its closure inside the space ℓ_2. This space is given by all $\Phi \in \ell_2$ such that $V_{\infty}^* \Phi = b$. Note that the matrix V_{∞} need not be split into its real and complex components as before.

The corresponding dual problem (which is equivalent to applying the projection theorem) is given by

$$\sup_\alpha |\alpha^* b|$$

subject to   (10.4)

$$\|V_\infty \alpha\|_2 \le 1.$$

An easy way to obtain the solution from this argument is to invoke the alignment condition before solving the dual problem, even though the dual is finite dimensional. The optimal primal solution is aligned with any solution of the dual problem, all of the form $V_\infty \alpha$. But two elements in ℓ_2 are aligned if and only if one is a scalar multiple of the other. Hence the optimal Φ has the form: $\Phi^o = \nu V_\infty \alpha^o$. From the interpolation conditions, $V_\infty^* \Phi^o = \nu V_\infty^* V_\infty \alpha^o = b$. Hence,

$$\nu \alpha^o = (V_\infty^* V_\infty)^{-1} b \tag{10.5}$$

from which it follows that Φ^o is given by

$$\Phi^o = V_\infty (V_\infty^* V_\infty)^{-1} b. \tag{10.6}$$

The optimal solution is the pseudo-inverse of V_∞ acting on b. It is immediate that ν is the minimum value and is given by

$$\nu = (b^* (V_\infty^* V_\infty)^{-1} b)^{\frac{1}{2}}. \tag{10.7}$$

Notice that the matrix $(V_\infty^* V_\infty)$ is given by

$$(V_\infty^* V_\infty) = \begin{pmatrix} \frac{1}{1-|a_1|^2} & \frac{1}{1-a_1\overline{a_2}} & \cdots & \frac{1}{1-a_1\overline{a_N}} \\ \vdots & \vdots & \vdots & \vdots \\ \frac{1}{1-a_N\overline{a_1}} & \frac{1}{1-a_N\overline{a_2}} & \cdots & \frac{1}{1-|a_N|^2} \end{pmatrix} \tag{10.8}$$

and will always have an inverse.

The optimal solution always exists and is unique. In fact, it is always rational, and has poles equal to the mirror images (with respect to the unit circle) of the interpolation points that lie inside the unit disc. A more direct approach to solving the $\mathcal{H}_2$ problem is shown in Problem 10.13.

10.9 THE $\mathcal{H}_\infty$ MODEL MATCHING PROBLEM

The $\mathcal{H}_\infty$ problem is to be stated as follows:

$$\inf_{\hat{Q} \in \mathcal{R}\mathcal{H}_\infty} \|\hat{H} - \hat{U}\hat{Q}\|_\infty. \tag{10.9}$$

We have posed this problem inside the space $\mathcal{R}\mathcal{H}_\infty$. It is clear that the infimum obtained is the same as the one obtained by minimizing over the space ℓ_1. There are different approaches to solving this problem. One particular approach is to employ duality theory. A maximization problem in the dual space of $\mathcal{H}_\infty$ can be formulated, from which properties of the optimal solution are derived. We will not follow this approach.

An elegant approach to solving the $\mathcal{H}_\infty$ problem is through the theory of Hankel operators developed earlier. In fact, under some mild assumptions, the $\mathcal{H}_\infty$ model matching problem is a special case of the model reduction problem presented in Chapter 4. To see this, first assume that $\hat{U}$ has no zeros on the unit circle. Then $\hat{U}$ can be written as:

$$\hat{U}(\lambda) = \hat{U}_i(\lambda)\hat{U}_o(\lambda)$$

$$= \prod_{i=1}^{N} \frac{\lambda - a_i}{1 - \bar{a}_i\lambda} \hat{U}_o(\lambda). \tag{10.10}$$

This is known as the inner-outer factorization of $\hat{U}$. Notice that $\hat{U}_i$ has constant magnitude on the unit circle, and $\hat{U}_o$ has no zeros inside the unit disc and thus has a stable inverse. Using this factorization, it follows that:

$$\inf_{\hat{Q} \in \mathcal{RH}_\infty} \|\hat{H} - \hat{U}\hat{Q}\|_\infty = \inf_{\hat{Q} \in \mathcal{RH}_\infty} \|\hat{H} - \hat{U}_i\hat{U}_o\hat{Q}\|_\infty$$

$$= \inf_{\hat{Q} \in \mathcal{RH}_\infty} \|\hat{U}_i^{\sim}\hat{H} - \hat{U}_o\hat{Q}\|_\infty, \quad \hat{U}_i^{\sim}(\lambda) = \hat{U}_i\left(\frac{1}{\lambda}\right) = \hat{U}_i^{-1}$$

$$= \inf_{\hat{Q}_1 \in \mathcal{RH}_\infty} \|\hat{U}_i^{\sim}\hat{H} - \hat{Q}_1\|_\infty, \quad \hat{Q}_1 = \hat{U}_o\hat{Q}$$

$$= \inf_{\hat{Q}_1 \in \mathcal{RH}_\infty} \|\hat{G} - \hat{Q}_1\|_\infty, \quad \hat{G} = \hat{U}_i^{\sim}\hat{H} \in \mathcal{L}_\infty.$$

If we write $\hat{G} = \hat{G}_- + \hat{G}_+$, where $\hat{G}_-$ is anti-causal, and $\hat{G}_+$ is causal, then

$$\inf_{\hat{Q}_1 \in \mathcal{RH}_\infty} \|\hat{G} - \hat{Q}_1\|_\infty = \inf_{\hat{Q}_1 \in \mathcal{RH}_\infty} \|\hat{G}_- + \hat{G}_+ - \hat{Q}_1\|_\infty$$

$$= \inf_{\hat{Q}_2 \in \mathcal{H}_\infty} \|\hat{G}_- - \hat{Q}_2\|_\infty, \quad \hat{Q}_2 = \hat{Q}_1 - \hat{G}_+$$

$$= \inf_{\hat{Q}_2^{\sim} \text{ anti-causal}} \|\hat{G}_-^{\sim} - \hat{Q}_2^{\sim}\|_\infty$$

$$= \inf_{\hat{X} \text{ anti-causal}} \|\hat{G}_c - \hat{X}\|_\infty, \quad \hat{G}_c = \hat{G}_-^{\sim} \text{ and } \hat{X} = \hat{Q}_2^{\sim}.$$

The last equality shows that the $\mathcal{H}_\infty$ model matching problem is equivalent to finding the anti-causal system that optimally approximates a causal system in the $\mathcal{L}_\infty$ sense. This problem has already been solved in the context of model reduction and the solution is explicitly stated in Theorem 4.5.3. We summarize the results in the following theorem.

Theorem 10.9.1.   Assume $\hat{U}$ has no zeros on the unit circle, and let $\hat{G}_c$ be the adjoint of the anti-causal part of $\hat{U}_i^{\sim}\hat{H}$. Also let σ_1 denote the first (largest) Hankel singular value of G_c, with the associated Schmidt pair w_1 and v_1. Then

1. $\inf_{\hat{Q} \in \mathcal{RH}_\infty} \|\hat{H} - \hat{U}\hat{Q}\|_\infty = \sigma_1$.
2. The optimal solution exists and is given by

$$(\hat{H} - \hat{U}\hat{Q})(\lambda) = \sigma_1 \hat{U}_i(\lambda) \frac{\hat{w}_1\left(\frac{1}{\lambda}\right)}{\hat{v}_1\left(\frac{1}{\lambda}\right)}.$$

3. The optimal solution is allpass, i.e., has magnitude σ_1 everywhere on the unit circle.
4. The optimal solution is unique.

Proof. From the previous analysis, we have

$$\inf_{\hat{Q} \in \mathcal{H}_\infty} \|\hat{H} - \hat{U}\hat{Q}\|_\infty = \inf_{\hat{X} \text{ anti-causal}} \|\hat{G}_c - \hat{X}\|_\infty.$$

The problem on the RHS is solved in Theorem 4.5.2 (Nehari's theorem) and its value is equal to σ_1, which proves part 1 of the theorem. The solution is given in Corollary 4.5.1:

$$(\hat{G}_c - \hat{X})(\lambda) = \sigma_1 \frac{\hat{w}_1(\lambda)}{\hat{v}_1(\lambda)}.$$

The result in part 2 follows by backward substitution. To show part 3, we note that

$$\hat{w}_1^\sim \hat{w}_1 = \hat{v}_1^\sim \hat{v}_1. \tag{10.11}$$

This can be verified by direct substitution and using the definition of the Schmidt pair (see Exercise 10.11). Finally Part 4 follows from Corollary 4.5.1. ∎

The theorem shows that the computation of the $\mathcal{H}_\infty$ solution follows the same steps as computing a reduced order model of the plant G_c. The optimal solution can be computed exactly, emphasizing that the infinite-dimensional optimization problem is equivalent to a finite-dimensional one.

Example 10.9.1

This example illustrates how to solve a simple sensitivity minimization problem using the ℓ_1, $\mathcal{H}_2$, and $\mathcal{H}_\infty$ system norms.

Consider the following unstable, nonminimum phase, SISO plant:

$$\hat{P} = \frac{\lambda - \frac{1}{2}}{\lambda + \frac{2}{3}}.$$

We would like to minimize the ℓ_1, $\mathcal{H}_2$, and $\mathcal{H}_\infty$ norms of the sensitivity function, $\Phi = (1 - PK)^{-1}$, where K is the controller. We start by computing a parametrization of all feasible closed-loop maps, Φ, as shown in Chapter 5. Let $\hat{P} = \hat{M}^{-1}\hat{N}$ where $\hat{M} = (\lambda + \frac{2}{3})$ and $\hat{N} = (\lambda - \frac{1}{2})$. Then, a coprime factorization is completed by solving

$$\hat{X}\hat{M} - \hat{Y}\hat{N} = 1.$$

Note that, since the plant is SISO, right and left coprime factors coincide. The above equation is satisfied with $\hat{X} = \hat{Y} = \frac{6}{7}$. Thus,

$$\hat{\Phi}(\lambda) = \hat{H} - \hat{U}\hat{Q} \tag{10.12}$$

$$= \frac{6}{7}\left(\lambda + \frac{2}{3}\right) - \left(\lambda - \frac{1}{2}\right)\left(\lambda + \frac{2}{3}\right)\hat{Q}.$$

ℓ_1 **Norm** We start with $\|\Phi\|_1$. In this case we have $a_1 = \frac{1}{2}$ and $a_2 = -\frac{2}{3}$, both real. Hence,

$$V_\infty = \begin{pmatrix} 1 & 1 \\ \frac{1}{2} & -\frac{2}{3} \\ \frac{1}{4} & \frac{4}{9} \\ \vdots & \vdots \end{pmatrix}.$$

So the zero interpolation conditions are given by

$$V_\infty^T \Phi = V_\infty^T H = \begin{pmatrix} \hat{H}(\frac{1}{2}) \\ \hat{H}(-\frac{2}{3}) \end{pmatrix} = \begin{pmatrix} 1 \\ 0 \end{pmatrix}.$$

Next, we compute an upper bound for the length of Φ^o. For $L = 1$, $|V_2^{-1}|_{\infty,1} = \frac{13}{7}$; hence Φ^o is at most first order, i.e., $L^* = 2$ and

$$\hat{\Phi}^o(\lambda) = \phi(0) + \phi(1)\lambda.$$

The solution can be obtained by solving the primal problem directly:

$$\min_{\phi^+(k),\phi^-(k)} \phi^+(0) + \phi^-(0) + \phi^+(1) + \phi^-(1)$$

subject to

$$\begin{pmatrix} 1 & -1 & \frac{1}{2} & -\frac{1}{2} \\ 1 & -1 & -\frac{2}{3} & \frac{2}{3} \end{pmatrix} \begin{pmatrix} \phi^+(0) \\ \phi^-(0) \\ \phi^+(1) \\ \phi^-(1) \end{pmatrix} = \begin{pmatrix} 1 \\ 0 \end{pmatrix},$$

$$\phi^+(k),\ \phi^-(k) \geq 0.$$

The results are $\nu^o = \frac{10}{7}$ and $\hat{\Phi}^o(\lambda) = \frac{4}{7} + \frac{6}{7}\lambda$.

For the purpose of comparison, we will compute all the other system norms for each solution. In this case, $\|\Phi^o\|_2 = 1.0302$ and, interestingly, $\|\hat{\Phi}^o\|_\infty = \nu^o = \frac{10}{7}$ (note that this is always the case when the impulse response does not change sign; hence the $\mathcal{H}_\infty$-norm is achieved at $\lambda = 1$).

$\mathcal{H}_2$ **Norm** First, the pseudo-inverse is computed:

$$(V_\infty^* V_\infty)^{-1} = \frac{1}{49} \begin{pmatrix} 48 & -20 \\ -20 & 35.556 \end{pmatrix}.$$

Then, from Equation (10.7), we obtain $\nu^o = 0.9897$ and

$$\Phi^o = \frac{48}{49}(V_\infty)^1 - \frac{20}{49}(V_\infty)^2$$

where $(V_\infty)^1$ and $(V_\infty)^2$ denote the first and second (infinite) columns of V_∞. The above sequence can be expressed in terms of the λ-transform:

$$\hat{\Phi}^o(\lambda) = \frac{28}{49} \frac{(1 + \frac{3}{2}\lambda)}{(1 - \frac{1}{2}\lambda)(1 + \frac{2}{3}\lambda)}.$$

Note how the reflection (i.e., inverse) of each interpolation zero shows as a pole of the optimal solution. The other norms are given by: $\|\Phi^o\|_1 = 1.8413$ and $\|\hat{\Phi}^o\|_\infty = 1.7143$.

$\mathcal{H}_\infty$ **Norm** Following the procedure of the previous section, we have that

$$\hat{U}_i(\lambda) = \frac{\left(\lambda - \frac{1}{2}\right)\left(\lambda + \frac{2}{3}\right)}{\left(1 - \frac{1}{2}\lambda\right)\left(1 + \frac{2}{3}\lambda\right)}.$$

Hence,

$$\hat{G}(\lambda) = \underbrace{-\frac{2}{7}\lambda}_{\hat{G}_+} + \underbrace{\frac{6}{7}\frac{1}{\lambda - \frac{1}{2}}}_{\hat{G}_-},$$

and

$$\hat{G}_c(\lambda) = \frac{6}{7}\frac{\lambda}{1 - \frac{1}{2}\lambda}.$$

Next, we compute the reachability and observability gramians of $\hat{G}_c$ to obtain $\mathbf{P} = \frac{4}{3}$ and $\mathbf{Q} = \frac{48}{49}$. Therefore, $\sigma_1 = \sqrt{\mathbf{PQ}} = 8/7$ is the value of the optimal $\mathcal{H}_\infty$ norm. To find the explicit form of the closed-loop map we compute the Schmidt pair corresponding to σ_1:

$$\hat{v}_1(\lambda) = \frac{2}{2\lambda - 1} \quad ; \quad \hat{w}_1(\lambda) = \frac{1}{1 - \frac{1}{2}\lambda}.$$

Finally,

$$\hat{\Phi}^o(\lambda) = \frac{8}{7}\frac{\lambda + \frac{2}{3}}{1 + \frac{2}{3}\lambda},$$

which is inner as expected. The other measures are: $\|\Phi^o\|_1 = \frac{8}{3}$ and, interestingly, $\|\Phi^o\|_2 = \sigma_1$ due to the all pass property.

10.10 COMPARISONS AND EXTREME EXAMPLES

Now that we have solved the ℓ_1, $\mathcal{H}_2$, and $\mathcal{H}_\infty$ SISO model matching problems, it is worthwhile to explore some of the differences between these solutions, in particular the $\mathcal{H}_\infty$ and the ℓ_1 problems, both being worst-case measures. Table 10.1 gives a summary of the various properties of these solutions for the SISO weighted sensitivity minimization problem that we have already derived.

Next, we present various examples illustrating contrasting behavior between the ℓ_1 and the $\mathcal{H}_\infty$ minimization problems. Before we do this, however, we stress the fact that these methods are optimal each under different assumptions on the exogenous signals and different measures on the regulated variables. Hence, direct comparisons are not possible. On the other hand, we are interested in quantifying the gap between such solutions. We are guided by the inequality proved in Theorem 4.3.1:

$$\|\hat{\Phi}\|_\infty \le \|\Phi\|_1 \le (2N + 1)\|\hat{\Phi}\|_\infty$$

where N is the McMillan degree of Φ. In all of the subsequent examples, we take Φ to be the weighted sensitivity function for some plant and weight. In Example 4.3.1, we have shown that there exist functions such that the above right-hand side inequality

TABLE 10.1   PROPERTIES OF SISO OPTIMAL SOLUTIONS.

Property	ℓ_1	$\mathcal{H}_2$	$\mathcal{H}_\infty$
Structure	FIR	rational	all pass
Degree	arbitrary	$\leq n$	$\leq n-1$
Poles	at infinity	mirror image of zeros and poles	new locations
Controller	stable or unstable	stable or unstable	unstable if No. of zeros $\geq$ 2 and plant is unstable
Existence	guaranteed if $|a_i| < 1$	guaranteed if $|a_i| < 1$	guaranteed if $|a_i| < 1$
Uniqueness	not unique	unique	unique

is tight. Such functions are allpass functions, with poles close to the unit circle. The following question arises at this point:

Do there exist plants where the optimal $\mathcal{H}_\infty$ solutions make the right-hand side inequality tight, yet the optimal ℓ_1 solutions are much smaller than that? In particular, are there examples where the value of the optimal ℓ_1 solution is independent of the McMillan degree?

We will answer this question by examples. These examples show that minimizing the $\mathcal{H}_\infty$ norm of a sensitivity function can make the ℓ_1 norm comparable to the McMillan degree of the solution, yet the optimal ℓ_1 norm has a very close value to the optimal $\mathcal{H}_\infty$ norm.

10.10.1 A Set of Plants with Separated Pole-Zero Configuration

For every η such that $0 < \eta < 1$, define the pole-zero configuration

$$\lambda_i^p = (1 - \eta^i), \qquad i = 1, \ldots, 2N - 1,$$
$$\lambda_i^z = -(1 - \eta^i), \qquad i = 1, \ldots, 2N - 1.$$

Let $\hat{P}(\lambda)$ be given by

$$\hat{P}(\lambda) = \prod_{i=1}^{N} \frac{\lambda - \lambda_i^z}{\lambda - \lambda_i^p}.$$

$\mathcal{H}_\infty$ optimal solution.   Choose the weight $\hat{W}(\lambda)$ so that the $\mathcal{H}_\infty$ optimal solution for the weighted sensitivity minimization problem is

$$\hat{\Phi}(\lambda) = \prod_{i=1}^{2N-1} \frac{\lambda_i^p - \lambda}{1 - \lambda_i^p \lambda}.$$

It follows that $\hat{W}(\lambda)$ is minimum phase and satisfies

$$\hat{W}(\lambda_k^p) \neq 0, \qquad k = 1, \ldots, N$$

$$\hat{W}(\lambda_k^z) = \prod_{i=1}^{2N-1} \frac{\lambda_i^p - \lambda_k^z}{1 - \lambda_i^p \lambda_k^z}, \qquad k = 1, \ldots, N.$$

It can be shown that a stable minimum phase weight satisfying the above conditions can always be constructed. We will skip the proof of this result (see references). From Example 4.3.1, we know that $\|\hat{\Phi}\|_{\mathcal{H}_\infty} = 1$ and $\|\Phi\|_{\ell_1} \to 4N - 1$ as $\eta \to 0$.

ℓ_1 optimal solution. Now we like to see what happens if we minimize the ℓ_1-norm instead of the $\mathcal{H}_\infty$-norm for the same system and weight. The primal problem is stated as follows:

$$\nu(\eta) = \min \|\Phi\|_1$$

subject to

$$A(\eta)\Phi = b(\eta),$$

$$\Phi = \{\phi(k)\}_{k=0}^\infty \in \ell_1,$$

where

$$A(\eta) = \begin{pmatrix} 1 & \lambda_1^p & (\lambda_1^p)^2 & \cdots & (\lambda_1^p)^{2k} & (\lambda_1^p)^{2k+1} & \cdots \\ 1 & -\lambda_1^p & (\lambda_1^p)^2 & \cdots & (\lambda_1^p)^{2k} & -(\lambda_1^p)^{2k+1} & \cdots \\ \vdots & \vdots & \vdots & \vdots & \vdots & \vdots & \vdots \\ 1 & \lambda_N^p & (\lambda_N^p)^2 & \cdots & (\lambda_N^p)^{2k} & -(\lambda_N^p)^{2k+1} & \cdots \\ 1 & -\lambda_N^p & (\lambda_N^p)^2 & \cdots & (\lambda_N^p)^{2k} & (\lambda_N^p)^{2k+1} & \cdots \end{pmatrix}, \quad b(\eta) = \begin{pmatrix} 0 \\ \hat{W}(\lambda_1^z) \\ \vdots \\ 0 \\ \hat{W}(\lambda_N^z) \end{pmatrix}.$$

The dual problem is given by

$$\nu(\eta) = \max \sum_{i=1}^N \alpha_{2i} \hat{W}(\lambda_i^z)$$

subject to

$$\left| \sum_{i=1}^N (\alpha_{2i-1} + \alpha_{2i})(1 - \eta^i)^{2k} \right| \leq 1$$

$$\left| \sum_{i=1}^N (\alpha_{2i-1} - \alpha_{2i})(1 - \eta^i)^{2k+1} \right| \leq 1$$

$$0 < \eta < 1, \ k = 0, 1, \ldots.$$

Notice that as η goes to zero, all the poles approach the value 1 and the zeros approach the value -1. If the solution of the problem is continuous with respect to η, then we can find the limiting value of $\nu(\eta)$ by solving the problem with N multiple poles at 1 and N multiple zeros at -1. We have shown earlier that having multiple poles and zeros at the boundary is equivalent to having simple poles and zeros. We can therefore replace the N poles at 1 and the N zeroes at -1 with a simple pole at 1 and a simple zero at -1. The problem in the limit is given by

$$\nu(0) = \max(\alpha_2)$$

subject to

$$|\alpha_1 + \alpha_2| \leq 1,$$

$$|\alpha_1 - \alpha_2| \leq 1.$$

By inspection we can see that $\nu(0) = 1$. However, we still have not shown that the limiting behavior is indeed so. To show this we need to verify that this solution is continuous.

We first observe that

$$\liminf_{\eta \to 0} \nu(\eta) \geq 1.$$

Indeed, since $\hat{W}(\lambda_i^z) \to 1$ as $\eta \to 0$, choosing $\alpha_2 = 1$, $\alpha_i = 0$, for all i, we see that for any $\epsilon > 0$, $\nu(\eta) \geq 1 - \epsilon$ whenever η is small enough.

It remains to show that $\nu(\eta)$ decreases to 1 as η goes to zero. That is, for any arbitrary $\epsilon > 0$ we can find a $\delta > 0$ such that for all $0 < \eta < \delta$ the solution $\nu(\eta) < 1 + \epsilon$. To show this, we construct a sequence of feasible solutions indexed by η in which each element has $2N$ nonzero components and such that their norms converge to 1 as η goes to zero. Consider the constraint equation in the primal problem, $A(\eta)\Phi = b(\eta)$. Notice that by adding the $(2k-1)^{th}$ row to the $(2k)^{th}$, $k = 1, \ldots, N$ and then again in the resulting matrix equation subtracting the $(2k)^{th}$ row from $(2k-1)^{th}$ we will have the equivalent set of constraints: $\tilde{A}(\eta)\Phi = \tilde{b}(\eta)$ where

$$\tilde{A}(\eta) = \begin{pmatrix} 1 & 0 & (\lambda_1^p)^2 & \cdots & (\lambda_1^p)^{2k} & 0 & \cdots \\ 0 & -\lambda_1^p & 0 & \cdots & 0 & -(\lambda_1^p)^{2k+1} & \cdots \\ \vdots & \vdots & \vdots & \vdots & \vdots & \vdots & \\ 1 & 0 & (\lambda_N^p)^2 & \cdots & (\lambda_N^p)^{2k} & 0 & \cdots \\ 0 & -\lambda_N^p & 0 & \cdots & 0 & -(\lambda_N^p)^{2k+1} & \cdots \end{pmatrix}, \quad \tilde{b}(\eta) = \begin{pmatrix} \hat{W}(\lambda_1^z)/2 \\ \hat{W}(\lambda_1^z)/2 \\ \vdots \\ \hat{W}(\lambda_N^z)/2 \\ \hat{W}(\lambda_N^z)/2 \end{pmatrix}.$$

We can now separate this in two problems as shown below:

$$\nu(\eta) = \left\{ \begin{array}{cc} \min(\|\Phi_1\|_1) & + \quad \min(\|\Phi_2\|_1) \\ \bar{A}(\eta)\Phi_1 = b_1(\eta) & \bar{A}(\eta)\Phi_2 = b_2(\eta) \end{array} \right\}$$

$$\Phi_1 = \{\phi(2k+1)\}_{k=0}^{\infty}, \qquad \Phi_2 = \{\phi(2k)\}_{k=0}^{\infty}$$

where

$$\bar{A}(\eta) = \begin{pmatrix} 1 & (\lambda_1^p)^2 & \cdots & (\lambda_1^p)^{2k} & \cdots \\ \vdots & \vdots & \vdots & \vdots & \vdots \\ 1 & (\lambda_N^p)^2 & \cdots & (\lambda_N^p)^{2k} & \cdots \end{pmatrix}, \quad b_1(\eta) = (1/2) \begin{pmatrix} \dfrac{-\hat{W}(\lambda_1^z)}{\lambda_1^p} \\ \vdots \\ \dfrac{-\hat{W}(\lambda_N^z)}{\lambda_N^p} \end{pmatrix}, \quad b_2 = (1/2) \begin{pmatrix} \hat{W}(\lambda_1^z) \\ \vdots \\ \hat{W}(\lambda_N^z) \end{pmatrix}.$$

The above two problems are independent standard ℓ_1 problems that have the same limiting behavior. Next, we choose a particular sequence of basic feasible solutions to the problem, each of length $2N$, and show that $v(\eta)$ converges to 1 as $\eta \to 0$.

Let $\bar{A}(\eta) = \{a_{ij}\}$. Choose the indices of the basic feasible solution as

$$I = \{k_1, k_2, \ldots, k_N\} = \{1, \{2\lfloor 1/\eta^{j+1/2} \rfloor\}_{j=1}^{N-1}\}$$

where $\lfloor . \rfloor$ denotes the greatest integer less than the number. Then the basis $B(\eta) = \{a_{ik_j}\}_{j=1}^{N}$, for all $i = 1, \ldots, N$. Note that

$$a_{ik_1} = 1,$$

$$a_{ik_j} = (1 - \eta^i)^{k_j}, \quad j = 2, \ldots, N, \quad i = 1, \ldots, N.$$

Hence by applying L'Hospital's rule, we can verify that:

$$\lim_{\eta \to 0} a_{ik_j} = \begin{cases} 1 & i \geq j \\ 0 & i < j \end{cases}.$$

As $\eta \to 0$ the basis $B(\eta)$, $b_1(\eta)$, and $b_2(\eta)$ go to

$$B_0 = \begin{pmatrix} 1 & 0 & 0 & \cdots \\ 1 & 1 & 0 & \cdots \\ \vdots & \vdots & \vdots & \vdots \\ 1 & 1 & \cdots & 1 \end{pmatrix}, \quad b_1 = -b_2 = -1/2 \begin{pmatrix} 1 \\ 1 \\ \vdots \\ 1 \end{pmatrix},$$

respectively. It follows that in the limit, $\|\Phi_1\|_1 = 1/2$, $\|\Phi_2\|_1 = 1/2$ (choose Φ_0 such that $\phi(0) = 1/2$, $\phi(1) = -1/2$, $\phi(k) = 0$ for all other k). Now notice that since B_0 is positive definite and $B(\eta) = \{a_{ik_j}\}_{i,j=1}^{N} \to B_0$ there exists some $\delta > 0$ such that $\Phi_1 = B(\eta)^{-1}b(\eta)$ exists and is continuous, similarly Φ_2. Hence, $\Phi \to \Phi_0$ and $v(\eta) \to 1$.

Comments. Notice that this example illustrates an extreme situation. It basically shows that minimizing the $\mathcal{H}_\infty$ norm of a closed-loop transfer function may cause the ℓ_1 norm to grow. In this example the ℓ_1 norm of the $\mathcal{H}_\infty$ optimal solution increases as $O(N)$ (order N); however, the ℓ_1 norm of the ℓ_1 optimal solution is exactly equal to the $\mathcal{H}_\infty$ norm of the $\mathcal{H}_\infty$ optimal solution (in the limit as $\eta \to 0$), which is a constant. Further, since the $\mathcal{H}_\infty$ norm of any system is bounded by its ℓ_1 norm, the $\mathcal{H}_\infty$ norm of the ℓ_1 optimal solution is also a constant and does not increase with the order. Moreover, this is the maximum possible ratio between the ℓ_1 norm and the $\mathcal{H}_\infty$ norm achievable for a stable system of a given McMillan degree.

The above results were derived for $\eta \to 0$. It would be interesting to investigate how the limiting values of the sensitivity norms are approached as η approaches zero. Figures 10.2 and 10.3 show, for $N = 3$, the values of $\|WS\|_1$ and $\|\hat{W}\hat{S}\|_\infty$ corresponding to the ℓ_1 optimal solution, and $\|WS\|_1$ corresponding to the $\mathcal{H}_\infty$ optimal solution (recall that $\|\hat{W}\hat{S}\|_\infty = 1$ in this case).

Finally, it is also possible to come up with examples for which minimizing the ℓ_1 norm of the sensitivity function may not reduce the value of the $\mathcal{H}_\infty$ norm, in comparison

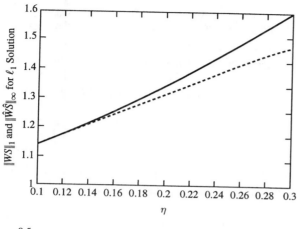

Figure 10.2 Example with separated poles and zeros and $N = 3$, showing $\|WS\|_1$ (solid line) and $\|\hat{W}\hat{S}\|_\infty$ (broken line) for ℓ_1 optimal solution.

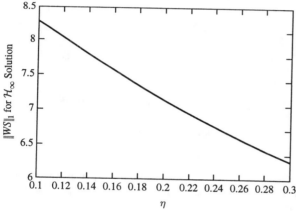

Figure 10.3 Example with separated poles and zeros and $N = 3$, showing $\|WS\|_1$ for $\mathcal{H}_\infty$ optimal solution.

to the optimal $\mathcal{H}_\infty$ norm. Thus, minimizing an upper bound of a function may not yield desirable results.

10.11 SUMMARY

In this chapter we have solved the SISO model matching problem for the ℓ_1, $\mathcal{H}_2$, and $\mathcal{H}_\infty$ norms. The power of duality theory was demonstrated in the solution of both the ℓ_1 and the $\mathcal{H}_2$ problems. The three problems were shown to be equivalent to finite-dimensional optimization problems and were solved exactly. These solutions were then compared and extreme examples were presented. These examples highlight important issues:

1. The solution resulting from a worst-case methodology depends heavily on the particular norm considered. The $\mathcal{H}_\infty$ methodology gives solutions that are allpass, which tend to have large ℓ_1 norms. The existence of the above extreme examples supports the argument that the gap between ℓ_1 and $\mathcal{H}_\infty$ can be large.

2. The only way to bound both the $\mathcal{H}_\infty$ and the ℓ_1 solutions is to perform mixed minimization problems. We will discuss such problems in later chapters.

3. If the optimization problem resulted from a desire to maximize the margin of uncertainty, the above examples indicate that the solution is sensitive to the way the uncertainty is modeled, i.e., ℓ_∞ or ℓ_2 stable. This has serious implications on the worst-case paradigm for robust control, which we will not discuss in this book.

EXERCISES

10.1. Given a plant $\hat{G}_{22}(\lambda) = \frac{\lambda(\lambda - .2)}{(\lambda - 2)(\lambda - .5)}$. Find the controllers that minimize the ℓ_1 norm, $\mathcal{H}_2$ norm, and the $\mathcal{H}_\infty$ norm of the weighted sensitivity function, with $\hat{W}(\lambda) = \frac{1}{1 - 0.6\lambda}$. Which controller is stable?

10.2. Let $\hat{G}_{22}$ be a stable, SISO discrete-time plant, with one nonminimum phase zero. Show that the controller that minimizes the $\mathcal{H}_\infty$ norm of the weighted sensitivity function will also minimize that ℓ_1 norm of the same function. Assume the weight is stable, minimum phase. Does this follow for the $\mathcal{H}_2$ problem?

10.3. For the SISO model matching problem $H - UQ$, suppose $\hat{U} = \lambda^N \hat{U}_1$ and $\hat{H} = \hat{H}_1 + \lambda^N \hat{H}_2$ where $\hat{H}_1$ is a polynomial in λ with degree less than N.
   (a) Show that

$$\min_{Q \in \ell_1} \|H - UQ\|_1 = \|H_1\|_1 + \min_{Q \in \ell_1} \|H_2 - U_1 Q\|.$$

   (b) If U_1 has a stable inverse, find all solutions Q satisfying

$$\|H - UQ\|_1 \le \gamma.$$

10.4. Let c be any complex number such that $|c| < 1$, and $\theta_0 \in [-\pi, \pi]$ (which is not a rational multiple of π). Show that there exists a function $F \in \ell_1$, with norm $\|F\|_1 < 1$ such that $\hat{F}(e^{i\theta_0}) = c$. Is this result true if θ_0 is a rational multiple of π?
Use this result to show that for SISO systems, $(I - M\Delta)^{-1}$ is stable for all $\Delta \in \ell_1$, with $\|\Delta\|_1 < 1$ if and only if $\|\hat{M}\|_\infty \le 1$.

10.5. Given the plant $\hat{G}_{22} = \lambda(\lambda - .5)$. Suppose the plant is subjected to a disturbance set

$$D = \left\{ d = We, \ e \in \ell_\infty, \ \hat{W} = \frac{1}{1 - \lambda} \right\}.$$

   (a) Are the admissible disturbances bounded in time?
   (b) Design a compensator that minimizes the worst possible ℓ_∞ norm of the output $y = (I - G_{22}K)^{-1}We$. (Be careful! If your compensator is only stabilizing, some of the possible outputs are unbounded. The compensator should do more.)

10.6. Let $\hat{U}$ have zeros both in the unit disc and the on the unit circle. Assume that every zero on the unit circle is a multiple of $\frac{2\pi}{N}$ for some N. Show how the dual problem can be computed to any desired degree of accuracy. Does your procedure extend if the zeros are not rational multiples of π? (Hint: The constraints due to zeros that are rational multiples of π are periodic.)

10.7. In general, the optimal $\hat{\Phi}^o$ (in the ℓ_1 norm) may be of a degree larger than the number of interpolation points. Show that the degree of $\hat{\Phi}^o$ equals the number of interpolation points minus one, $N - 1$, if

$$\sum_{i=0}^{N-1} |c_i| < 1,$$

where $\prod_{j=1}^{N} (\lambda - a_j) = \lambda^N + c_{N-1} \lambda^{N-1} + \cdots + c_1 \lambda + c_0$. (See references.)

10.8. (ℓ_∞ minimization) It is useful to compare a worst-case methodology to one based on fixed exogenous inputs. The $\mathcal{H}_2$ and $\mathcal{H}_\infty$ problems give such a comparison for ℓ_2 signals. In this problem, we will study the solution of minimizing the ℓ_∞ norm of the output due to a fixed disturbance w_f. We will assume that w_f has a rational λ-transform.

    (a) Obtain a parametrization of all stabilizing controllers with the property that the output goes to zero as t goes to infinity.

    (b) Show that the output z is parametrized as follows:

$$z = (I - PK)^{-1} w_f = \tilde{H} - \tilde{U} Q, \quad Q \in \ell_1,$$

    where $\tilde{H}$ and $\tilde{U}$ are stable.

    (c) Define the optimization problem

$$\nu^o = \inf_{Q \in \ell_1} \| \tilde{H} - \tilde{U} Q \|_{\ell_\infty},$$

    where $\| \cdot \|_{\ell_\infty}$ denotes the ℓ_∞ norm of the sequence. Show that the infimum does not change if we search over all $Q \in c_0$, i.e.,

$$\nu^o = \inf_{Q \in c_0} \| \tilde{H} - \tilde{U} Q \|_{\ell_\infty}.$$

    (d) Assume $\tilde{U}$ has no zeros on the unit circle. Write down the dual formulation of the above problem. In what space does it lie? Is the dual problem finite dimensional?

    (e) Does there exist a solution to the primal problem in c_0? How about ℓ_∞?

    (f) Discuss how the dual problem can be computed to any desired degree of accuracy. Show how suboptimal solutions to the primal problem can be computed.

10.9. Solve the above problem for

$$\hat{P} = \frac{\lambda(\lambda - .5)}{(\lambda - .6)}$$

and each of the following inputs:

    (a) $w_f(k) = 1$ for all $k \geq 0$.

    (b) $w_f(k) = (-1)^k$ for all $k \geq 0$.

Comment on both of these solutions. Compare them to the optimal ℓ_1 solution minimizing $\| \tilde{H} - \tilde{U} Q \|_1$.

10.10. The following procedure is proposed to avoid the problems arising from interpolation points on the unit circle in solving the ℓ_1 problem. For $r > 1$, define the exponentially weighted norm

$$\| H \|_{(r)} = \sum_{t=0}^{\infty} r^t |H(t)|.$$

It is clear that not every function in ℓ_1 has a bounded $\| \cdot \|_{(r)}$ norm. Let the space $\ell_1(r)$ denote the subspace of ℓ_1 that has elements of bounded $\| \cdot \|_{(r)}$ norm, and equipped with the $\| \cdot \|_{(r)}$ norm. Elements of this space have to decay faster than r^{-t} (hence, the associated λ-transform is analytic in the disc of radius r).

(a) Show that the dual space of $\ell_1(r)$ is given by $\ell_\infty(r^{-1})$, which contains all sequences (not necessarily bounded) satisfying

$$\sup_t |r^{-t} g(t)| < \infty.$$

Verify that the above measure is exactly equal to the induced norm over $\ell_1(r)$.

(b) Consider the minimization problem

$$\nu_r = \inf_{Q \in \ell_1(r)} \|H - UQ\|_{(r)}.$$

Assume that $\hat{U}$ has no zeros on the circle of radius r. $\hat{U}$ may have zeros on the unit circle. Denote all the zeros in the disc of radius r by $a_1, \ldots a_N$, and assume for simplicity that they are real and distinct. Define the feasible space S as

$$S = \{R \in \ell_1(r) | R = UQ, \quad Q \in \ell_1(r)\}.$$

Show that the annihilator subspace $S^\perp$ is given by

$$S^\perp = \text{Span}\{\underline{a}_i, \ i = 1, \ldots N\}.$$

What is the norm defined on this subspace?

(c) Write down the dual problem of the above primal problem. Show that the problem is equivalent to a finite-dimensional linear program.

(d) Show that

$$\lim_{r \to 1} \nu_r = \nu_1$$

where ν_1 denotes the solution of the standard ℓ_1 problem. Show this by means of an example if you cannot prove it.

10.11. Verify Equation 10.11. (Hint: Start with the Lyapunov equation defining the observability gramian, add the term $1/\lambda I$ in the appropriate place, and construct $\hat{w}_1^\sim \hat{w}_1$. Do the same for the reachability gramian.)

10.12. We have shown that the optimal $\mathcal{H}_\infty$ solution of $\hat{H} - \hat{U}\hat{Q}$ is allpass. Suppose $\hat{U}$ has N-zeros, z_i, in the unit disc.

(a) Show that the optimal solution $\hat{\Phi}^o = \hat{H} - \hat{U}\hat{Q}^o$ has at most $N - 1$ poles and zeros.

(b) Let $\hat{r}(\lambda)$ be a polynomial of degree $N - 1$ with zeros all outside the unit disc. Show that $\hat{\Phi}_0$ has the form

$$\hat{\Phi}^o = \alpha \lambda^{N-1} \frac{\hat{r}(1/\lambda)}{\hat{r}(\lambda)}$$

satisfying the following constraints:

$$\hat{\Phi}^o(z_i) = \hat{H}(z_i), \quad i = 1, 2, ..., N.$$

(c) Set up a generalized eigenvalue problem in terms of the coefficients of $r(\lambda)$, and α. (Hint: Use the form of $\hat{\Phi}^o$ and plug in the constraints.)

10.13. Compute the minimizing solution of the $\mathcal{H}_2$ problem directly in the frequency domain. In other words, start with an inner-outer factorization of $\hat{U}$, and apply the projection theorem to $\hat{U}_i^\sim \hat{H}$.

10.14. (Equivalence of $\mathcal{H}_\infty$ and $\mathcal{H}_2$ optimal problems) It is generally argued that any controller produced by an $\mathcal{H}_\infty$ minimization of a weighted SISO problem can be produced by an $\mathcal{H}_2$

minimization of the same problem with possibly a different weight. In particular, one can find a stable weight $\hat{W}(\lambda)$ such that $\hat{Q}$ that solves the following problem:

$$\min_{\hat{Q} \in \mathcal{RH}_\infty} \|(\hat{H} - \hat{U}\hat{Q})\hat{W}\|_{\mathcal{H}_2}$$

also solves:

$$\min_{\hat{Q} \in \mathcal{RH}_\infty} \|\hat{H} - \hat{U}\hat{Q}\|_{\mathcal{H}_\infty}.$$

We claim that we have already proved this result, and, in fact, we have constructed the precise $\hat{W}$ that does it. Verify this claim, i.e., find $\hat{W}$ and prove the equivalence. (Hint: Think of the role of the Schmidt pair in solving the $\mathcal{H}_\infty$ problem.)

10.15. In weighted sensitivity minimization, show that the controller that minimizes the $\mathcal{H}_\infty$ norm is always unstable if the system has at least two nonminimum phase zeros.

10.16. Let $\hat{G}_{22}$ be a minimum phase, unstable plant with only one unstable pole. Assume that the actual plant lies in the set $\Omega = \{\hat{G}_{22} + \hat{\Delta}\}$, where $\hat{\Delta} \in \mathcal{H}_\infty$. Show that there exists a compensator that robustly stabilizes all plants in Ω if and only if $\|\hat{\Delta}\|_\infty \leq \sigma$, where σ is the smallest singular value associated with the Hankel operator $\Gamma_{\hat{G}_{22}^\sim}$. (Hint: Write $\hat{G}_{22} = \hat{N}/\hat{M}$, $\hat{M} = (\lambda - a)/(1 - a\lambda)$.)

Comment: This is a special case of the general result: The maximum radius of uncertainty is equal to the smallest nonzero singular value of $\Gamma_{\hat{G}_{22}^\sim}$. In the above problem, there is only one such Hankel singular value.

10.17. Let

$$\nu_{TI} = \min_{\hat{Q} \in \mathcal{RH}_\infty} \|\hat{H} - \hat{U}\hat{Q}\|_\infty,$$

and

$$\nu_{NL} = \min_{Q \text{ bounded } \ell_2(\mathbf{z})-\text{gain}} \|H - UQ\|_{\ell_2(\mathbf{z})-\text{gain}}.$$

Show that

$$\nu_{TI} = \nu_{NL}.$$

(Hint: Show that the second problem is also bounded from below by $\|\Pi_{\mathcal{H}_2^\perp}\hat{U}_i^\sim\hat{H}\Pi_{\mathcal{H}_2}\|$. Relate the latter to ν_{TI}.)

10.18. (A continuous-time example) Let the continuous-time plant $\hat{G}$ be given by

$$\hat{G}(s) = \frac{s - z}{s - p},$$

where both z and p lie in the open right-half plane. The objective is to find a controller that minimizes the $\mathcal{L}_1$ norm of the weighted sensitivity function $\Phi = W(I - PK)^{-1}$.

(a) Show that Φ is feasible if and only if $\hat{\Phi}$ is analytic in the right-half plane and

$$\hat{\Phi}(z) = \hat{W}(z), \quad \text{and} \quad \hat{\Phi}(p) = 0.$$

(b) Show that the minimization problem can be written as

$$\min_{R \in \mathcal{S}} \|H - R\|_1, \quad \mathcal{S} = \{R \in \mathcal{L}_1 \mid \hat{R}(z) = 0, \text{ and } \hat{R}(p) = 0\}$$

with some feasible $H \in \mathcal{L}_1$.

(c) Show that

$$S^\perp = \mathrm{span}\{e^{-zt},\ e^{-pt}\} \subset \mathcal{L}_\infty.$$

(d) Apply duality theory to the above problem and show that it is equivalent to a linear program with two variables and an infinite number of constraints.

(e) Argue that if a solution to the primal problem exists, then it has the form

$$\hat{\Phi}(s) = b_0 + b_1 e^{-s\delta}.$$

(f) Solve the above problem for $z = 1$ and $p = 2$ and any stable proper $\hat{W}$.

NOTES AND REFERENCES

The solution of the ℓ_1 model matching problem follows [DP87a] in which duality was employed to give a complete solution to the one-block problem (Independent results on ℓ_1 were reported in [BG84].) The two examples showing the properties of long FIR solutions and nonuniqueness are taken from [Mey88]. Various results discussing continuity of the solution as a function of the problem data, as well as the formulation of superoptimal minimization problems, can be found in [DD90]. The case where the interpolation conditions include points on the unit circle was completely analyzed in [Vid91]. Extreme examples are presented in [SD93]. The parallel results for continuous time plants were reported in [BS93, DP87b, DO92], and the case of boundary interpolations was analyzed in [Vid91]. Problems concerning fixed input minimization and their relation to ℓ_1 minimization were discussed in [DP88b]. The proof of Exercise 10.7 can be found in [DV90].

The solution of the $\mathcal{H}_2$ model matching problem is quite standard using pseudoinverses [YJB76a, Lue69]. Solution of the $\mathcal{H}_\infty$ model matching problem was given in [FZ84, Sar67, ZF83] for the SISO case, and in [FHZ84, CP84] for the general one-block problem. The solution presented here is a special case of the general model reduction solution presented in Chapter 4; see [Glo89, Fra87]. For a thorough bibliography on the $\mathcal{H}_\infty$ problem see [Fra87, Vid85].

CHAPTER 11 ———————————————

MIMO ℓ_1 Model Matching Problem

In Chapter 6 we have shown that the closed-loop achievable maps are characterized by a set of interpolation conditions: zero conditions and rank conditions. In the SISO case, only zero interpolation conditions exist, resulting in an ℓ_1 model matching problem that is equivalent to a finite-dimensional linear program, as was discussed in the previous chapter. In the MIMO case, there will be several cases to consider depending on whether the rank interpolation conditions are absent or not. In this chapter we present the basic results regarding the primal-dual formulation and existence of ℓ_1 optimal solutions to the standard control problem in its most general form. At the same time, the equivalence of these problems and infinite linear programs is established explicitly. The solution of the various cases is left to the next chapter.

11.1 PROBLEM DEFINITION

The ℓ_1 model matching problem is defined as follows: Find ν^o and Q (if it exists) as in Figure 11.1 such that

$$\nu^o := \inf_{Q \in \ell_1^{n_u \times n_y}} \| H - UQV \|_1. \tag{11.1}$$

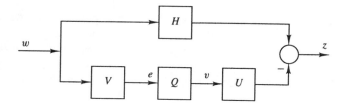

Figure 11.1 Model matching problem.

255

To be able to transform the above formulation into a minimum distance problem in ℓ_1, we make the following definition.

Definition 11.1.1. The subspace of $\ell_1^{n_z \times n_w}$, $\mathcal{S}$, is defined as

$$\mathcal{S} := \{R \in \ell_1^{n_z \times n_w} \mid R = UQV \text{ for some } Q \in \ell_1^{n_u \times n_y}\}. \qquad (11.2)$$

∎

Problem (11.1) can be restated as a minimum distance problem:

$$\nu^o := \inf_{R \in \mathcal{S}} \|H - R\|_1. \qquad (11.3)$$

In this chapter, the following issues are discussed:

1. Characterization of the subspace $\mathcal{S}$ in terms of interpolation conditions.
2. A primal-dual pair of Linear Programs equivalent to Problem (11.3).
3. Existence of optimal solutions in both the primal and dual problems.

11.2 FEASIBILITY CONDITIONS

With Theorem 6.2.2 we have established a compact algebraic description of the set $\mathcal{S}$. We need to interpret these results in the context of the duality theorems discussed in Chapter 9. We have the following possibilities:

1. Identify the annihilator subspace of $\mathcal{S}$ in ℓ_∞ and apply Theorem 9.3.4.
2. Identify a subspace M of $c_0^{n_z \times n_w}$ for which $\mathcal{S}$ is its annihilator and apply Theorem 9.3.5.
3. Represent the subspace $\mathcal{S}$ as the kernel of some operator, i.e.,

$$R \in \mathcal{S} \Longleftrightarrow \mathcal{A}R = 0$$

and apply the results in Section 9.6.2.

Of course, these methods are not different from one another, and setting the problem in any one method gives us the rest of them. For instance, item 3 above is really a consequence of item 1, and items 2 and 3 are related as in Theorem 9.5.4. The advantage of the formulation in item 2 is that existence in the primal problem follows directly. Clearly, not every problem can be formulated as in item 2, and in the sequel we will show that this is possible only if all the zero interpolation conditions are in the open unit disc. In the following development, we will characterize the feasible space as in items 1 and 3. Formulations involving item 2 will be left to the existence discussion.

Notice that Theorem 6.2.2 shows that the feasible subspace is characterized in terms of two classes of linear operators (which include bounded linear functionals):

1. Operators due to zeros inside the disc. Each zero defines (at least) a bounded linear functional on the space $\ell_1^{n_w \times n_z}$. The feasible subspace lies in its null-space.

2. Operators due to rank interpolation conditions. These are convolution operators that can be represented as multiplication by Toeplitz matrices, each having an infinite-dimensional range.

In the sequel, we will find explicit descriptions of such operators. First, note that the zero interpolation conditions (i.e., first set of conditions in Theorem 6.2.2) can be rewritten as follows:

$$(\hat{\alpha}_i \hat{R} \hat{\beta}_j)^{(k)}(\lambda_0) = \text{trace} \left[\hat{R} \hat{\beta}_j \hat{\alpha}_i \right]^{(k)} (\lambda_0)$$

$$= \left[\sum_{p=1}^{n_z} \sum_{q=1}^{n_w} \hat{r}_{pq} \hat{\beta}_{qj} \hat{\alpha}_{ip} \right]^{(k)} (\lambda_0), \qquad (11.4)$$

where α_{ip} denotes the pth column of α_i, β_{qj} denotes the qth row of β_j, and $\hat{r}_{pq}$ denotes the pq^{th} term of $\hat{R}$.

Furthermore, the algebraic product of λ-transforms is equivalent to convolution of sequences. That is,

$$(\hat{\alpha}_{ip} \hat{\beta}_{qj} \hat{r}_{pq})(\lambda) = \sum_{t=0}^{\infty} (\alpha_{ip} * \beta_{qj} * r_{pq})(t) \lambda^t. \qquad (11.5)$$

Then, by expanding the convolution operations we have

$$(\alpha_{ip} * \beta_{qj} * r_{pq})(t) = \left[\sum_{s=0}^{\infty} \alpha_{ip}(t-s) \beta_{qj}(s) \right] * r_{pq}$$

$$= \sum_{l=0}^{\infty} \sum_{s=0}^{\infty} \alpha_{ip}(t-l-s) \beta_{qj}(s) r_{pq}(l). \qquad (11.6)$$

By substituting Equation (11.6) into (11.5) and then into (11.4), and changing the order of the infinite sums, the following expression is obtained:

$$(\hat{\alpha}_i \hat{R} \hat{\beta}_j)^{(k)}(\lambda_0) = \sum_{p=1}^{n_z} \sum_{q=1}^{n_w} \sum_{l=0}^{\infty} \left[\sum_{t=0}^{\infty} \sum_{s=0}^{\infty} \alpha_{ip}(t-l-s) \beta_{qj}(s) (\lambda^t)^{(k)} \right]_{\lambda = \lambda_0} r_{pq}(l).$$

Clearly, the sequence that annihilates R in $\mathcal{S}$ is given, by definition, by the term in brackets. That is,

$$(\hat{\alpha}_i \hat{R} \hat{\beta}_j)^{(k)}(\lambda_0) = \left\langle R , \left. \sum_{t=0}^{\infty} \sum_{s=0}^{\infty} \alpha_i(t-l-s) \beta_j(s) (\lambda^t)^{(k)} \right|_{\lambda = \lambda_0} \right\rangle.$$

Furthermore, when λ_0 is complex, both the real and imaginary parts of the sequence in brackets annihilate the real sequence R independently. Therefore, for all (i, j, k) in the ranges established in Theorem 6.2.2 and all $\lambda_0 \in \Lambda_{UV}$, define

$$F_{ijk\lambda_0}(l) := \sum_{t=0}^{\infty} \sum_{s=0}^{\infty} \alpha_i(t - l - s)\beta_j(s)(\lambda^t)^{(k)}\bigg|_{\lambda=\lambda_0} \tag{11.7}$$

and

$$RF_{ijk\lambda_0} := \Re(F_{ijk\lambda_0}), \quad IF_{ijk\lambda_0} := \Im(F_{ijk\lambda_0}), \tag{11.8}$$

where $\Re(\cdot)$ and $\Im(\cdot)$ denote the real and imaginary parts, respectively. Then,

$$\langle R, RF_{ijk\lambda_0}\rangle = 0, \quad \langle R, IF_{ijk\lambda_0}\rangle = 0.$$

Notice that $RF_{ijk\lambda_0}$ and $IF_{ijk\lambda_0}$ are inside the space $\ell_\infty^{n_z \times n_w}$. If λ_0 is not on the boundary of the disc, then these sequences will be inside $c_0^{n_z \times n_w}$. In addition, any sequence in the linear span of the $RF_{ijk\lambda_0}$'s and the $IF_{ijk\lambda_0}$'s also annihilates every element $R \in S$. Note that there is only a finite number of sequences; thus the subspace spanned by the zero interpolation sequences is finite dimensional.

Next, we look at the rank interpolation conditions (i.e., the second set of conditions in Theorem 6.2.2). Again, these algebraic conditions can be viewed as convolution of matrix sequences:

$$(\hat{\alpha}_i \hat{R})(\lambda) \equiv 0 \iff (\alpha_i * R)(t) = \sum_{l=0}^{t} \alpha_i(t - l)R(l) = (0 \cdots 0) \quad \text{for all } t \geq 0,$$

$$(\hat{R}\hat{\beta}_j)(\lambda) \equiv 0 \iff (R * \beta_j)(t) = \sum_{l=0}^{t} R(l)\beta_j(t - l) = \begin{pmatrix} 0 \\ \vdots \\ 0 \end{pmatrix} \quad \text{for all } t \geq 0.$$

For $n_u + 1 \leq i \leq n_z$, $1 \leq q \leq n_w$ and $t \in \mathbf{Z}_+$, define the following sequence of $n_z \times n_w$ matrices:

$$G_{\alpha_i qt}(l) := \begin{pmatrix} \vdots & & \vdots & \overbrace{\vdots}^{q\text{th column}} & \vdots & & \vdots \\ 0 & \cdots & 0 & \alpha_i^T(t-l) & 0 & \cdots & 0 \\ \vdots & & \vdots & \vdots & \vdots & & \vdots \end{pmatrix}. \tag{11.9}$$

Similarly, for $n_y + 1 \le j \le n_w$, $1 \le p \le n_z$ and $t \in \mathbf{Z}_+$, define

$$
G_{\beta_j pt}(l) := \begin{pmatrix}
\cdots & 0 & \cdots \\
& \vdots & \\
\cdots & 0 & \cdots \\
\cdots & \beta_j^T(t-l) & \cdots \\
\cdots & 0 & \cdots \\
& \vdots & \\
\cdots & 0 & \cdots
\end{pmatrix} \}\text{pth row.} \qquad (11.10)
$$

Therefore, $\langle R, G_{\alpha_i qt} \rangle = 0$ and $\langle R, G_{\beta_j pt} \rangle = 0$ for all $t \ge 0$ and all $R \in \mathcal{S}$. Note that, in contrast with the zero interpolation sequences, the linear span of the $G_{\alpha_i qt}$'s and $G_{\beta_j pt}$'s is infinite dimensional since for every (i, q, p), t ranges over all nonnegative integers (i.e., $t \in \mathbf{Z}_+$). This observation turns out to be of great importance in connection with the computation of optimal solutions of multiblock problems.

The reader should realize at this point that the construction of the above sequences is simply a way to rewrite the algebraic conditions of Theorem 6.2.2 in a form compatible with the definition of how linear functionals act on sequences.

From the previous discussion it is clear that a given Φ is feasible (i.e., there exists a stable Q such that $\Phi = H - UQV$) if and only if

$$
\begin{cases}
\langle \Phi, RF_{ijk\lambda_0} \rangle = \langle H, RF_{ijk\lambda_0} \rangle \\
\langle \Phi, IF_{ijk\lambda_0} \rangle = \langle H, IF_{ijk\lambda_0} \rangle
\end{cases}
\text{for}
\begin{cases}
\lambda_0 \in \Lambda_{UV} \\
i = 1, \ldots, n_u \\
j = 1, \ldots, n_y \\
k = 0, \ldots, \sigma_{U_i}(\lambda_0) + \sigma_{V_j}(\lambda_0) - 1
\end{cases}
, \qquad (11.11)
$$

and

$$
\begin{cases}
\langle \Phi, G_{\alpha_i qt} \rangle = \langle H, G_{\alpha_i qt} \rangle \\
\langle \Phi, G_{\beta_j pt} \rangle = \langle H, G_{\beta_j pt} \rangle
\end{cases}
\text{for}
\begin{cases}
i = n_u + 1, \ldots, n_z \\
j = n_y + 1, \ldots, n_w \\
q = 1, \ldots, n_w \\
p = 1, \ldots, n_z \\
t = 0, 1, 2, \ldots
\end{cases}
. \qquad (11.12)
$$

Each of these linear functionals is equivalent to a linear equality constraint on the sequence Φ.

Based on this characterization, we define the linear operator $\mathcal{A}_{\text{zero}}$ mapping $\ell_1^{n_z \times n_w}$ to $\mathbb{R}^{c_z}$, to be the collected action of (i.e., stacking of) all functionals in Equation (11.11), and c_z is the total number of equality constraints associated with the zero interpolation conditions, and is given by

$$
c_z = \sum_{\lambda_0 \in \Lambda_{UV}} \sum_{i=1}^{n_y} \sum_{j=1}^{n_u} \sigma_{U_i}(\lambda_0) + \sigma_{V_j}(\lambda_0). \qquad (11.13)
$$

A note should be made on the way c_z is computed. If a given $\lambda_0 \in \Lambda_{UV}$ is complex then $\bar{\lambda}_0 \in \Lambda_{UV}$ too, since $\hat{U}$ and $\hat{V}$ are real-rational. However, for the purpose of constructing functionals, only one of each pair of complex-conjugate zeros should be considered since the other one would generate redundant functionals. But, for the purpose of counting the number of independent functionals (i.e., computing c_z), both zeros should be included in Λ_{UV}, since a complex-conjugate pair of zeros generates twice as many functionals as a real zero.

Similarly, define the linear operator $\mathcal{A}_{\text{rank}}$ mapping $\ell_1^{n_z \times n_w}$ to ℓ_1 to be the collected action of all functionals in Equation (11.12) on Φ and H. Then, $\Phi - H \in \mathcal{S}$ is equivalent to

$$\begin{aligned} \mathcal{A}_{\text{zero}}\Phi &= \mathcal{A}_{\text{zero}}H =: b_{\text{zero}} \in \mathbb{R}^{c_z}, \\ \mathcal{A}_{\text{rank}}\Phi &= \mathcal{A}_{\text{rank}}H =: b_{\text{rank}} \in \ell_1. \end{aligned} \tag{11.14}$$

Again, it is noted that while the range of $\mathcal{A}_{\text{zero}}$ is finite dimensional, the range of $\mathcal{A}_{\text{rank}}$ is not.

11.3 DUAL PROBLEM FORMULATION

Duality Theorem 9.3.4 takes the following form when applied to the problem of ℓ_1 optimization:

$$\nu^o = \inf_{R \in \mathcal{S}} \|H - R\|_1 = \max_{\substack{G \in \mathcal{S}^\perp \\ \|G\|_\infty \leq 1}} \langle H, G \rangle, \tag{11.15}$$

where $\mathcal{S}$ is given by

$$\mathcal{S} = \mathcal{N}\begin{pmatrix} \mathcal{A}_{\text{zero}} \\ \mathcal{A}_{\text{rank}} \end{pmatrix}. \tag{11.16}$$

To follow the development in Section 9.6.2, we note that

$$\mathcal{R}\begin{pmatrix} \mathcal{A}_{\text{zero}} \\ \mathcal{A}_{\text{rank}} \end{pmatrix} \tag{11.17}$$

is closed. This is a consequence of two basic properties of these operators: The first is that $\mathcal{A}_{\text{zero}}$ has a finite-dimensional range, and the second is that $\mathcal{A}_{\text{rank}}$ is constructed from operators involving unimodular matrices, i.e., having stable inverses. It follows that $\mathcal{S}^\perp \subset \ell_\infty^{n_z \times n_w}$ is given by

$$\mathcal{S}^\perp = \mathcal{R}\begin{pmatrix} \mathcal{A}_{\text{zero}}^* & \mathcal{A}_{\text{rank}}^* \end{pmatrix}.$$

From the explicit representation we derived earlier, it follows that $\mathcal{S}^\perp$ is the linear span of the sequences (11.8), (11.9), and (11.10). Hence, G is any element in this subspace with infinity norm not greater than one, i.e.,

$$G \in \text{span}\{RF_{ijk\lambda_0}, IF_{ijk\lambda_0}, G_{\alpha_i qt}, G_{\beta_j pt}\} \tag{11.18}$$

with the appropriate index ranges.

We can rewrite Problem (11.15) explicitly in terms of the closed-loop pulse response Φ:

$$\nu^o = \inf_{\Phi - H \in \mathcal{S}} \|\Phi\|_1 = \max_{\substack{G \in \mathcal{S}^\perp \\ \|G\|_\infty \leq 1}} \langle H, G \rangle. \tag{11.19}$$

The next definition and theorem formalize the discussion above.

Definition 11.3.1.   In the context of Problem (11.19), a sequence Φ is said to be feasible to the primal problem if $\Phi - H \in \mathcal{S}$. Similarly, a sequence G is said to be feasible to the dual problem if $G \in \mathcal{S}^\perp$ and $\|G\|_\infty \leq 1$.   ∎

Theorem 11.3.1.   Given any feasible sequences Φ and G, it follows that $\|\Phi\|_1 \geq \langle H, G \rangle$. Further, if equality is attained, then such sequences are optimal solutions of the primal and dual problems, respectively, and thus are aligned.

Proof. Using the fact that $\langle R, G \rangle = 0$ we have:

$$\langle H, G \rangle = \langle H - R, G \rangle$$
$$= \langle \Phi, G \rangle$$
$$\leq \|\Phi\|_1 \|G\|_\infty$$
$$\leq \|\Phi\|_1.$$

The rest follows from Theorem 9.3.4.   ∎

Next, it will be shown that Problem (11.19) is equivalent to a primal-dual pair of infinite dimensional linear programs, that can be written in standard form.

11.4 A PRIMAL-DUAL PAIR OF LINEAR PROGRAMS: THE MIMO CASE

At this point we could continue developing the theory in terms of the tensor notation employed so far in this chapter, or change to a more familiar and "computer-ready" matrix notation. This second approach will also prove to be convenient in analyzing and exploiting the internal structure of the problem. For these reasons we introduce the following notation.

Define an "ordering" operator $\mathcal{O}$ mapping $\ell_1^{n_z \times n_w}$ onto ℓ_1 as follows: It takes the rows of $\Phi(t)$ for $t = 0, 1, \ldots$, and stacks them up into an infinite column vector denoted by ϕ; that is,

$$\phi := \mathcal{O}\Phi := \begin{pmatrix} \phi_{11}(0) \\ \vdots \\ \phi_{1n_w}(0) \\ \phi_{21}(0) \\ \vdots \\ \phi_{2n_w}(0) \\ \vdots \\ \phi_{n_z 1}(0) \\ \vdots \\ \phi_{n_z n_w}(0) \\ \phi_{11}(1) \\ \vdots \end{pmatrix} \in \ell_1. \tag{11.20}$$

The linear operator $\mathcal{O}$ is one-to-one and onto, and its inverse is equal to its adjoint, i.e., $\Phi = \mathcal{O}^* \phi$. Denote by A_{zero} and A_{rank} the matrix representations of the linear operators $\mathcal{A}_{\text{zero}} \mathcal{O}^*$ and $\mathcal{A}_{\text{rank}} \mathcal{O}^*$, respectively. With this notation, feasibility of the sequence Φ reduces to the following:

$$\Phi - H \in \mathcal{S} \iff A_{\text{zero}} \phi = b_{\text{zero}} \quad \text{and} \quad A_{\text{rank}} \phi = b_{\text{rank}}. \tag{11.21}$$

Finally, the constraints $\|(\Phi)_i\|_1 \le \nu$ needs to be rewritten in matrix form. Recall the definition of the linear operator $\mathcal{A}_{\ell_1}$ introduced in Chapter 3. It maps $\ell_1^{n_z \times n_w}$ to $\mathbb{R}^{n_z}$ as follows: for $X \in \ell_1^{n_z \times n_w}$, $(\mathcal{A}_{\ell_1} X)_i := \sum_{j=1}^{n_w} \sum_{t=0}^{\infty} x_{ij}(t)$. Let A_{ℓ_1} denote the (infinite) matrix representation of $\mathcal{A}_{\ell_1} \mathcal{O}^*$. It follows that

$$\|\Phi\|_1 \le \nu \iff A_{\ell_1}(\phi^+ + \phi^-) \le \mathbf{1}\nu, \tag{11.22}$$

where $\mathbf{1}$ is a vector of ones of dimension n_z. As before, ϕ^+ and ϕ^- are nonnegative infinite vectors such that $\phi = \phi^+ - \phi^-$, and we assume that for every positive integer k, either $\phi^+(k)$ or $\phi^-(k)$ is zero. Note that A_{ℓ_1} is nothing but a matrix with n_z rows and infinite columns, with its entries being ones or zeros such that $A_{\ell_1}(\phi^+ + \phi^-)$ captures the ℓ_1 norm of each row of Φ (see Chapter 3).

Primal linear program.   In terms of the new matrix notation, Problem (11.19) takes the following form:

$$\nu^o = \inf_{\nu, \phi^+, \phi^-} \nu$$

subject to

$$\begin{aligned} A_{\ell_1}(\phi^+ + \phi^-) &\le \mathbf{1}\nu, \\ A_{\text{zero}}(\phi^+ - \phi^-) &= b_{\text{zero}}, \\ A_{\text{rank}}(\phi^+ - \phi^-) &= b_{\text{rank}}, \\ \phi^+ \ge 0 \,, \ \phi^- &\ge 0. \end{aligned} \tag{11.23}$$

The discussion in Section 9.6.2 shows that this problem has a dual linear program. It is sometimes convenient to convert all of the inequality constraints to equality constraints. For this purpose, define a vector of slack variables $\xi \in \mathbb{R}^{n_z}$ such that $A_{\ell_1}(\phi^+ + \phi^-) \leq \mathbf{1}\nu$ if and only if $A_{\ell_1}(\phi^+ + \phi^-) + \xi = \mathbf{1}\nu$ for some $\xi \geq 0$, and collect all relevant variables into a single (infinite) column vector $x := (\nu \ \ \xi \ \ \phi^+ \ \ \phi^-)^T$. Then,

$$\nu^o = \inf_x \langle x, c \rangle$$

subject to

$$Ax = b, \tag{11.24}$$
$$x \geq 0,$$
$$x \in \mathbb{R} \times \mathbb{R}^{n_z} \times \ell_1 \times \ell_1,$$

where

$$A := \begin{pmatrix} -\mathbf{1} & I & A_{\ell_1} & A_{\ell_1} \\ 0 & 0 & A_{\text{zero}} & -A_{\text{zero}} \\ 0 & 0 & A_{\text{rank}} & -A_{\text{rank}} \end{pmatrix}, \qquad b := \begin{pmatrix} 0 \\ b_{\text{zero}} \\ b_{\text{rank}} \end{pmatrix},$$

and $c := (1 \ 0 \ 0 \ \cdots)^T$. Problem (11.24) is finally in standard form.

A few observations are in order. First, Problem (11.24) shows that ℓ_1 optimal control problems are equivalent to linear programming problems with infinitely many variables (dimension of x) and infinitely many constraints (row dimension of A). Note, however, that if the problem is one-block, then A_{rank} is not present and A retains a finite range (i.e., finite number of equality constraints). Indeed, in such cases, the number of equality constraints equals $c_z + n_z$. Further, if Λ_{UV} is empty, then H can be matched exactly with a feasible R and the optimal solution is trivial (i.e., $\nu^o = 0$).

Dual linear program.   To complete this discussion, it remains to show that the right-hand side of Equation (11.15) is also equivalent to a linear programming problem. In fact, it can be shown that such a problem corresponds to the standard dual formulation of Problem (11.24). To illustrate this fact, we simply write the dual form of (11.24) and compare it with (11.15):

$$\nu^o = \max_\gamma \langle b, \gamma \rangle$$

subject to

$$A^T \gamma \leq c, \tag{11.25}$$
$$\gamma \in \ell_\infty,$$

where γ is an element in the dual space, ℓ_∞, and A^T is the adjoint (transpose) of A that maps ℓ_∞ to ℓ_∞. To get more insight into the structure of this problem, let us partition the dual vector γ according to the natural partitioning of A^T; that is, $\gamma =: (-\gamma_0 \ \gamma_1 \ \gamma_2)^T$, where $\gamma_0 \in \mathbb{R}^{n_z}$, $\gamma_1 \in \mathbb{R}^{c_z}$, and $\gamma_2 \in \ell_\infty$ (it is convenient to have the sign of γ_0 changed). Then, it is easy to show that the set of inequality constraints $A^T \gamma \leq c$ is equivalent to the following set:

$$\gamma_0 \geq 0, \quad \sum_{i=1}^{n_z} \gamma_0(i) \leq 1, \tag{11.26}$$

$$-A_{\ell_1}^T \gamma_0 \leq A_{\text{zero}}^T \gamma_1 + A_{\text{rank}}^T \gamma_2 \leq A_{\ell_1}^T \gamma_0, \tag{11.27}$$

while the objective function takes the form

$$\max_{\gamma} \quad \langle b_{\text{zero}}, \gamma_1 \rangle + \langle b_{\text{rank}}, \gamma_2 \rangle.$$

When we compare these equations with the dual form of Problem (11.15), the following equivalences become apparent:

1. γ_1 and γ_2 are nothing but the coefficients that combine the linear functionals associated with the zero interpolation conditions (columns of A_{zero}^T) and the rank interpolation conditions (columns of A_{rank}^T), respectively, to obtain G.

2. The set of inequalities (11.26, 11.27) is equivalent to $\|G\|_{\infty} \leq 1$. Recall the definition of $\| \cdot \|_{\infty}$ as a dual norm:

$$\|G\|_{\infty} = \sum_{i=1}^{n_z} \max_{1 \leq j \leq n_w} \|g_{ij}\|_{\infty}.$$

Then, Equation (11.27) bounds G pointwise, where the maximum of the ith row, given by $\gamma_0(i)$, is captured in the product $A_{\ell_1}^T \gamma_0$. Hence, Equation (11.26) completes the computation of $\|G\|_{\infty}$ and bounds it with one.

3. The objective function results from expanding $\langle H, G \rangle$, where G is expressed as a linear combination of the elements in the generator of $S^{\perp}$ with coefficients (γ_1, γ_2).

Having established the equivalence of ℓ_1 optimal control problems and (infinite dimensional) linear programs, both in their primal and dual formulations, we are ready to move on to the problem of finding a solution. This is the subject of the next chapter.

11.5 REDUNDANCY

In this section we explore the occurrence of redundant constraints in the setting of the standard linear program (see Equation (11.24)). We seek to show that the interpolation conditions of Theorem 6.2.2 are not only necessary and sufficient for the solvability of $\hat{R} = \hat{U}\hat{Q}\hat{V}$ in ℓ_1, but also are minimal in the sense that they generate a set of linearly independent equality constraints. This is a desirable characteristic from the point of view of algorithm efficiency. In fact, the set of rank interpolation conditions, as stated in Theorem 6.2.2, is not minimal when the problem is four block; however, these conditions can be easily restated without any redundancy, as we shall see in the sequel.

First consider one-block problems, in which the set of equality constraints is given by $A_{\text{zero}}(\phi^+ - \phi^-) = b_{\text{zero}}$ (Equation (11.23)). The simplest possible case is a SISO problem with n simple and real interpolation conditions (i.e., real zeros with algebraic multiplicity equal to one that are not common to U and V). It is easy to see that the matrix A_{zero} takes the following form (see Section 10.3.2):

$$A_{\text{zero}} = \begin{pmatrix} 1 & \lambda_1 & \lambda_1^2 & \lambda_1^3 & \cdots \\ 1 & \lambda_2 & \lambda_2^2 & \lambda_2^3 & \cdots \\ \vdots & & & & \vdots \\ 1 & \lambda_n & \lambda_n^2 & \lambda_n^3 & \cdots \end{pmatrix},$$

where $\Lambda_{UV} = \{\lambda_1, \lambda_2, \ldots, \lambda_n\}$. This matrix can be recognized as a (infinite) Vandermonde matrix, known to have full rank. Indeed, the determinant formed with the first n columns is equal to $\prod_{i>j}(\lambda_i - \lambda_j) \neq 0$. Thus, the generated constraints are linearly independent. This extends easily to the complex zero case.

Next, consider the SISO case with multiple or common real zeros. Say that λ_0 is such that $\sigma_U(\lambda_0) + \sigma_V(\lambda_0) = m > 1$. This zero, according to Theorem 6.2.2, generates m linearly independent constraints of the form

$$\begin{pmatrix} 1 & \lambda_0 & \lambda_0^2 & \cdots & \lambda_0^{m-1} & \cdots \\ 0 & 1 & 2\lambda_0 & \cdots & (m-1)\lambda_0^{m-2} & \cdots \\ \vdots & & & & & \vdots \\ 0 & 0 & 0 & \cdots & (m-1)! & \cdots \end{pmatrix}.$$

In fact, the determinant of the first m columns is independent of λ_0 and equal to

$$1! \, 2! \cdots (n-2)! \, (n-1)! \neq 0.$$

It turns out that the combined set of constraints (due to simple and multiple interpolations) is also linearly independent.

The MIMO case incorporates the directional properties of the zeros, characterized by the $\hat{\alpha}_i$'s and $\hat{\beta}_j$'s or by the equivalent left and right set of null chains. To avoid the complexity of the most general case, consider a MIMO problem with only left simple interpolations ($\hat{V}$ has no zeros in the disc), and further, consider the constraints on the first column of $\hat{R}$, denoted $\hat{r}$ (constraints on the different columns of $\hat{R}$ are independent). Let λ_0 be an unstable left zero of $\hat{U}$, then

$$\hat{\alpha}_{i_k}(\lambda_0)\hat{r}(\lambda_0) = 0 \quad \text{for} \quad k = 1, \ldots, m$$

are the zero interpolation conditions on the first column of $\hat{R}$, where m is the geometric multiplicity of λ_0. This results in the following matrix, representation of the constraints:

$$\begin{pmatrix} \hat{\alpha}_{i_1}(\lambda_0) & \lambda_0\hat{\alpha}_{i_1}(\lambda_0) & \lambda_0^2\hat{\alpha}_{i_1}(\lambda_0) & \cdots \\ \hat{\alpha}_{i_2}(\lambda_0) & \lambda_0\hat{\alpha}_{i_2}(\lambda_0) & \lambda_0^2\hat{\alpha}_{i_2}(\lambda_0) & \cdots \\ \vdots & & & \vdots \\ \hat{\alpha}_{i_m}(\lambda_0) & \lambda_0\hat{\alpha}_{i_m}(\lambda_0) & \lambda_0^2\hat{\alpha}_{i_m}(\lambda_0) & \cdots \end{pmatrix}.$$

Clearly, the rows of the above matrix are linearly independent since the set of vectors $\{\hat{\alpha}_{i_1}(\lambda_0), \ldots, \hat{\alpha}_{i_m}(\lambda_0)\}$ are linearly independent. This argument, combined with the properties of the Vandermonde matrix, can be extended for the general one-block MIMO case.

Next, we consider the rank interpolation conditions. Such conditions have the following algebraic form:

$$\begin{pmatrix} \hat{\alpha}_{n_u+1} \\ \vdots \\ \hat{\alpha}_{n_z} \end{pmatrix} \hat{R} \equiv 0 \quad \text{and} \quad \hat{R} \begin{pmatrix} \hat{\beta}_{n_y+1} & \cdots & \hat{\beta}_{n_w} \end{pmatrix} \equiv 0. \tag{11.28}$$

Recall the following expression from the proof of Theorem 6.2.2:

$$\hat{L}_U^{-1} \hat{R} \hat{R}_V^{-1} = \begin{pmatrix} \hat{\bar{\mathcal{E}}}_U \hat{\bar{Q}} \hat{\bar{\mathcal{E}}}_V & \mathbf{0} \\ \mathbf{0} & \mathbf{0} \end{pmatrix}. \tag{11.29}$$

The rank interpolations are responsible for generating the appropriate blocks of zeros in the equation above. Note, however, that the block of zeros on the diagonal of Equation (11.29) is "forced" twice, from the left and the right rank interpolation conditions. Thus, there is a certain degree of redundancy introduced by the set of algebraic constraints (11.28). To avoid it, consider the following partition of $\hat{R}$:

$$\hat{R} = \begin{pmatrix} \hat{R}_{11} & \hat{R}_{12} \\ \hat{R}_{21} & \hat{R}_{22} \end{pmatrix},$$

where $\hat{R}_{11}$ is $n_u \times n_y$. There are two ways of eliminating the aforementioned redundancy. These are given by the following conditions:

$$\begin{pmatrix} \hat{\alpha}_{n_u+1} \\ \vdots \\ \hat{\alpha}_{n_z} \end{pmatrix} \hat{R} \equiv 0 \quad \text{and} \quad \begin{pmatrix} \hat{R}_{11} & \hat{R}_{12} \end{pmatrix} \begin{pmatrix} \hat{\beta}_{n_y+1} & \cdots & \hat{\beta}_{n_w} \end{pmatrix} \equiv 0, \tag{11.30}$$

or, equivalently,

$$\begin{pmatrix} \hat{\alpha}_{n_u+1} \\ \vdots \\ \hat{\alpha}_{n_z} \end{pmatrix} \begin{pmatrix} \hat{R}_{11} \\ \hat{R}_{21} \end{pmatrix} \equiv 0 \quad \text{and} \quad \hat{R} \begin{pmatrix} \hat{\beta}_{n_y+1} & \cdots & \hat{\beta}_{n_w} \end{pmatrix} \equiv 0. \tag{11.31}$$

In this way, each block of zeros in (11.29) is "forced" only once and the redundancy is eliminated. These considerations are unnecessary for two-block problems.

The remaining question is whether the above reduced set of rank interpolation conditions generates an infinite set of linearly independent constraints. To be able to answer this question we need to review some of the properties of unimodular matrices. From the definition of unimodular matrices, it follows that the polynomial row and column vectors that make up a unimodular matrix form a polynomial basis for the space they generate (i.e., they are linearly independent over the field of rational functions). This is equivalent to the invertibility of such matrices at all points of the complex plane.

This observation implies that the set $\{\hat{\alpha}_{n_u+1}, \ldots, \hat{\alpha}_{n_z}\}$ is linearly independent and forms a basis for the left null-space of $\hat{U}$. Similarly, the set $\{\hat{\beta}_{n_y+1}, \ldots, \hat{\beta}_{n_w}\}$ forms a basis for the right null-space of $\hat{V}$. In this sense, the number of algebraic equations representing the rank interpolation conditions is minimal. Therefore, it follows that the set of constraints over each column or row of the sequence R, generated by such algebraic conditions, is linearly independent.

11.6 EXISTENCE

From Theorem 9.3.5, it is clear that the characterization of $\mathcal{S}$ as some annihilator space $M^\perp$ is enough to show existence. Section 11.2 suggests the possibility of splitting the subspace $\mathcal{S}$ into two larger subspaces: $\mathcal{S}_{(i)}$ containing all R's that satisfy the zero interpolation conditions, and $\mathcal{S}_{(ii)}$ containing all R's that satisfy the rank interpolation conditions. Then, $\mathcal{S} = \mathcal{S}_{(i)} \cap \mathcal{S}_{(ii)}$, and the problem reduces to establishing the equivalence $\mathcal{S}_{(i)} = M_{(i)}^\perp$ and $\mathcal{S}_{(ii)} = M_{(ii)}^\perp$. With this in mind we present the main existence theorem.

Theorem 11.6.1.   If every $\lambda_0 \in \Lambda_{UV}$ is in the open unit disk, then there exists a $R^o \in \ell_1^{n_z \times n_w}$ such that

$$\|H - R^o\|_1 = \inf_{R \in \mathcal{S}} \|H - R\|_1.$$

Proof. First we show that $\mathcal{S}_{(i)} = M_{(i)}^\perp$ for some $M_{(i)} \subset c_0^{n_z \times n_w}$. It is easy to see that, with the stated condition on λ_0 (i.e., $\lambda_0 \in D$), Equations (11.8) represent sequences in $c_0^{n_z \times n_w}$ that are annihilated by every element of $\mathcal{S}_{(i)}$. Let $M_{(i)}$ be the span of such sequences. Then, from Definition 9.3.3 and the fact that $M_{(i)}$ is finite dimensional, we conclude that $M_{(i)}^\perp = \mathcal{S}_{(i)}$. Moreover, $\mathcal{S}_{(i)}$ is $weak^*$-closed in $\ell_1^{n_z \times n_w}$ (Theorem 9.5.3).

Next, consider the sequences defined by Equations (11.9) and (11.10). It is not as easy to identify $M_{(ii)}$ since the linear span of such sequences is infinite dimensional. For simplicity, consider left rank interpolation conditions only (the proof for the right rank interpolation conditions is identical). Define a bounded linear operator $T : \ell_1^{n_z \times n_w} \to \ell_1^{(n_z - n_u) \times n_w}$ as follows: Given $R \in \ell_1^{n_z \times n_w}$,

$$TR := \begin{pmatrix} \alpha_{n_u+1} \\ \vdots \\ \alpha_{n_z} \end{pmatrix} * R.$$

Then, $\mathcal{N}(T) = \mathcal{S}_{(ii)}$. Now, define a bounded linear operator $W : c_0^{(n_z - n_u) \times n_w} \to c_0^{n_z \times n_w}$ in the following way: Given $x \in c_0^{(n_z - n_u) \times n_w}$,

$$Wx(t) := \sum_{l=0}^{\infty} T^T(l - t)x(l).$$

It follows by direct computation that $T = W^*$; that is, $\langle Wx, R \rangle = \langle x, TR \rangle$. Therefore, from Theorem 9.5.5, we have that $\mathcal{S}_{(ii)} = \mathcal{N}(T) = \mathcal{N}(W^*) = \mathcal{R}(W)^\perp$. Thus, $M_{(ii)} = \mathcal{R}(W)$ is exactly the linear span of the sequences defined in (11.9) and (11.10) but with the sequence of coefficients in c_0. Note that a direct consequence of this is that $\mathcal{S}_{(ii)}$ is $weak^*$-closed in $\ell_1^{n_z \times n_w}$, which then implies the $weak^*$-closedness of the whole subspace $\mathcal{S}$. We will refer to this fact in the future.   ∎

As a final note to this section, it should be pointed out that although solutions exist (given that the λ_0's are not on the boundary of $\overline{\mathcal{D}}$) they are in general not unique. Examples of nonunique solutions were presented in the previous chapter.

11.7 SUMMARY

We have discussed in detail how to rewrite an ℓ_1 MIMO model matching problem as an infinite dimensional linear program, in primal or dual form. By writing explicit formulae we were able to show the particular infinite matrix structure of such linear programs. This will be useful for finding solutions in the next chapter. We also showed that the interpolation conditions are both necessary and sufficient, and minimal in the sense that are equivalent to an infinite but nonredundant set of linear constraints. Finally, the issue of existence was addressed. As in the SISO case, the absence of zero interpolations on the unit circle guarantees the existence of optimal solutions.

EXERCISES

11.1. Suppose Φ_i is the closed loop map from w_i to z_i, $i = 1, 2$. Show how one can set up the constraint optimization problem

$$\min_{\text{K stablizing}} \quad \{\|\Phi_1\|_1 | \, \|\Phi_2\|_1 \leq \epsilon\}$$

as an infinite dimensional LP.

11.2. Derive the dual formulation of the problem above. Is there a duality gap? Relate this problem to the standard problem

$$\min_{\text{K stablizing}} \left\| \begin{pmatrix} \Phi_1 \\ \frac{1}{\epsilon}\Phi_2 \end{pmatrix} \right\|_1$$

and its dual. Discuss existence of both primal and dual solutions.

11.3. As we have shown in Exercise 10.18, solutions to continuous-time $\mathcal{L}_1$ problems can be irrational for rational plants and weights. Hence, controllers have to be approximated by rational ones for implementation purposes. In this exercise, we will highlight a method that allows us to use standard discrete-time techniques to find rational $\mathcal{L}_1$ suboptimal controllers.

Let G denote the continuous-time system. The Euler approximation of this system with a sampling interval τ corresponds to the discrete-time system G^E such that $\hat{G}^E(\lambda) := \hat{G}(\frac{\lambda^{-1}-1}{\tau})$ where $\hat{G}(s)$ is the Laplace transform of G. Two facts hold for this approximation:

Fact 1: For any given system G with a realization (A, B, C, D), and any $\tau > 0$, if G^E is stable, then G is stable and

$$\|G\|_{\mathcal{L}1} \leq \|G^E\|_{\ell 1}.$$

Fact 2: If G is stable with $\|G\|_{\mathcal{L}1} = \nu$, then

$$\lim_{r \to 0} \|G^E\|_{\ell 1} = \nu.$$

(a) Let G denote the continuous-time 2-input 2-output system. Show that if G has the state-space realization:

$$\left[\begin{array}{c|cc} A & B_1 & B_2 \\ \hline C_1 & D_{11} & D_{12} \\ C_1 & D_{21} & D_{22} \end{array} \right],$$

then G^E corresponds to

$$\left[\begin{array}{c|cc} I + \tau A & \tau B_1 & \tau B_2 \\ \hline C_1 & D_{11} & D_{12} \\ C_1 & D_{21} & D_{22} \end{array} \right].$$

(b) Show that for any rational controller $\hat{K}(s)$ and any closed loop map $\Phi = F_\ell(G, K)$, we have

$$\Phi^E = F_\ell(G, K)^E = F_\ell(G^E, K^E).$$

(c) Show that every rational discrete-time controller K designed for G^E corresponds to a rational continuous-time controller K^c such that

$$\|F_\ell(G, K^c)\|_{\mathcal{L}1} \leq \|F_\ell(G^E, K)\|_{\ell 1}.$$

Find $\hat{K}^c(s)$ as a function of $\hat{K}(\lambda)$.

(d) Propose a scheme for synthesizing a continuous-time suboptimal controller with any desired accuracy by combining this method and the standard discrete-time design based on the zero-order hold equivalence discussed in Chapter 4. Notice that the latter provides lower bounds on the optimal solution.

(e) Apply this scheme to Exercise 10.18 and compare answers.

NOTES AND REFERENCES

The extension of the SISO ℓ_1 problem to the one-block MIMO problem was reported in [DP87a]. The setup for the multiblock problem was suggested in [DP88a] in which closed-loop maps were characterized by both zeros and relations, and was further developed in [MP91], which resulted in a systematic way for setting up the problem based on the coprime factorizations method suggested in Exercise 6.3; see also [Dah93]. Further improvements and results were reported in [Sta90] in which the issue of redundancy was raised. Setting up the problem directly from the Smith-McMillan decomposition (the least redundant method) was proposed in [DBD93]. The conversion of a general ℓ_1 problem to a linear program (infinite-dimensional) was reported in [Men89]. The operator form notation is similar to that in [DBD93, Sta90]. The setup of the dual problem and the proof of existence was reported in [DP88a, MP91].

CHAPTER 12 ——————

Solution of the MIMO ℓ_1 Model Matching Problem

From an engineering point of view, it is not as interesting to prove the existence of optimal solutions as it is to device computable methods for finding (sub-)optimal solutions. Because the problems at hand are infinite dimensional, in general we will look for finite-dimensional approximations that will enable us to compute suboptimal solutions. The accuracy of such approximations will be often proportional to the computational burden. This chapter addresses the issue of finding such optimal or suboptimal solutions in a systematic way.

12.1 THE ONE-BLOCK PROBLEM

In the previous chapter we have established the equivalence between the primal-dual formulation of a general ℓ_1-optimal multiblock problem (Problem (11.19)) and an infinite dimensional but otherwise standard primal-dual pair of linear programs (Problems (11.24) and (11.25)).

This section provides a complete treatment of the one-block problem. In essence, it is a detailed discussion of the following observation: The dual Linear Program is finite dimensional. In this sense, this section is a straightforward extension of the SISO Model Matching Problem.

To avoid problems of existence, we make the following assumption valid for the rest of the chapter:

Assumption 1:   Every $\lambda_0 \in \Lambda_{UV}$ is inside the open unit disk.

12.1.1 Dual Problem

The one-block problem has a very specific interpolation structure, namely no rank interpolation conditions. From a primal formulation point of view (see Problem (11.24)), this simplifies the problem significantly by bringing the number of equality constraints down to a finite value, $c_z + n_z$. There remains, however, an infinite number of variables represented by x in ℓ_1. Nevertheless, we will show by looking at the structure of the dual problem, that the underlying problem is finite dimensional. Indeed, the dual formulation has an infinite number of inequality constraints but retains a finite number of variables:

$$\nu^o = \max_{\gamma_0, \gamma_1} \langle b_{\text{zero}}, \gamma_1 \rangle$$

subject to

$$\gamma_0 \geq 0, \quad \sum_{i=1}^{n_z} \gamma_0(i) \leq 1, \tag{12.1}$$

$$-A_{\ell_1}^T \gamma_0 \leq A_{\text{zero}}^T \gamma_1 \leq A_{\ell_1}^T \gamma_0.$$

Recall that $A_{\ell_1}^T$ is the matrix representation of an operator mapping $\mathbb{R}^{n_z}$ to ℓ_∞ while A_{zero}^T is the matrix representation of an operator mapping $\mathbb{R}^{c_z}$ to ℓ_∞. However, with Assumption 1 holding, the actual range of A_{zero}^T is in c_0 since each of the columns of A_{zero}^T is in c_0 and there are only finitely many of them.

The operators A_{ℓ_1} and A_{zero}.   Before going into the next result, it is useful to look at the matrices A_{ℓ_1} and A_{zero} in more detail. Recall that the matrix A_{ℓ_1} simply collects and sums the appropriate elements of the infinite vector ϕ to construct the ℓ_1 norm of each row of Φ. So in general A_{ℓ_1} has n_z rows and infinite columns with the following structure: The first row is given by

$$(A_{\ell_1})_1 = (\mathbf{1}\ \mathbf{0}\ \mathbf{1}\ \mathbf{0}\ \cdots),$$

where $\mathbf{1}$ is a row vector of n_w ones and $\mathbf{0}$ is a row vector of $n_w(n_z - 1)$ zeros. Then, the ith row, $(A_{\ell_1})_i$, is equal to the first row shifted $n_w(i - 1)$ times, and the product $A_{\ell_1}^T \gamma_0$ is given by

$$A_{\ell_1}^T \gamma_0 = (\overbrace{\underbrace{\gamma_0(1) \cdots \gamma_0(1)}_{}\ \gamma_0(2) \cdots \gamma_0(2)\ \cdots\ \gamma_0(n_z) \cdots \gamma_0(n_z)}^{n_w}\ \gamma_0(1) \cdots \gamma_0(1)\ \cdots)^T.$$

$$\underset{n_z \times n_w}{}$$

Based on this structure we define a collection of $n_z n_w$ auxiliary submatrices of A_{zero}. Let M_{ij}^T be an infinite matrix mapping $\mathbb{R}^{c_z}$ to c_0, formed by collecting the subsequence of rows of A_{zero}^T that multiply $\phi_{ij}(k)$, $k \geq 0$, in the product $A_{\text{zero}}\phi$. That is,

$$(M_{ij}^T)_k = (A_{\text{zero}}^T)_{n_z n_w k + n_w(i-1) + j} \quad \text{for} \quad \begin{cases} i = 1, \ldots, n_z \\ j = 1, \ldots, n_w \\ k = 0, 1, \ldots \end{cases}.$$

It follows that the set of inequalities $|(M_{ij}^T \gamma_1)(k)| \leq \gamma_0(i)$ for the appropriate range of indices is equivalent to the set $|(A_{\text{zero}}^T \gamma_1)(k)| \leq (A_{\ell_1}^T \gamma_0)(k)$ for $k = 0, 1, \ldots$ and can effectively replace them. Therefore, the dual problem (12.1) can be written as

$$\nu^o = \max_{\gamma_0, \gamma_1} \langle b_{\text{zero}}, \gamma_1 \rangle$$

subject to

$$\gamma_0 \geq 0 \ , \ \sum_{i=1}^{n_z} \gamma_0(i) \leq 1,$$

$$-\gamma_0(i) \leq (M_{ij}^T \gamma_1)(k) \leq \gamma_0(i) \quad \text{for} \quad \begin{cases} i = 1, \ldots, n_z \\ j = 1, \ldots, n_w \\ k = 0, 1, \ldots \end{cases}.$$

In the sequel, this notation is used to expose more details of the primal-dual pair of infinite-dimensional linear programs corresponding to the one-block problem. As in the SISO case, this will reveal the underlying finite-dimensional structure of the problem.

12.1.2 Primal-Dual LP Solution

The next lemma is a generalization of the result presented in Section 10.3.2:

Lemma 12.1.1. Let M^T be a full column rank infinite matrix mapping $\mathbb{R}^n$ to c_0. Then there exists a positive integer N such that

$$\|(I - P_N)M^T x\|_\infty < \|P_N M^T x\|_\infty$$

for all nonzero $x \in \mathbb{R}^n$.

Proof. For some $\tilde{N} \geq n$, let D be the $n \times n$ matrix formed with n linearly independent rows from the first $\tilde{N}$ rows of M^T. Then, for any x, let $y = Dx$ and therefore $x = D^{-1}y$. Thus,

$$\|(I - P_k)M^T x\|_\infty \leq \|(I - P_k)M^T\|_1 \|D^{-1}\|_1 \|y\|_\infty$$

for all $k = 0, 1, \ldots$ But, since the columns of M^T decay to zero, there exists a positive integer $N > \tilde{N}$ such that $\|(I - P_N)M^T\|_1 \|D^{-1}\|_1 < 1$. Then,

$$\|(I - P_N)M^T x\|_\infty < \|y\|_\infty \leq \|P_{\tilde{N}} M^T x\|_\infty \leq \|P_N M^T x\|_\infty.$$

Note, in particular, that N is independent of x and is only a function of M. ∎

In other words, the above lemma states that given a matrix mapping a finite-dimensional space to c_0, it is always possible to bound the index at which the infinity norm of any sequence in the range is achieved.

The following theorem proves a very interesting result by exploiting the structure of the matrices M_{ij}'s.

Theorem 12.1.1.   The exact solution of a one-block ℓ_1-optimal control problem is given by the following finite-dimensional (dual) linear program,

$$\nu^o = \max_{\gamma_0, \gamma_1} \langle b_{\text{zero}}, \gamma_1 \rangle$$

subject to

$$\gamma_0 \geq 0 \ , \quad \sum_{i=1}^{n_z} \gamma_0(i) \leq 1, \tag{12.2}$$

$$-\gamma_0(i) \leq (M_{ij}^T \gamma_1)(k) \leq \gamma_0(i) \quad \text{for} \quad \begin{cases} i = 1, \ldots, n_z \\ j = 1, \ldots, n_w \\ k = 0, \ldots, N_{ij} < \infty \end{cases}.$$

Proof. Form matrices M_{ij}^T as defined before. Assume they have full column rank (if not, reduce the number of columns). Apply Lemma 12.1.1 to each M_{ij}^T and let N_{ij} denote the corresponding index bound. Then, we claim that for every feasible solution of Problem (12.1) all inequalities of the form $|(M_{ij}^T \gamma_1)(k)| \leq \gamma_0(i)$ for $k > N_{ij}$ are inactive constraints (i.e., the inequality is strict) and they can be ignored in the solution. Indeed, by Lemma 12.1.1, if there is an active constraint for $k > N_{ij}$, then there must have been a violation of a constraint for some $k < N_{ij}$ since the ℓ_∞ norm of the sequence $M_{ij}^T \gamma_1$ is attained before N_{ij} and is always bounded by $\gamma_0(i)$. ■

This fact has an immediate and important implication on the primal linear programming formulation of one-block problems. Owing to the alignment conditions, if a dual optimal solution is such that all inequality constraints are inactive for $k > N$, then the primal optimal solution vanishes for $k > N$. The next corollary exploits this fact in the context of one-block problems.

Corollary 12.1.1.   For any one-block problem, the ℓ_1-optimal closed-loop response, Φ^o, has finite support (i.e., finite pulse response).

Proof. Simply write the primal form of (12.2) by noting that $|(M_{ij}^T \gamma_1)(k)| < \gamma_0(i)$ implies that both $\phi_{ij}^+(k)$ and $\phi_{ij}^-(k)$ are zero in the optimal solution:

$$\nu^o = \min_{\nu, \xi, \tilde{\phi}_{ij}^+, \tilde{\phi}_{ij}^-} \nu$$

subject to

$$\xi(i) + \sum_{j=1}^{n_w} \sum_{k=0}^{N_{ij}} \tilde{\phi}_{ij}^+(k) + \tilde{\phi}_{ij}^-(k) = \nu \quad \text{for} \ i = 1, \ldots, n_z, \tag{12.3}$$

$$\sum_{i=1}^{n_z} \sum_{j=1}^{n_w} \tilde{M}_{ij}(\tilde{\phi}_{ij}^+ - \tilde{\phi}_{ij}^-) = b_{\text{zero}},$$

$$\xi, \tilde{\phi}_{ij}^+, \tilde{\phi}_{ij}^- \geq 0,$$

where $\tilde{\phi}_{ij}$ is a finite vector formed with the first N_{ij} elements of ϕ_{ij}, and $\tilde{M}_{ij}$ is a finite matrix formed with the first N_{ij} columns of M_{ij}. Then, $\Phi^o(k) = 0$ for k greater than $\max_{ij}(N_{ij})$. ■

Note that the N_{ij}'s provide a priori bounds on the lengths of the optimal ϕ_{ij}'s. Moreover, these bounds are independent of H and only depend on the zero interpolation structure of the problem.

Another interesting aspect of the primal optimal solution is that, if such solution is basic, it will have at most as many nonzero variables as the number of equality constraints. So, generically, the number of nonzero elements in Φ^o is bounded by $c_z + n_z - 1$ owing to the fact that ν^o is nonzero in any nontrivial case. Furthermore, if the solution is nondegenerate in ϕ and $\xi^o = 0$, then the bound is tight.

We conclude this section with the following corollary that suggests that most one-block problems have optimal solutions with all row-norms equal to ν^o.

Corollary 12.1.2.   Given a one-block problem and some $i \in \{1, \ldots, n_z\}$, if for any $j \in \{1, \ldots, n_w\}$ the matrix M_{ij}^T has full column rank, then $\|(\Phi^o)_i\|_1 = \nu^o$.

Proof. Assume $\|(\Phi^o)_i\|_1 < \nu^o$, then $\xi(i) > 0$. By the alignment conditions, this implies that $\gamma_0(i) = 0$, and in view of Equation (12.2) and the rank condition on M_{ij}^T, γ_1 must be zero. But this implies that $\nu^o = 0$, which is a contradiction. ∎

It should be noted that there are some pathological cases where the rank condition on M_{ij}^T is violated. For instance, if the given one-block problem is a combination of two or more totally decoupled subproblems, then some M_{ij}^T's will have entire columns of zeros. In most cases, however, the solution is such that the norm of each row of Φ^o is equal to ν^o.

12.2 THE MULTIBLOCK PROBLEM

In solving the one-block problem, we have exploited the fact that the primal linear programming problem has only finitely many equality constraints (or, equivalently, the dual problem has finitely many variables). For multiblock problems, however, both primal and dual problems have an infinite number of variables and constraints (see Equations (11.24) and (11.25)). So in principle, one can attempt to get approximate solutions by an appropriate truncation of the original problem. There are basically three approximation methods:

1. The first one results from constraining the closed-loop response $\hat{\Phi}(\lambda)$ to be a matrix polynomial of some given order, thus providing a suboptimal polynomial feasible solution to the problem. We will call this the *Finitely Many Variables* (FMV) approximation.

2. In the second approach, the dual variables are approximated by a finite-length vector, which is equivalent to solving the primal problem with only finitely many equality constraints retained. The solution to this problem is superoptimal but infeasible. Its value is complementary to the first approach in the sense that it

generates lower bounds of the optimal norm, ν^o. We will call this approach the *Finitely Many Equations* (FME) approximation.

3. A third approximation scheme can be obtained by embedding the problem into a one-block problem through augmenting the operators U and V with delays. It will be shown that this method carries more information about the structure of the optimal solution than the previously described methods. We will call this approach the *Delay Augmentation* (DA) approximation.

The next few sections provide a more detailed description of these methods along with their main characteristics.

12.2.1 The FMV and FME Approximation Methods

Before getting into the specifics of these methods, let us extend the notation developed for the one-block problem to the multiblock case. Analogous to the definition of M_{ij}^T, define a collection of $n_z n_w$ infinite matrices mapping ℓ_∞ to ℓ_∞ and denoted by $\bar{M}_{ij}^T$, by collecting the appropriate rows of A_{rank}^T, as the rows of A_{zero}^T were collected in M_{ij}^T. (An example, where these matrices are explicitly written down, is shown later in this chapter.) Then, in the context of the dual LP formulation, the following sets of constraints are equivalent:

$$-(A_{\ell_1}^T \gamma_0)(k) \le (A_{\text{zero}}^T \gamma_1 + A_{\text{rank}}^T \gamma_2)(k) \le (A_{\ell_1}^T \gamma_0)(k) \quad \text{for} \quad k = 0, 1, \ldots$$

$$\Updownarrow$$

$$-\gamma_0(i) \le (M_{ij}^T \gamma_1 + \bar{M}_{ij}^T \gamma_2)(k) \le \gamma_0(i) \quad \text{for} \quad \begin{cases} i = 1, \ldots, n_z \\ j = 1, \ldots, n_w \\ k = 0, 1, \ldots \end{cases},$$

and in the context of the primal LP formulation,

$$A_{\ell_1}(\phi^+ + \phi^-) + \xi = \mathbf{1}\nu,$$
$$A_{\text{zero}}(\phi^+ - \phi^-) = b_{\text{zero}},$$
$$A_{\text{rank}}(\phi^+ - \phi^-) = b_{\text{rank}},$$

$$\Updownarrow$$

$$\xi(i) + \sum_{j=1}^{n_w} \sum_{k=0}^{\infty} \phi_{ij}^+(k) + \phi_{ij}^-(k) = \nu \quad \text{for} \quad i = 1, \ldots, n_z,$$

$$\sum_{i=1}^{n_z} \sum_{j=1}^{n_w} M_{ij}(\phi_{ij}^+ - \phi_{ij}^-) = b_{\text{zero}},$$

$$\sum_{i=1}^{n_z} \sum_{j=1}^{n_w} \bar{M}_{ij}(\phi_{ij}^+ - \phi_{ij}^-) = b_{\text{rank}}.$$

The FMV method. With this notation, given a certain order of approximation N, the FMV primal formulation is given by the following linear program:

$$\bar{\nu}_N := \min_{\nu, \xi, \phi_{ij}^+, \phi_{ij}^-} \nu$$

subject to

$$\xi(i) + \sum_{j=1}^{n_w} \sum_{k=0}^{N} \phi_{ij}^+(k) + \phi_{ij}^-(k) = \nu \quad \text{for} \quad i = 1, \ldots, n_z,$$

$$\sum_{i=1}^{n_z} \sum_{j=1}^{n_w} M_{ij}(\phi_{ij}^+ - \phi_{ij}^-) = b_{\text{zero}}, \tag{12.4}$$

$$\sum_{i=1}^{n_z} \sum_{j=1}^{n_w} \bar{M}_{ij}(\phi_{ij}^+ - \phi_{ij}^-) = b_{\text{rank}},$$

$$\phi_{ij}^+(k) = \phi_{ij}^-(k) = 0 \quad \text{for} \quad k > N,$$

$$\xi, \phi_{ij}^+, \phi_{ij}^- \geq 0.$$

Note that without the constraints $\phi_{ij}^+(k) = \phi_{ij}^-(k) = 0$ for $k > N$, Problem (12.4) is equivalent to the full (nontruncated) optimization problem. Clearly, the added constraints will make

$$\bar{\nu}_N \geq \nu^o$$

in general. It is yet unclear, however, if the resulting problem is finite dimensional or not, since we still carry an infinite number of constraints. A closer look at the matrices $\bar{M}_{ij}$ will answer this question.

FMV as a finite-dimensional linear program. Recall that these matrices represent the rank interpolation conditions of the form (see Theorem 6.2.2):

$$\begin{pmatrix} \alpha_{n_u+1} \\ \vdots \\ \alpha_{n_z} \end{pmatrix} * \Phi = \begin{pmatrix} \alpha_{n_u+1} \\ \vdots \\ \alpha_{n_z} \end{pmatrix} * H$$

and

$$\Phi * \begin{pmatrix} \beta_{n_y+1} & \cdots & \beta_{n_w} \end{pmatrix} = H * \begin{pmatrix} \beta_{n_y+1} & \cdots & \beta_{n_w} \end{pmatrix},$$

where the results from the right-hand side convolutions are collected in the infinite vector b_{rank}. The matrix representation of the convolution of the α_i's and β_j's on the different entries of Φ, say ϕ_{ij}, is precisely given by $\bar{M}_{ij}$. Therefore, such infinite matrices will have a band structure inherited from the fact that the $\hat{\alpha}_i(\lambda)$'s and $\hat{\beta}_j(\lambda)$'s are polynomials. To illustrate this point, consider the following simple two-block row example, where

$$\begin{pmatrix} \phi_{11} & \phi_{12} \end{pmatrix} = \begin{pmatrix} h_{11} & h_{12} \end{pmatrix} - q\begin{pmatrix} v_1 & v_2 \end{pmatrix}, \quad q \in \ell_1,$$

and assume (without loss of generality) that $\hat{v}_1$ and $\hat{v}_2$ are coprime polynomials (i.e., $\hat{V}$ has no zeros). Then, the rank interpolation conditions take the following algebraic form:

$$\hat{\phi}_{11}\hat{v}_2 - \hat{\phi}_{12}\hat{v}_1 = \hat{h}_{11}\hat{v}_2 - \hat{h}_{12}\hat{v}_1$$

since $\hat{\beta}_1 = (\hat{v}_2 \quad -\hat{v}_1)$. It is easy to see that the matrices $\bar{M}_{11}$ and $\bar{M}_{12}$ have a band structure. For instance,

$$\bar{M}_{11} = \begin{pmatrix} v_2(0) & 0 & 0 & \cdots \\ v_2(1) & v_2(0) & 0 & \cdots \\ \cdots & v_2(1) & v_2(0) & \\ \vdots & \vdots & \vdots & \\ v_2(n) & \cdots & v_2(1) & \ddots \\ 0 & v_2(n) & & \ddots & \ddots \\ 0 & 0 & v_2(n) & \ddots \\ \vdots & \vdots & & \ddots \end{pmatrix}$$

where n is the order of the polynomial $\hat{v}_2(\lambda)$.

In view of this particular structure, forcing $\phi_{ij}(k) = 0$ for $k > N$ will make the product $(\bar{M}_{ij}\phi_{ij})(k)$ eventually vanish for $k > N + constant$, where the constant depends on the order of the polynomial $\hat{\alpha}(\lambda)$'s and $\hat{\beta}(\lambda)$'s. If, however, the infinite vector b_{rank} is not zero at that point, then the equality constraints will be violated for any Φ, implying that the added constraints have transformed the feasible set into an empty set and that the linear program has no solution. Furthermore, this will always be the case if b_{rank} has infinite support, no matter how large N is chosen to be. This leads to the following theorem and corollary (see Exercise 6.4).

Theorem 12.2.1.   Given a multiblock problem, there exists a finitely supported feasible solution, Φ, if and only if $\alpha_i * H$ and $H * \beta_j$ are finitely supported for $i = n_u + 1, \ldots, n_z$ and $j = n_y + 1, \ldots, n_w$.

Proof. For simplicity we consider problems with no right-rank interpolations. The general case is a straightforward extension of the argument in this proof. For the "only if" part of the proof, consider the following:

$$\alpha_i * \Phi = \alpha_i * H - \alpha_i * R = \alpha_i * H.$$

But $\alpha_i * \Phi$ is finitely supported since both factors are; then $\alpha_i * H$ must be finitely supported.

The "if" part requires a little more work. Define the following partitions of $\hat{L}_U^{-1}$:

$$\hat{L}_{U,1}^{-1} := \begin{pmatrix} \hat{\alpha}_1 \\ \vdots \\ \hat{\alpha}_{n_u} \end{pmatrix}, \qquad \hat{L}_{U,2}^{-1} := \begin{pmatrix} \hat{\alpha}_{n_u+1} \\ \vdots \\ \hat{\alpha}_{n_z} \end{pmatrix}.$$

Then, consider the following transformed two-block column problem:

$$\begin{pmatrix} \Phi_1' \\ \Phi_2' \end{pmatrix} = \begin{pmatrix} L_{U,1}^{-1} H \\ L_{U,2}^{-1} H \end{pmatrix} - \begin{pmatrix} L_{U,1}^{-1} R \\ 0 \end{pmatrix}.$$

Clearly, Φ_2' is finitely supported since $L_{U,2}^{-1} H$ is (by assumption). Also, note that the first block row defines a one-block problem: $\Phi_1' = H_1' - R'$, where $H_1' := L_{U,1}^{-1} H$ and $R' := L_{U,1}^{-1} R$. Then, there exists an R' that satisfies the zero interpolation conditions, such that Φ_1' is finitely supported (e.g., the ℓ_1-optimal solution). For this choice of R', let

$$R = L_U \begin{pmatrix} R' \\ 0 \end{pmatrix} \implies L_{U,1}^{-1} R = R',$$

then R is feasible such that

$$\begin{pmatrix} \Phi_1' \\ \Phi_2' \end{pmatrix} = L_U^{-1}(H - R)$$

is finitely supported. Therefore, $\Phi = H - R$ is finitely supported since $\hat{L}_U$ is a polynomial matrix. ∎

Corollary 12.2.1. Given a positive integer N, the FMV Problem (12.4) has a nonempty feasible set and therefore a solution, if and only if $(\alpha_i * H)(k) = 0$ and $(H * \beta_j)(k) = 0$ for $k > N + constant$, $i = n_u + 1, \ldots, n_z$ and $j = n_y + 1, \ldots, n_w$, where the constant depends on the order of $\hat{\alpha}_i$ and $\hat{\beta}_j$. ∎

It is clear from the above results that there is a class of multiblock problems for which the FMV method fails regardless of the order of approximation N. Also, given any multiblock problem, there is in general a lower-bound for N under which the FMV method fails. Without overlooking these limitations, we are going to assume for the rest of this subsection that the problems at hand allow polynomial feasible solutions and that N is large enough to capture at least one such solution.

Under these assumptions, all but finitely many constraints in Problem (12.4) are satisfied trivially, so that the problem is in effect a finite-dimensional linear program.

Convergence of FMV. Assuming the existence of finitely supported feasible solutions, the FMV method has a nice convergence property.

Theorem 12.2.2. In the FMV method, $\bar{\nu}_N \to \nu^o$ as $N \to \infty$.

Proof. Consider Problem (12.4). Increasing N is equivalent to removing constraints on the sequence Φ; therefore, $\{\bar{\nu}_N\}$ forms a nonincreasing sequence bounded from below by ν^o. The result follows from the fact that the set of finitely supported sequences are dense in ℓ_1 (i.e., every sequence in ℓ_1 can be approximated by an FIR to any degree of accuracy by simply truncating the tail of the sequence). Let H_p be a finitely supported (i.e., polynomial) feasible solution (assumed to exist), then the optimization problem can always be written as

$$v^o = \inf_{Q \in \ell_1^{n_u \times n_y}} \| H_p - UQV \|_1.$$

Next, by Corollary 4.1.1, write UQV as $L_U \mathcal{E}_U \tilde{Q} \mathcal{E}_V R_V$ where $\tilde{Q} := (\Psi_U)_R^{-1} Q(\Psi_V)_L^{-1}$. Let $\tilde{Q}^o$ denote the optimal free parameter such that $\Phi^o = H_p - L_U \mathcal{E}_U \tilde{Q}^o \mathcal{E}_V R_V$. Then, it is always possible to pick an approximation of $\tilde{Q}^o$ with finite support, say $\tilde{Q}_p^o$, such that the finitely supported matrix sequence $\Phi_p^o := H_p - L_U \mathcal{E}_U \tilde{Q}_p^o \mathcal{E}_V R_V$ is arbitrarily close to Φ^o.   ∎

Drawbacks.   Besides the necessary assumptions regarding the existence of polynomial feasible solutions, the FMV approximation method suffers from two other significant drawbacks:

1. Although it provides an upper bound for v^o and a feasible solution that achieves it, it gives no information about how far away from optimal the solution is.

2. The compensators obtained with this method suffer from order inflation (i.e., the order of the controller increases with N).

These aspects of the solutions will be illustrated through several examples later in this chapter.

The first drawback is solved by introducing a second optimization problem, the FME approximation method.

The FME method.   Such method further exploits the structure of the matrices $\bar{M}_{ij}$ to get lower bounds on v^o. The name stems from the fact that only finitely many equality constraints associated with the rank interpolation conditions are included in the optimization problem. The rest are simply ignored. Therefore, the solution obtained will in general fail to satisfy those constraints that were left out, rendering it infeasible to the nontruncated problem. A formal statement of the FME approximation problem (in its primal form) is as follows:

$$\underline{v}_N := \min_{v, \xi, \phi_{ij}^+, \phi_{ij}^-} v$$

subject to

$$\xi(i) + \sum_{j=1}^{n_w} \sum_{k=0}^{\infty} \phi_{ij}^+(k) + \phi_{ij}^-(k) = v \text{ for } i = 1, \ldots, n_z,$$

$$\sum_{i=1}^{n_z} \sum_{j=1}^{n_w} M_{ij}(\phi_{ij}^+ - \phi_{ij}^-) = b_{\text{zero}}, \tag{12.5}$$

$$\left(\sum_{i=1}^{n_z} \sum_{j=1}^{n_w} \bar{M}_{ij}(\phi_{ij}^+ - \phi_{ij}^-) \right)(k) = b_{\text{rank}}(k) \text{ for } k = 0, \ldots, N,$$

$$\xi, \phi_{ij}^+, \phi_{ij}^- \geq 0.$$

The dual form of FME is

$$\underline{\nu}_N = \max_{\gamma_0, \gamma_1, \gamma_2} \langle b_{\text{zero}}, \gamma_1 \rangle + \langle b_{\text{rank}}, \gamma_2 \rangle$$

subject to

$$\gamma_0 \geq 0 \;, \quad \sum_{i=1}^{n_z} \gamma_0(i) \leq 1,$$

$$-\gamma_0(i) \leq (M_{ij}^T \gamma_1 + \bar{M}_{ij}^T \gamma_2)(k) \leq \gamma_0(i) \quad \text{for} \quad \begin{cases} i = 1, \ldots, n_z \\ j = 1, \ldots, n_w \;, \\ k = 0, 1, \ldots \end{cases}$$

$$\gamma_2(l) = 0 \text{ for } l > N.$$

(12.6)

It is clear that the FME method is obtained by truncating the variables of the infinite-dimensional dual problem.

FME as a finite-dimensional linear program.   This truncation scheme transforms the original problem into one with a finite number of constraints but still an infinite number of variables. An argument similar to the one used for the one-block problem shows that the above infinite-dimensional linear program is indeed equivalent to a finite-dimensional one. Let $\bar{M}_{ij,N}$ denote the truncated $\bar{M}_{ij}$ (i.e., the first N rows of it). Since $\bar{M}_{ij,N}$ has only a finite number of rows, then the combined matrix

$$\left(M_{ij}^T \;\; \bar{M}_{ij,N}^T \right)$$

maps a finite-dimensional space to ℓ_∞. Moreover, owing to the band structure of $\bar{M}_{ij}$, all the columns of the combined matrix are in c_0 and thus the range is in c_0. Therefore, by Lemma 12.1.1 and Theorem 12.1.1, the FME problem is equivalent to a finite-dimensional linear program whose solution has finite support.

Convergence of FME.   The sequence of linear programs in Problem (12.5) is such that the number of constraints increases with N. Therefore, $\underline{\nu}_N$ forms a nondecreasing sequence bounded from above by ν^o. The next theorem shows that it actually converges to ν^o.

Theorem 12.2.3.   In the FME method, $\underline{\nu}_N \to \nu^o$ as $N \to \infty$.

Proof. The proof is immediate after writing the dual form of FME. Clearly, the support of the dual vector $\gamma = (\gamma_0 \;\; \gamma_1 \;\; \gamma_2)^T$ is finite and increases with N. But the set of finitely supported sequences is dense in c_0. Then the result follows as in Theorem 12.2.2. ∎

Summary.   Based on these convergence properties, a multiblock problem can be solved iteratively to any degree of accuracy by solving two finite-dimensional linear programs, corresponding to the FMV and FME truncation schemes, at each iteration. Given N, it was shown that

$$\underline{\nu}_N \leq \nu^o \leq \bar{\nu}_N.$$

(12.7)

If there exists finitely supported feasible solutions to the problem, then

$$\lim_{N \to \infty} |\bar{\nu}_N - \underline{\nu}_N| = 0. \tag{12.8}$$

The stopping criterion is based on the last equation.

12.2.2 Delay Augmentation Method

This subsection presents the Delay Augmentation (DA) approximation. This method provides a conceptually attractive and computationally efficient way of solving general multiblock problems, with the added benefit of not requiring assumptions on the existence of polynomial feasible solutions and with the capacity of generating suboptimal controllers without order inflation.

The main idea is very simple:

1. Augment U and V with pure delays (i.e., right shifts) such that the augmented problem is one-block;
2. Apply all the machinery developed for one-block problems to the augmented system;
3. Reduce it back to the original system and compute the controller.

In more precise terms, partition the original system as follows:

$$\begin{pmatrix} \Phi_{11} & \Phi_{12} \\ \Phi_{21} & \Phi_{22} \end{pmatrix} = \begin{pmatrix} H_{11} & H_{12} \\ H_{21} & H_{22} \end{pmatrix} - \begin{pmatrix} U_1 \\ U_2 \end{pmatrix} Q \begin{pmatrix} V_1 & V_2 \end{pmatrix}, \tag{12.9}$$

where $U_1 \in \ell_1^{n_u \times n_u}$ and $V_1 \in \ell_1^{n_y \times n_y}$. Then, augment U and V with Nth order shifts and augment the free parameter Q accordingly:

$$\begin{pmatrix} \Phi_{11,N} & \Phi_{12,N} \\ \Phi_{21,N} & \Phi_{22,N} \end{pmatrix} := \begin{pmatrix} H_{11} & H_{12} \\ H_{21} & H_{22} \end{pmatrix} - \begin{pmatrix} U_1 & 0 \\ U_2 & S_N \end{pmatrix} \begin{pmatrix} Q_{11} & Q_{12} \\ Q_{21} & Q_{22} \end{pmatrix} \begin{pmatrix} V_1 & V_2 \\ 0 & S_N \end{pmatrix} \tag{12.10}$$

or, equivalently,

$$\Phi_N := H - U_N Q_N V_N =: H - R_N, \tag{12.11}$$

where U_N, Q_N and V_N have the obvious definitions.

DA as a one-block problem.   Clearly, Problem (12.11) is of the one-block class since $U_N \in \ell_1^{n_z \times n_z}$ and $V_N \in \ell_1^{n_w \times n_w}$. By expanding Equation (12.10) we have

$$\Phi_N = H - U Q_{11} V - S_N \tilde{R}_N \tag{12.12}$$

and

$$\tilde{R}_N := \begin{pmatrix} 0 & U_1 Q_{12} \\ Q_{21} V_1 & Q_{21} V_2 + U_2 Q_{12} + S_N Q_{22} \end{pmatrix}.$$

This follows since all operators are time invariant. With this notation we are ready to define the Delay Augmentation problem of order N as the following optimization problem:

$$\underline{\eta}_N := \inf_{Q_N \in \ell_1^{n_z \times n_w}} \|\Phi_N\|_1. \tag{12.13}$$

It follows from the above definition that $\underline{\eta}_N$ is a lower bound for v^o since

$$\underline{\eta}_N \leq \inf_{\substack{Q_{11} \in \ell_1^{n_u \times n_z} \\ Q_{12}=Q_{21}=Q_{22}=0}} \|\Phi_N\|_1 = \inf_{Q_{11} \in \ell_1^{n_u \times n_z}} \|\Phi\|_1 = v^o.$$

In other words, the extra degree of freedom in the free parameter Q_N (as compared to Q) makes the construction of superoptimal solutions possible. Such solutions, however, are clearly infeasible to the unaugmented problem. Also, it is interesting to note that the extra parameters (namely Q_{12}, Q_{21}, and Q_{22}) have no effect on the solution $\Phi_N(k)$ for $k < N$ due to the presence of the shift operator in Equation (12.12). And even more interesting, the term Φ_{11} is not affected at all by the added parameters (note the block of zeros in $\tilde{R}_N$). This observation will let us construct a suboptimal feasible solution directly from the solution of (12.13).

Upper and lower bounds.   Given some positive integer N, let

$$\underline{\eta}_N = \|\Phi_N^o\|_1 = \|H - U Q_{11}^o V - S_N \tilde{R}_N^o\|_1$$

then,

$$v^o = \inf_{Q \in \ell_1^{n_u \times n_y}} \|H - UQV\|_1 \leq \|H - U Q_{11}^o V\|_1 =: \bar{\eta}_N. \tag{12.14}$$

Equivalently, the solution obtained by making the extra free parameters zero after solving Problem (12.13) is feasible and suboptimal to the unaugmented problem. The following lemma summarizes these results.

Lemma 12.2.1.   Given a positive integer N and definitions (12.13) and (12.14), then

$$\underline{\eta}_N \leq v^o \leq \bar{\eta}_N$$

where $\bar{\eta}_N$ is achieved with Q_{11}^o.   ∎

Before addressing the convergence properties of this method, a word on existence is in order. Recall that existence is assured if there are no zero interpolations on the boundary of the unit disc. Now, it may happen that a multiblock problem that satisfies this condition augments into a one-block problem that does not. Indeed, notice that the zeros of $\hat{U}_N$ are given by the zeros of $\hat{U}_1$ plus a multiple zero at the origin (due to the block of delays, $\lambda^N I$, resulting from the λ-transform of S_N). Clearly, the zeros of $\hat{U}$ are also zeros of $\hat{U}_1$. However, $\hat{U}_1$ may have more zeros, possibly on the boundary of the disc. The same applies to the zeros of $\hat{V}$. In many instances this situation may be remedied by a proper reordering of inputs and outputs, such that the resulting $\hat{U}_1$ and $\hat{V}_1$ have no zeros on the boundary, respectively. In any case, this limitation has few practical implications since it is always possible to find approximate rational solutions to Problem (12.13) that are arbitrarily close to $\underline{\eta}_N$. In view of this, we will make the following simplifying assumption:

Assumption 2: $\hat{U}_1(\lambda)$ and $\hat{V}_1(\lambda)$ have no zeros on the unit circle, respectively.

Note that under this assumption the results of Theorem 6.2.2 are applicable. Furthermore, in the analysis that follows we will be able to exploit the existence of optimal solutions for any N and thus avoid approximate arguments.

Convergence of DA method.   The previous discussion suggests that the way inputs and outputs are ordered in the problem setup has an impact on the DA approximation method. Note that this is not the case with FMV and FME methods, where the solutions remain the same after inputs and outputs are reordered. In fact, ordering in the DA approximation method has important convergence implications on top of the existence issue mentioned before. To motivate this discussion observe the following:

1. In the DA method, the interpolation zeros are given by the nonminimum phase zeros of $\hat{U}_1$ and $\hat{V}_1$, which strongly depend on the way inputs and outputs are ordered. As an example, consider a two-block row problem, with $\hat{U} = 1$ and $\hat{V} = (1 \quad \lambda - 0.5)$. A DA approximation of such a problem will generate no zero interpolation constraints other than those at $\lambda = 0$. However, if inputs are reordered, then $\hat{V} = (\lambda - 0.5 \quad 1)$, and there will be an extra interpolation condition at $\lambda = 0.5$. Hence, it is not surprising that convergence is affected by the ordering of inputs and outputs.

2. In Corollary 12.1.2 we have shown that most one-block problems have optimal solutions with all row norms equal to ν^o. With multiblock problems, however, this is not the case. To illustrate this point, consider the following SISO example:

$$\phi_1 = h_1 - u_1 q$$

where all operators are in ℓ_1 and $\hat{u}_1(\lambda)$ has no zeros on the unit circle. Let ϕ_1^o denote an ℓ_1-optimal solution to such (one-block) problem, that is achieved with q^o. Next, add a new row to the problem,

$$\begin{pmatrix} \phi_1 \\ \phi_2 \end{pmatrix} = \begin{pmatrix} h_1 \\ h_2 \end{pmatrix} - \begin{pmatrix} u_1 \\ u_2 \end{pmatrix} q$$

such that $\|h_2 - u_2 q^o\|_1 < \nu^o$ (this is always possible simply by choosing a scalar weight on the second row of small enough value). Then, it is clear that the optimal solution to the new two-block column problem is still given by q^o and that $\|\phi_2^o\|_1 < \|\phi_1^o\|_1 = \nu^o$. In other words, the new row does not affect the optimal solution, which is given by the first row alone. In contrast with a one-block problem with two outputs, a two-block problem with two outputs has to minimize both outputs with just one scalar-free parameter, q. The "shortage" of degrees of freedom is what makes this situation quite common in multiblock problems. In fact, this is analogous to the finite-dimensional matrix case.

The second observation motivates the following terminology.

Definition 12.2.1. Given a multiblock problem with closed-loop map Φ of dimension $n_z \times n_w$, the ith row of Φ is said to be *active* in the ℓ_1 optimization problem if $\|(\Phi)_i^o\|_1 = \nu^o$ for all optimal solutions Φ^o. ∎

Based on the above definition, we summarize the convergence results of the DA approximation:

1. The lower bound always converges, i.e.,
$$\lim_{N \to \infty} \underline{\eta}_N = \nu^o.$$

2. The upper bound converges to ν^o if the first n_u rows of Φ^o are active, i.e., if the rows corresponding to Φ_{11}^o achieve the optimal norm.

3. For every sequence of optimal solutions to the DA problem, Φ_N^o, there exists a subsequence that converges to an optimal solution of the original problem, Φ^o, on finite subsets of $\mathbf{Z}_+$. In other words, there exists a subsequence with index set $\{N_j\}$, such that
$$P_L \Phi_{N_j}^o \overset{j \to \infty}{\longrightarrow} P_L \Phi^o$$
for any finite $L \in \mathbf{Z}_+$.

4. Furthermore, if the first n_u rows of Φ are active (i.e., if the upper bound converges to ν^o), then there exists a subsequence of Φ_N^o that converges to Φ^o. That is,
$$\Phi_{N_j}^o \overset{j \to \infty}{\longrightarrow} \Phi^o$$

In the sequel, we will describe how one arrives at these results.

Convergence proofs. By definition, Problem (12.13) is equivalent to
$$\underline{\eta}_N = \min_{R_N \in \mathcal{S}_N} \|H - R_N\|_1 = \sup_{\substack{G_N \in {}^\perp \mathcal{S}_N \\ \|G_N\|_\infty \leq 1}} \langle G_N, H \rangle. \tag{12.15}$$

It is easy to see that, as N increases, the subspace $\mathcal{S}_N$ gets smaller and such that
$$\mathcal{S}_N \supseteq \mathcal{S}_{N+1} \supseteq \cdots \supseteq \mathcal{S}. \tag{12.16}$$

This is a consequence of the fact that the only change in the interpolation structure is due to a higher multiplicity of the zero at the origin. Therefore, $\underline{\eta}_N$ forms a nondecreasing sequence, bounded from above by ν^o. Also, in the dual problem we have the inclusion property:
$${}^\perp \mathcal{S}_N \subseteq {}^\perp \mathcal{S}_{N+1} \subseteq \cdots \subseteq {}^\perp \mathcal{S}.$$

The next theorem states an interesting convergence result.

Theorem 12.2.4.   Given the sequence Φ_N^o, there exists a subsequence that converges *weak** to some Φ^o. If the optimal solution is unique then the whole sequence converges to it.

Proof. The sequence $\{\Phi_N^o,\ N \geq 0\}$ is bounded in $\ell_1^{n_z \times n_w}$; hence it follows that there exists a *weak**-convergent subsequence $\Phi_{N_s}^o$ by Theorem 9.5.2. Let Φ^{w^*} denote such limit point. As mentioned before, Φ_N^o is infeasible to the original (unaugmented) problem. However, we will show that Φ^{w^*} is in fact feasible. From Equation (12.12), after taking the *weak** limit, we have

$$\Phi^{w^*} = H - (U Q_{11,N_s}^o V)^{w^*} - (S_{N_s} \tilde{R}_{N_s}^o)^{w^*} = H - U(Q_{11,N_s}^o)^{w^*} V,$$

where the superscript w^* denotes *weak** limit. The last term drops since $\{\tilde{R}_N^o\}$ is uniformly bounded in N. For if $\{\tilde{R}_N^o\}$ were unbounded, then $\{Q_{11,N}^o\}$ would necessarily be unbounded to keep $\underline{\eta}_N$ bounded. But this contradicts the fact that $\underline{\eta}_N$ is larger than $\|H_{11} - U_1 Q_{11,N}^o V_1\|_1$; therefore, Φ^{w^*} is feasible. To show that Φ^{w^*} is actually an optimal solution, we need to view Φ_N^o as a bounded linear operator from $c_0^{n_z \times n_w}$ to $\mathbb{R}$ (i.e., bounded linear functional on $c_0^{n_z \times n_w}$) with strong operator limit Φ^{w^*}. In such context we have the following inequality from Theorem A.7.2:

$$\|\Phi^{w^*}\|_1 \leq \liminf_{s \to \infty} \|\Phi_{N_s}^o\|_1 \leq \|\Phi^o\|_1.$$

Therefore, since Φ^{w^*} is feasible, all inequalities above are equalities and $\Phi^{w^*} = \Phi^o$. If the solution is unique then the whole sequence converges to Φ^o *weak**.   ∎

The last claim in the above lemma simply reflects the fact that if there are several optimal solutions, Φ^o, then a sequence of DA problems can be such that Φ_N^o (in the limit) "jumps" from one optimal solution to the other, therefore not converging as a whole. Then, a subsequence that "keeps track" of a single optimal solution will converge *weak** to it. This technicality is unnecessary when the optimal solution is unique.

An immediate corollary to Theorem 12.2.4 is the following:

Corollary 12.2.2.   The sequence of lower bounds, $\underline{\eta}_N$, converge to ν^o as $N \to \infty$.   ∎

The next convergence result, set in the dual space, will prove useful in studying the support structure of optimal solutions, as we shall see later in this chapter. In the context of Theorem 9.3.4, the DA problem can be stated in its dual form as

$$\underline{\eta}_N = \max_{\substack{G_N \in \mathcal{S}_N^{\perp} \\ \|G_N\|_\infty \leq 1}} \langle H, G_N \rangle =: \langle H, G_N^o \rangle, \tag{12.17}$$

while the original optimization problem is

$$\nu^o = \max_{\substack{G \in \mathcal{S}^{\perp} \\ \|G\|_\infty \leq 1}} \langle H, G \rangle =: \langle H, G^o \rangle. \tag{12.18}$$

Note that G_N^o as well as G^o may not be unique. In this framework we state the following lemma.

Lemma 12.2.2. Given the sequence $\{G_N^o\}$, there exists a subsequence $\{G_{N_s}^o\}$ that converges $weak^*$ in $\ell_\infty^{n_z \times n_w}$ to an optimal solution G^o. Furthermore, if the solution G^o is unique, then the whole sequence converges $weak^*$ to it.

Proof. The sequence $\{G_N^o, N \geq 0\}$ is bounded by one. Hence, from Definition 9.5.4 and Theorem 9.5.2, there exists a subsequence that converges $weak^*$ in $\ell_\infty^{n_z \times n_w}$. Also, from Equation (12.16) we have that

$$\mathcal{S}_N^\perp \subseteq \mathcal{S}_{N+1}^\perp \subseteq \cdots \subseteq \mathcal{S}^\perp,$$

or, equivalently, G_N^o is feasible to the original (dual) problem for all N. Further, since the feasible subspace $\mathcal{S}^\perp$ is $weak^*$-closed (by Theorem 9.5.3), then $G_{N_s}^o$ converges $weak^*$ to a feasible limit point, say G^{w^*}. Therefore,

$$\underline{\eta}_{N_s} = \langle H, G_{N_s}^o \rangle \to \langle H, G^{w^*} \rangle.$$

But, by Corollary 12.2.2, $\underline{\eta}_{N_s} \to \nu^o$, thus, $\nu^o = \langle H, G^{w^*} \rangle$. This implies that G^{w^*} is in fact an optimal dual solution, G^o, for it achieves the optimal value and is feasible.

If the solution G^o is unique, then all subsequences must converge $weak^*$ to it so the whole sequence converges $weak^*$ to it. ∎

Next, we focus our attention on the sequence of suboptimal solutions that attain the upper bound $\bar{\eta}_N$. Let $\bar{\Phi}_N := H - U Q_{11,N}^o V$, then $\bar{\eta}_N = \|\bar{\Phi}_N\|_1$ by definition. It is easy to see that $\{\bar{\Phi}_N\}$ is a uniformly bounded sequence in $\ell_1^{n_z \times n_w}$ (if not $\{\Phi_{11,N}^o\}$ and thus $\underline{\eta}_N$ would be unbounded). Therefore, by Theorem 9.5.2, there exists a subsequence that converges $weak^*$ in $\ell_1^{n_z \times n_w}$. Also, $\bar{\Phi}_N$ is clearly feasible to the original problem for any N, and since $\mathcal{S}$ is $weak^*$-closed (shown in Theorem 11.6.1) then all $weak^*$ limit points are feasible. The question is whether the subsequence $\bar{\eta}_{N_s} = \|\bar{\Phi}_{N_s}\|_1$ converges to ν^0 in general. Here is where the notion of active row plays a central role.

Theorem 12.2.5. Given a general multiblock problem, let $\bar{\Phi}_{N_s}$ converge $weak^*$ to an optimal solution $\Phi^o = H - U Q^o V$ such that $\|(\Phi^o)_i\|_1 = \nu^o$ for $i \in \{1, \ldots, n_u\}$. Then, $\bar{\Phi}_{N_s}$ converges strongly to Φ^o as $N \to \infty$, and further, $\bar{\eta}_N \to \nu^o$.

Proof. We seek to apply Example 9.5.1. However, such a result is valid only for scalar and row-vector sequences in ℓ_1 (it is easy to think of a counter example in the general matrix case). Therefore, we can apply it to each individual row of Φ_{N_s} to conclude the following: $(\Phi_{N_s}^o)_i$ converges strongly (i.e., in the norm) to $(\Phi^o)_i$ for all $i \in \{1, \ldots, n_u\}$ such that $\|(\Phi^o)_i\|_1 = \nu^o$.

At the same time, it follows from Assumption 2 that $\hat{U}_1$ and $\hat{V}_1$ have full normal rank. Hence, the map from $Q_{11,N}^o$ to $\Phi_{11,N}^o$ is continuous with continuous inverse, that is

$$\hat{Q}_{11,N}^o = \hat{U}_1^{-1}(\hat{H}_{11} - \hat{\Phi}_{11,N}^o)\hat{V}_1^{-1}.$$

Then, using the fact that $\|(\Phi^o)_i\|_1 = \nu^o$ for $i \in \{1, \ldots, n_u\}$, we conclude that Φ^o_{11,N_s} converges strongly to Φ^o_{11} which in turn implies that Q^o_{11} converges strongly to Q^o and the result follows.   ∎

The above result indicates that the construction of the feasible solution that attains the upper bound, $\bar{\Phi}_N$, can be viewed as an attempt to compute the *weak** limit of the sequence $\Phi^o_{N_s}$ by "throwing away its tail" contained in the term $S_N \tilde{R}^o_N$.

Exact solutions.   Theorem 12.2.5 states that a sufficient condition for the convergence of the upper bound is that the first n_u rows of Φ be active. It should be stressed at this point that most multiblock problems have optimal solutions where at least n_u of the n_z rows achieve the optimal norm (a natural extension of how optimal solutions of one-block problems behave). Furthermore, those rows that are not active can be left out of the optimization problem without affecting the overall solution, so eventually the problem can be reduced up to a two-block row problem. If the original problem is two-block column with only n_u active rows, then an exact solution could be computed by reducing the problem to a one-block. In general, however, a balanced control problem will tend to have none of its rows "redundant," so that $\bar{\eta}_N$ converges to ν^o without further considerations. In this context we have the following corollary valid for two-block column problems of the form

$$\begin{pmatrix} \Phi_{11} \\ \Phi_{21} \end{pmatrix} = \begin{pmatrix} H_{11} \\ H_{21} \end{pmatrix} - \begin{pmatrix} U_1 \\ U_2 \end{pmatrix} QV.$$

Corollary 12.2.3.   Given a two-block column problem, if $\|\Phi^o_{21}\|_1 < \nu^o$ then $\bar{\Phi}_N$ is the exact optimal solution for any N.

Proof. Follows immediately from the fact that the first block-row $H_{11} - U_1 Q_{11} V$ is independent of the extra free parameter. That is,

$$\Phi^o_{11,N} = H_{11} - U_1 Q^o_{11} V,$$

$$\Phi^o_{21,N} = H_{21} - U_2 Q^o_1 V - S_N Q^o_{21} V.$$

Then, for any N we have

$$\|\Phi^o_{11,N}\|_1 \geq \nu^o \geq \underline{\eta}_N = \max(\|\Phi^o_{11,N}\|_1, \|\Phi^o_{21,N}\|_1) \geq \|\Phi^o_{11,N}\|_1.$$

Thus, equality is attained throughout and the result follows, i.e., $Q^o_{11,N} = Q^o$.   ∎

An important question arises—namely how to find a priori the rows of the problem that are not going to achieve the optimal norm. A straightforward answer to this is simply to solve all possible one-block problems formed by taking n_u rows out of the given n_z rows. If any solution is such that all the rows that were left out have smaller norms than the corresponding ν^o, then those rows are the inactive ones and should be ordered in U_2. However, this approach may require a considerable amount of work. We will return to this question later in the chapter where a different approach is presented.

Two-block row problems, although less understood, show a similar behavior. Indeed, such problems may have columns that are inactive in the optimization process in the sense that they can be removed without affecting the solution. (Why this happens in general remains an open question.) Note that in the previous case, the phenomenon of inactive rows was intimately related with the fact that the ℓ_1 norm on matrices takes the maximum row norm, which allowed us to construct an example easily. This structure, however, is absent in the two-block row problem with inactive columns. If the DA method is applied to a two-block row problem such that the columns associated with V_2 are inactive, then again the solution $\bar{\Phi}_N$ is exact for any N. However, $\underline{\eta}_N$ will not give the exact optimal norm (although it will tend to it) since the extra parameter contributes in reducing the norm of $\Phi_{12,N}^o$.

Finally, let us point out that both forms of redundancy (row and column) can occur in a multiblock problem simultaneously. This discussion motivates the following definitions.

Definition 12.2.2.   Given a general multiblock problem, a *one-block partition* is defined by taking n_y inputs and n_u outputs of the full problem, such that the reduced problem corresponds to a one-block problem with full rank U and V.   ■

Definition 12.2.3.   In a multiblock problem, a one-block partition is *Totally Dominant* (TD) if the optimal free parameter Q^o obtained from its solution also solves the original multiblock optimization problem.   ■

It follows from these definitions that, if there is a TD one-block partition corresponding to the partitions U_1 and V_1, then the DA method provides the exact answer for any N. The following sections illustrate some of these properties.

We conclude this section with a corollary regarding the connection between the FMV and DA methods. The proof is simple and is left to the reader.

Corollary 12.2.4.   Given an FMV approximation of order N_{FMV} of a multiblock problem, and the DA approximation to the same problem where the support of Φ is restricted to be N_{FMV}, then there exists a finite $N_{DA} > N_{FMV}$ such that the DA solution is equivalent to the FMV solution.   ■

12.3 AN ILLUSTRATIVE EXAMPLE

In view of the amount of notation introduced so far, and in order to clarify the exposition, this section includes the detailed FMV, FME, and DA solutions to a four-block example. All matrices involved in the setup of the linear program are shown explicitly in an effort to illustrate the simplicity of the procedures, despite the apparent notational complexity.

Consider the following 2-input 2-output four-block model matching problem:

$$\hat{\Phi} = \begin{pmatrix} 1 & \frac{1}{\lambda-3/2} \\ \frac{\lambda-1/2}{\lambda-3/2} & \frac{1}{\lambda-2} \end{pmatrix} - \begin{pmatrix} (\lambda-1/2)^2(\lambda-2/3) \\ 2(\lambda-1/2), \end{pmatrix} \hat{Q} \left(\lambda - 2/3 \ \lambda \right), \tag{12.19}$$

where $Q \in \ell_1$. The matrices $\hat{U}$ and $\hat{V}$ were chosen to be polynomial just to make the procedure tractable on paper. In general, however, the same solution methods apply to rational matrices.

The FMV and FME methods

1. First, the zero and rank interpolation conditions are stated according to Theorem 6.2.2. To this end, we compute the Smith-McMillan decompositions of $\hat{U}$ and $\hat{V}$.

$$
\hat{U} = \hat{L}_U \hat{M}_U = \begin{pmatrix} (\lambda - 1/2)(\lambda - 2/3) & 1 \\ 2 & 0 \end{pmatrix} \begin{pmatrix} \lambda - 1/2 \\ 0 \end{pmatrix},
$$

$$
\hat{V} = \hat{M}_V \hat{R}_V = \begin{pmatrix} 1 & 0 \end{pmatrix} \begin{pmatrix} \lambda - 2/3 & \lambda \\ 1 & 1 \end{pmatrix}.
$$

Hence, $\hat{U}$ has a simple zero at $\lambda_0 = 1/2$ and $\hat{V}$ has no zeros. From the inverse of the unimodular matrices $\hat{L}_U$ and $\hat{R}_V$, the polynomial vectors $\hat{\alpha}_i$'s and $\hat{\beta}_j$'s are computed:

$$
\hat{\alpha}_1 = (0 \ \ 1), \qquad \hat{\alpha}_2 = \left(1 \ \ -\tfrac{1}{2}(\lambda - 1/2)(\lambda - 2/3)\right),
$$

$$
\hat{\beta}_1^T = (-1 \ \ 1), \qquad \hat{\beta}_2^T = \left(\lambda \ \ -(\lambda - 2/3)\right).
$$

Note that the vectors have been rescaled to simplify the arithmetic. Then, the zero interpolation conditions are given by

$$
(\hat{\alpha}_1 \hat{\Phi} \hat{\beta}_1)(\lambda_0) = (\hat{\alpha}_1 \hat{H} \hat{\beta}_1)(\lambda_0), \tag{12.20}
$$

and the rank interpolations are

$$
\hat{\alpha}_2 \begin{pmatrix} \hat{\phi}_{11} \\ \hat{\phi}_{21} \end{pmatrix} = \hat{\alpha}_2 \begin{pmatrix} \hat{h}_{11} \\ \hat{h}_{21} \end{pmatrix}, \quad \hat{\Phi}\hat{\beta}_2 = \hat{H}\hat{\beta}_2. \tag{12.21}
$$

2. Next, the infinite matrix A_{zero} and vector b_{zero} are constructed based on Equation (12.20). As there is only one simple zero, the matrix A_{zero} reduces to an infinite row vector while b_{zero} reduces to a scalar:

$$
A_{\text{zero}} = \left(0 \ \ 0 \ \ -1 \ \ 1 \ \ 0 \ \ 0 \ \ -\tfrac{1}{2} \ \ \tfrac{1}{2} \ \ 0 \ \ 0 \ \ -(\tfrac{1}{2})^2 \ \ (\tfrac{1}{2})^2 \ \cdots\right), \tag{12.22}
$$

and

$$
b_{\text{zero}} = -\hat{h}_{21}(1/2) + \hat{h}_{22}(1/2) = -\frac{2}{3}. \tag{12.23}
$$

The infinite matrices M_{ij} defined at the beginning of this chapter can be read directly off A_{zero}:

$$
M_{11} = M_{12} = \left(0 \ \ 0 \ \cdots\right),
$$

$$
M_{22} = -M_{21} = \left(1 \ \ \tfrac{1}{2} \ \ (\tfrac{1}{2})^2 \ \cdots\right).
$$

3. Now, A_{rank} and b_{rank} are constructed. By combining and reordering the rank interpolation conditions, Equation (12.21) can be rewritten as follows: Define the auxiliary polynomial matrix

$$\hat{T}(\lambda) := \begin{pmatrix} 1 & 0 & (-\frac{1}{6} + \frac{7}{12}\lambda - \frac{1}{2}\lambda^2) & 0 \\ \lambda & (\frac{2}{3} - \lambda) & 0 & 0 \\ 0 & 0 & \lambda & (\frac{2}{3} - \lambda) \end{pmatrix},$$

then,

$$\hat{T}\begin{pmatrix} \hat{\phi}_{11} \\ \hat{\phi}_{12} \\ \hat{\phi}_{21} \\ \hat{\phi}_{22} \end{pmatrix} = \hat{T}\begin{pmatrix} \hat{h}_{11} \\ \hat{h}_{12} \\ \hat{h}_{21} \\ \hat{h}_{22} \end{pmatrix}. \tag{12.24}$$

The right-hand side of the equation above can be computed explicitly, since both factors are known. This gives rise to the infinite vector b_{rank}. Indeed,

$$\hat{T}\begin{pmatrix} \hat{h}_{11} \\ \hat{h}_{12} \\ \hat{h}_{21} \\ \hat{h}_{22} \end{pmatrix} = \begin{pmatrix} 1 - \frac{1}{2}\frac{(\lambda-1/2)^2(\lambda-2/3)}{\lambda-3/2} \\ \lambda - \frac{\lambda-2/3}{\lambda-3/2} \\ \lambda\frac{\lambda-1/2}{\lambda-3/2} - \frac{\lambda-2/3}{\lambda-2} \end{pmatrix}$$

and b_{rank} is simply the infinite impulse response of the above:

$$b_{\text{rank}}^T = \begin{pmatrix} \frac{17}{18} & -\frac{4}{9} & -\frac{1}{3} & \frac{29}{108} & \frac{37}{27} & \frac{2}{3} & \cdots \end{pmatrix}. \tag{12.25}$$

In fact, a closer look at this response indicates that the sequence b_{rank} is infinitely supported (i.e., decays asymptotically to zero). Hence, by Theorem 12.2.1, the FMV problem has no feasible solution for any order of truncation. One could compute, however, an "approximate" FMV solution by truncating the impulse response of $\hat{H}$. If $\hat{H}_{\text{approx}}$ is taken to be an Nth order polynomial approximation of $\hat{H}$, then a feasible point is guaranteed to exist for all FMV problems of order greater or equal to N (e.g., take $Q = 0$), and an approximate FMV solution can be computed. However, such approximation will deteriorate the rate of convergence of the method in general, as the tail of H is ignored. This is particularly true if $\hat{H}$ has slowly decaying modes. The construction of A_{rank} is immediate from the matrix $\hat{T}$ (i.e., simply the Toeplitz representation of $\hat{T}$),

$$A_{\text{rank}} = \begin{pmatrix} T(0) & 0 & 0 & 0 & 0 & \cdots \\ T(1) & T(0) & 0 & 0 & 0 & \cdots \\ T(2) & T(1) & T(0) & 0 & 0 & \cdots \\ 0 & T(2) & T(1) & T(0) & 0 & \cdots \\ \vdots & & \ddots & \ddots & \ddots & \end{pmatrix}, \tag{12.26}$$

where

$$T(0) = \begin{pmatrix} 1 & 0 & -\frac{1}{6} & 0 \\ 0 & \frac{2}{3} & 0 & 0 \\ 0 & 0 & 0 & \frac{2}{3} \end{pmatrix},$$

$$T(1) = \begin{pmatrix} 0 & 0 & \frac{7}{12} & 0 \\ 1 & -1 & 0 & 0 \\ 0 & 0 & 1 & -1 \end{pmatrix},$$

$$T(2) = \begin{pmatrix} 0 & 0 & -\frac{1}{2} & 0 \\ 0 & 0 & 0 & 0 \\ 0 & 0 & 0 & 0 \end{pmatrix}.$$

Notice that A_{rank} has the expected band structure. The matrices $\bar{M}_{ij}$ are nothing but the Toeplitz representation of each column of $\hat{T}$. For instance, the second column corresponds to $\bar{M}_{12}$:

$$\bar{M}_{12} = \begin{pmatrix} 0 & 0 & 0 & \cdots \\ \frac{2}{3} & 0 & 0 & \cdots \\ 0 & 0 & 0 & \cdots \\ 0 & 0 & 0 & \cdots \\ -1 & \frac{2}{3} & 0 & \cdots \\ 0 & 0 & 0 & \cdots \\ 0 & 0 & 0 & \cdots \\ 0 & -1 & \frac{2}{3} & \cdots \\ \vdots & & & \ddots \end{pmatrix}.$$

4. Finally, we construct the matrix A_{ℓ_1}:

$$A_{\ell_1} = \begin{pmatrix} 1 & 1 & 0 & 0 & 1 & 1 & 0 & 0 & \cdots \\ 0 & 0 & 1 & 1 & 0 & 0 & 1 & 1 & \cdots \end{pmatrix}. \tag{12.27}$$

5. Combining Equations (12.22), (12.23), (12.26), (12.25), and (12.27) we get the full infinite dimensional primal linear program as in Equation (11.24).

6. The solution of the FMV problem of order N is obtained by constraining the support of Φ to be no greater than $N + 1$, i.e., by including the first $4(N + 1)$ columns of A_{ℓ_1}, A_{zero}, and A_{rank} in the setup (see Figure 12.1). As pointed out before, this example has no such feasible solutions since a column truncation of A_{rank} results in a finite number of nonzero rows, whereas b_{rank} is infinitely supported.

7. The solution of the FME problem of order N is obtained by including the first N rows of A_{rank} and ignoring the rest (see Figure 12.1). Clearly, all the rows left in A_{rank} are in c_0, which implies that the problem has a finite-dimensional optimal solution (see Figure 12.1). Numerical results for this example will be presented after the setup of the DA method.

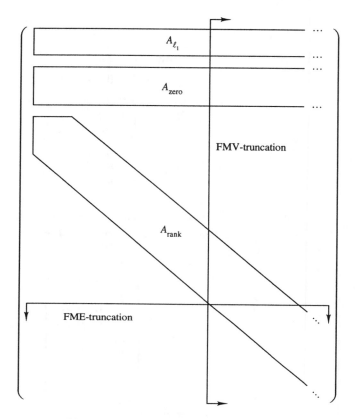

Figure 12.1 FMV and FME truncation schemes.

The DA method

1. We start by augmenting $\hat{U}$ and $\hat{V}$ with N delays:

$$\hat{U}_N = \begin{pmatrix} (\lambda - \frac{1}{2})^2(\lambda - \frac{2}{3}) & 0 \\ 2(\lambda - \frac{1}{2}) & \lambda^N \end{pmatrix}, \quad \hat{V}_N = \begin{pmatrix} (\lambda - \frac{2}{3}) & \lambda \\ 0 & \lambda^N \end{pmatrix}. \tag{12.28}$$

2. The DA method consists of solving the resulting one-block problem exactly. In this sense, we need to identify the zero interpolation frequencies of the augmented problem. It follows that

$$\Lambda_{U_N V_N} = \left\{ 0, \frac{1}{2}, \frac{2}{3} \right\}.$$

3. Next, the zero interpolation conditions are constructed based on the extended set of null chains corresponding to each zero frequency, as stated in Corollary 6.6.1. This can be done quite easily following the procedure described in Subsection 6.6.2, which involves the computation of null-spaces of certain Toeplitz matrices.

To simplify the arithmetic, we will borrow the polynomial vectors $\hat{\alpha}_2$ and $\hat{\beta}_2$ from the previous setup, as candidates for the null chains corresponding to $\lambda_0 = 0$. It should be noted that an explicit Smith-McMillan decomposition of $\hat{U}_N$ and $\hat{V}_N$ is not necessary in general. It is convenient in this case, since such vectors provide a numerically "simple" characterization of the necessary null-spaces. (Note that the extended set of null chains is not unique.)

Then, for $\lambda_0 = 0$, and say $N = 4$, we have

$$x^1 = \left(0 \ \tfrac{2}{3} \ 1 \ -1 \ 0 \ 0 \ 0 \ 0\right)^T, \quad x^2 = \left(1 \ 0\right)^T$$

of 4th and 0th order, respectively, and

$$y^1 = \left(1 \ -\tfrac{1}{6} \ 0 \ \tfrac{7}{12} \ 0 \ -\tfrac{1}{2} \ 0 \ 0\right)^T, \quad y^2 = \left(0 \ 1\right)^T$$

of 4th and 0th order, respectively. Similarly, for $\lambda_0 = 0.5$, we have

$$x^1 = \left(1 \ 0\right)^T, \quad x^2 = \left(0 \ 1\right)^T$$

both of 0th order, and

$$y^1 = \left(1 \ 0 \ 0 \ 0\right)^T, \quad y^2 = \left(0 \ 1\right)^T$$

of 2nd and 0th order, respectively. Finally, for $\lambda_0 = \tfrac{2}{3}$, the following extended set of right null chains can be found:

$$x^1 = \left(1 \ 0\right)^T, \quad x^2 = \left(0 \ 1\right)^T$$

of 1st and 0th order, respectively. The set of left null chains is identical to the above.

4. The actual constraints are obtained by applying Corollary 6.6.1; that is,

$$(\hat{y}^i_{\lambda_0} \hat{\Phi} \hat{x}^j_{\lambda_0})^{(k)}(\lambda_0) = (\hat{y}^i_{\lambda_0} \hat{H} \hat{x}^j_{\lambda_0})^{(k)}(\lambda_0) \quad \text{for} \quad \begin{cases} i = 1, 2 \\ j = 1, 2 \\ k = 0, \ldots, \sigma_{U_{Ni}}(\lambda_0) + \sigma_{V_{Nj}}(\lambda_0) - 1 \end{cases}.$$

As an example, we write explicitly the constraints corresponding to $\lambda_0 = \tfrac{2}{3}$:

$$\text{for } i = 1 \text{ and } j = 1 \ ; \quad \begin{cases} \hat{\phi}_{11}(\tfrac{2}{3}) = \hat{h}_{11}(\tfrac{2}{3}) \\ \hat{\phi}_{11}^{(1)}(\tfrac{2}{3}) = \hat{h}_{11}^{(1)}(\tfrac{2}{3}) \end{cases},$$

$$\text{for } i = 1 \text{ and } j = 2 \ ; \quad \hat{\phi}_{12}(\tfrac{2}{3}) = \hat{h}_{12}(\tfrac{2}{3}),$$

$$\text{for } i = 2 \text{ and } j = 1 \ ; \quad \hat{\phi}_{21}(\tfrac{2}{3}) = \hat{h}_{21}(\tfrac{2}{3}).$$

Note that for $i = j = 2$ there is no condition since the left and the right null chains have zero order.

The above conditions can be easily written in matrix form. The corresponding portions of A_{zero} and b_{zero} are given by:

$$\begin{pmatrix} 1 & 0 & 0 & 0 & \tfrac{2}{3} & 0 & 0 & 0 & (\tfrac{2}{3})^2 & 0 & 0 & 0 \cdots \\ 0 & 0 & 0 & 0 & 1 & 0 & 0 & 0 & 2(\tfrac{2}{3}) & 0 & 0 & 0 \cdots \\ 0 & 1 & 0 & 0 & 0 & \tfrac{2}{3} & 0 & 0 & 0 & (\tfrac{2}{3})^2 & 0 & 0 \cdots \\ 0 & 0 & 1 & 0 & 0 & 0 & \tfrac{2}{3} & 0 & 0 & 0 & (\tfrac{2}{3})^2 & 0 \cdots \end{pmatrix}$$

and

$$\begin{pmatrix} 1 & 0 & -1.2 & -0.2 \end{pmatrix}^T.$$

The rest of the constraints are computed in a similar fashion.

5. Once the setup is complete, the DA problem can be solved exactly for each N. The resulting cost, $\underline{\eta}_N$, is a lower bound of the ℓ_1 norm of the optimal solution of the original problem. An upper bound can be computed from the superoptimal solution Φ_N^o, by solving for $Q_{11,N}^o$ from $\Phi_{11,N}^o$. This parameter can be viewed as a suboptimal one for the original problem. Hence, $\|H - U Q_{11,N}^o V\|_1$ produces the desired upper bound $\bar{\eta}_N$. Furthermore, the corresponding suboptimal controller can be computed directly from $Q_{11,N}^o$ and the factors in the parameterization of stabilizing controllers.

Values for this particular example are shown in Figure 12.2, where the lower bounds resulting from the FME and DA methods are plotted as a function of the total number of constraints. Clearly, the DA approach generates a more "efficient" approximation of the infinite-dimensional problem, evidenced by the better rate of convergence. Interestingly, $Q_{11,N}^o = 0$ for all the DA solutions shown in Figure 12.2 (i.e., for $N = 0, 1, \ldots, 6$), implying that $\bar{\eta}_N = \|H\|_1 = 3$ for all such N's.

It is also interesting to point out to the staircase behavior of the FME lower bound. This is a direct consequence of the fact that the auxiliary matrix $\hat{T}(\lambda)$ has three rows. In other words, from Equation (12.26), it is clear that going from $\Phi(k)$ to $\Phi(k+1)$ requires the addition of one more block-row, i.e., three more constraints. The benefits of having an exact representation up to time index $k + 1$, as opposed to just k, is reflected in a "jump" of the lower bound. This phenomenon is not present in the DA solution, where the number of constraints increases in a more natural way as N increases.

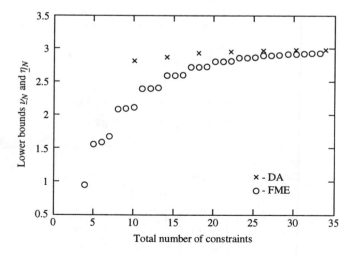

Figure 12.2 Convergence of the FME and DA methods.

12.4 COMPARISON OF METHODS

This section provides a general comparison of the approximation methods presented, based on a few simple multiblock examples. Particular attention will be paid to two aspects of the solutions: first, the support characteristics of the sequence of solutions, and second, the order of the suboptimal controller they generate.

12.4.1 Example I

Consider the following two-block column problem: Given the SISO plant P, minimize the ℓ_1 norm of the weighted sensitivity and complementary sensitivity,

$$\Phi = \begin{pmatrix} \phi_1 \\ \phi_2 \end{pmatrix} = \begin{pmatrix} W_1(1 - PK)^{-1} \\ W_2 PK(1 - PK)^{-1} \end{pmatrix}$$

where

$$\hat{P}(\lambda) = \frac{\lambda(\lambda - 0.5)}{(\lambda - 0.1)(1 - 0.5\lambda)}$$

and

$$\hat{W}_1(\lambda) = \frac{0.02}{1 - 0.2\lambda}, \quad \hat{W}_2(\lambda) = \frac{0.004\rho}{1 - 0.6\lambda}.$$

Note that a variable scalar weight on ϕ_2, denoted ρ, has included. By adjusting ρ, we will be able to generate two interesting cases: case (a) where ϕ_1 is TD (for "small" ρ) and case (b) where both rows are active in the optimization. The workings of Theorem 12.2.5 will be illustrated by reordering the outputs and forcing the TD row to be in the "wrong" place.

The results are presented in tables that show, for each N, the DA lower bound $(\underline{\eta}_N)$, the DA upper bound $(\bar{\eta}_N)$, and the FMV upper bound $(\bar{\nu}_N)$. The FME lower bound is omitted because it is equal to $\underline{\eta}_N$ in this particular case. Recall that in general $\underline{\eta}_N$ converges faster than $\underline{\nu}_N$, since the Delay Augmentation method generates more constraints than the FME method for any given N. In this particular case, however, both bounds are equal owing to the fact that the problem is two-block and the unstable zeros of $\hat{U}_1(\lambda)$ are also zeros of $\hat{U}_2(\lambda)$. Also included are the support characteristics of Φ_N^o and of the FMV solution (for reasons to be explained later) along with the order of the suboptimal controllers that achieve the corresponding upper bounds.

To describe the support characteristics we define a function, len($\cdot$), mapping $\ell_1^{n \times m}$ to $\mathbf{Z}_+^{n \times m}$ in the following way: Given $\Phi \in \ell_1^{n \times m}$, then $[\text{len}(\Phi)]_{ij}$ is a nonnegative integer equal to the maximum k for which $\phi_{ij}(k)$ is not zero, plus one. If ϕ_{ij} is infinitely supported then len$(\phi_{ij}) := \infty$. Also, we denote the order of a controller K by ord(K).

Case (a):   Let $\rho = 1$ and keep the same ordering of outputs as above (i.e., sensitivity first). The results are shown in Table 12.1. Clearly, the solution given by the Delay Augmentation method is exact since the upper and lower bounds are equal for any N. Then, by Corollary 12.2.3, the first row corresponding to the weighted sensitivity is

TABLE 12.1   EXAMPLE I, CASE (A) WHERE THE FIRST ROW IS TD

N	DA				FMV		
	$\underline{\eta}_N$	$\bar{\eta}_N$	$\text{len}(\Phi_N^o)^T$	$\text{ord}(K)$	$\bar{\nu}_N$	$\text{len}(\Phi_N^o)^T$	$\text{ord}(K)$
1	0.78222	0.78222	(3 2)	2	-	-	-
2	0.78222	0.78222	(3 3)	2	-	-	-
3	0.78222	0.78222	(3 4)	2	1.31912	(4 4)	3
4	0.78222	0.78222	(3 5)	2	0.97459	(5 5)	4
5	0.78222	0.78222	(3 6)	2	0.87547	(6 6)	5
6	0.78222	0.78222	(3 7)	2	0.83292	(7 7)	6

TD. Indeed, a simple computation shows that $\|\phi_2^o\|_1 = 0.2040 < \|\phi_1^o\| = 0.7822$. Note how the support of the second row of the augmented optimal solution increases with N while the first row remains constant. Since the controller is computed from the first row only, it is also exact and constant as N increases. In contrast, the FMV solution has increasing support on both rows, thus generating a suboptimal controller of increasing order that approximates the second-order optimal controller. Note that for some N's, the FMV problem has no solution (indicated with a dash) since the feasible set is empty.

Next, consider the same problem but with the outputs reordered (i.e., the complementary sensitivity in the first row). Table 12.2 shows how violating the condition of Theorem 12.2.5 (i.e., the fact that the first row is not active) affects the convergence of the upper bound. The lower bound, however, does converge as shown in Theorem 12.2.4. Although the upper bound does not converge, it is interesting to note that for $N \geq 2$ the length of $\phi_{2,N}^o$ (i.e., the weighted sensitivity) locks at a value of 3, which coincides with the length of the optimal solution. This seems to be a general characteristic of the DA method as we shall see later. At the same time, there is a clear order inflation on the suboptimal controller due to the constant increase in the length of $\phi_{1,N}^o$. (Note: FMV results are not included in Table 12.2 since this method is not affected by reordering.)

TABLE 12.2   EXAMPLE I, CASE (A) WHERE THE SECOND ROW IS TD

N	$\underline{\eta}_N$	$\bar{\eta}_N$	$\text{len}(\Phi_N^o)^T$	$\text{ord}(K)$
1	0.22000	1.1602	(3 2)	2
2	0.29095	1.9939	(4 3)	4
3	0.42826	3.1464	(5 3)	5
4	0.55995	3.9859	(6 3)	6
5	0.65664	4.5189	(7 3)	7
6	0.71550	4.8077	(8 3)	8
7	0.74789	4.9504	(9 3)	9
8	0.76483	5.0171	(10 3)	10
⋮	⋮	⋮	⋮	⋮
15	0.78159	5.1878	(15 3)	15

Case (b):   Let $\rho = 6$ and place the sensitivity back in the first row. For this weighting, both rows are active in the optimization as shown by the gradual convergence of the upper and lower bounds (see Table 12.3). Note that, even though the controller order growth is comparable in both methods, the support characteristics are quite different. Most interestingly, the length of $\phi_{2,N}^o$ remains equal to 4 for $N > 5$, suggesting the possibility that, by changing the order of the outputs, a low-order suboptimal controller can be computed. This is in fact the case, as shown in Table 12.4. (This procedure does not apply to the FMV method since the suboptimal solutions obtained by this method are such that all entries of $\Phi(k)$ are supported at $k = N$.) It is interesting how in both cases (a) and (b), a proper ordering of the outputs results in a much better approximation of the solution (exact, if one row is TD). Indeed, after some N, the sequence of suboptimal controllers is of fixed order and asymptotically approaches the optimal one. This is not an isolated case. A large class of multiblock problems behave in this way when solved by the DA method. In other words, given a general multiblock problem, there seems to be a one-block partition that preserves a polynomial optimal solution, and further, such support structure is eventually captured by the Delay Augmentation method for a

TABLE 12.3   EXAMPLE I, CASE (B) WHERE NO ROW IS TD

N	DA				FMV		
	$\underline{\eta}_N$	$\bar{\eta}_N$	$\text{len}(\Phi_N^o)^T$	$\text{ord}(K)$	$\bar{\nu}_N$	$\text{len}(\Phi_N^o)^T$	$\text{ord}(K)$
1	0.78222	1.2243	(3 2)	2	-	-	-
2	0.79333	1.2547	(4 3)	3	-	-	-
3	0.90230	1.5255	(5 4)	5	1.3191	(4 4)	3
4	0.99522	1.0389	(5 4)	5	1.0564	(5 5)	4
5	1.0082	1.0321	(5 6)	5	1.0121	(6 6)	6
6	1.0024	1.0043	(7 4)	6	1.0044	(7 7)	7
7	1.0026	1.0030	(8 4)	7	1.0030	(8 8)	8
8	1.0026	1.0027	(9 4)	8	1.0027	(9 9)	9

TABLE 12.4   EXAMPLE I, CASE (B) WITH THE OUTPUTS REORDERED

N	$\underline{\eta}_N$	$\bar{\eta}_N$	$\text{len}(\Phi_N^o)^T$	$\text{ord}(K)$
1	0.95745	1.1602	(3 2)	2
2	0.95745	1.1602	(3 3)	2
3	0.98658	1.0586	(4 4)	3
4	0.99889	1.0157	(4 5)	3
5	1.0019	1.0053	(4 6)	3
6	1.0022	1.0031	(4 7)	3
7	1.0026	1.0027	(4 8)	3
8	1.0026	1.0026	(4 9)	3

large enough N. Then, a proper ordering of inputs and outputs that places the one-block partition in the first n_u rows and n_y columns of Φ (corresponding to U_1 and V_1) will generate a sequence of suboptimal controllers without order inflation.

The DA algorithm.   These observations suggest that an efficient algorithm for computing low-order suboptimal controllers can be as follows: Given a general multi-block problem,

Step 1:   Pick a positive integer N.

Step 2:   Solve the corresponding Delay Augmentation problem.

Step 3:   Compute $\text{len}(\Phi_N^o)$ and reorder inputs and outputs such that the set of $n_u \times n_y$ input-output pairs of minimum length correspond to Φ_{11}.

Step 4:   If reordering was necessary in Step 3, solve the reordered system for the same N. Then, check the difference between the upper and lower bounds, i.e., $\bar{\eta}_N - \underline{\eta}_N$. If this difference is small enough, stop; otherwise, increase N by one (or more) and go to Step 2.

To illustrate the workings of such algorithm we include a four-block example.

12.4.2 Example II

Consider the following 2-input-2-output four-block problem where the regulated signals are the output of the plant and the control sequence (weighted with the scalar ρ), and the input disturbances are a disturbance at the plant output with frequency weighting $\hat{W}_1(\lambda)$ and measurement noise with frequency weighting $\hat{W}_2(\lambda)$. That is,

$$\Phi = \begin{pmatrix} (1 - PK)^{-1}W_1 & PK(1 - PK)^{-1}W_2 \\ \rho K(1 - PK)^{-1}W_1 & \rho K(1 - PK)^{-1}W_2 \end{pmatrix}$$

where

$$\hat{W}_1(\lambda) = \frac{0.4}{1 - 0.6\lambda}, \qquad \hat{W}_2(\lambda) = \frac{1 - 0.75\lambda}{1 - 0.25\lambda}.$$

Let $\rho = 0.1$ and $\hat{P}(\lambda)$ be the same as in Example 12.4.1. Then, the results in Table 12.5 are obtained by applying the above algorithm starting with $N = 3$. For $N = 10$, the suboptimal controller is of order five and achieves a norm that is within half a percent of the optimal. In contrast, it can be shown that the FMV method has no polynomial feasible solution for any N (due to the way $\hat{W}_1$ and $\hat{W}_2$ enter the problem). This example shows how the Delay Augmentation algorithm can generate low-order suboptimal controllers even when the FMV method has no solution.

TABLE 12.5   EXAMPLE II: DELAY AUGMENTATION ALGORITHM

	DA					FMV	
N	$\underline{\eta}_N$	$\bar{\eta}_N$	$\text{len}(\Phi_N^o)$	$\text{ord}(K)$	Comments	$\bar{\nu}_N$	$\text{ord}(K)$
3	60.453	102.34	$\begin{pmatrix} 5 & 3 \\ 5 & 3 \end{pmatrix}$	4	Reorder inputs	-	-
3	60.400	81.161	$\begin{pmatrix} 3 & 5 \\ 3 & 5 \end{pmatrix}$	2	Keep order	-	-
4	64.702	81.161	$\begin{pmatrix} 3 & 6 \\ 3 & 7 \end{pmatrix}$	2	,,	-	-
5	68.284	81.161	$\begin{pmatrix} 3 & 7 \\ 5 & 9 \end{pmatrix}$	2	,,	-	-
6	70.721	72.850	$\begin{pmatrix} 6 & 7 \\ 6 & 11 \end{pmatrix}$	5	,,	-	-
7	70.754	71.874	$\begin{pmatrix} 6 & 8 \\ 8 & 13 \end{pmatrix}$	5	,,	-	-
8	70.888	71.500	$\begin{pmatrix} 6 & 9 \\ 10 & 15 \end{pmatrix}$	5	,,	-	-
9	71.040	71.615	$\begin{pmatrix} 6 & 11 \\ 10 & 17 \end{pmatrix}$	5	,,	-	-
10	71.089	71.408	$\begin{pmatrix} 6 & 12 \\ 12 & 19 \end{pmatrix}$	5	,,	-	-
11	71.110	71.146	$\begin{pmatrix} 12 & 13 \\ 13 & 21 \end{pmatrix}$	12	,,	-	-
12	71.113	71.122	$\begin{pmatrix} 13 & 14 \\ 19 & 23 \end{pmatrix}$	14	,,	-	-

12.5 SUPPORT STRUCTURE OF OPTIMAL SOLUTIONS

Here we explore the support characteristics of the optimal solution in more general terms. The numerical examples in the previous section suggest that it may be possible to infer the support of the optimal solution by observing how the superoptimal solution, Φ_N^o, evolves as N increases. Here we make an important step in this direction by showing that such support structure is "hinted to" by the support of the sequence of superoptimal and suboptimal solutions.

More precisely, assume that all optimal dual solutions, G^o, for a given multiblock problem are such that $|g_{ij}^o(t)| < \gamma_0^o(i)$ for some $i \in \{1, \ldots, n_z\}$, $j \in \{1, \ldots, n_w\}$ and all $t > T$ (i.e., the dual optimal functional does not "hit" the bound after some T). This implies, by the alignment conditions, that ϕ_{ij}^o will not be supported beyond T. The question then is: How does the sequence of superoptimal solutions, Φ_N^o, behave?

Under these circumstances, it will be shown that if N is large enough, $\phi^o_{ij,N}$ will not be supported between T and some large integer L, a function of N. Furthermore, while L increases with N, T remains constant, suggesting that the actual ϕ^o_{ij} is supported only up to T.

We have already shown that given a multiblock problem, there exists a subsequence of superoptimal dual solutions, $G^o_{N_s}$, whose $weak^*$ limit point, G^o, is feasible and optimal (Theorem 12.2.2) (i.e., $G^o_{N_s}$ converges to G^o on finite subsets of $\mathbf{Z}_+$). By exploiting this result and Theorem 9.3.3 (concerning alignment), we will show that the finitely supported part of the optimal solution is eventually "captured" by the sequence of superoptimal solutions.

Support proofs.   For the purpose of proving this result, we need the following well-known lemma.

Lemma 12.5.1.   If a sequence $G_N \in \ell^{n \times m}_\infty$ converges $weak^*$ to G, then for any positive integer $L < \infty$, $\|P_L(G_N - G)\|_\infty \to 0$ as $N \to \infty$.   ∎

Note that the above lemma implies that each individual entry of G_N also enjoys this convergence property, i.e., $\|P_L(g_{ij,N} - g_{ij})\|_\infty \to 0$ as $N \to \infty$, for all $i = 1, \ldots, n$ and $j = 1, \ldots, m$.

Next, let us review the alignment properties of the optimal solutions. By Theorem 9.3.4, each optimal solution to the primal problem must be aligned with every optimal solution to the dual problem. In particular, if an optimal dual solution, G^o, is such that

$$|g^o_{ij}(t)| < \max_{1 \le j \le n_w} \|g^o_{ij}\|_\infty \text{ for all } t > T,$$

then all optimal primal solutions are such that $\phi^o_{ij}(t) = 0$ for all $t > T$ (see Theorem 9.3.3 part (1)). Note that, according to the notation developed at the end of Chapter 3, $\max_{1 \le j \le n_w} \|g^o_{ij}\|_\infty$ is nothing but $\gamma^o_0(i)$. The next theorem puts all these pieces together.

Theorem 12.5.1.   Given a multiblock problem, if all optimal dual solutions are such that $|g^o_{ij}(T)| = \gamma^o_0(i)$ for some $T \in \mathbf{Z}_+$ and $|g^o_{ij}(t)| < \gamma^o_0(i)$ for all $t > T$; then, for every $L > T$ there exists a positive integer N^* such that $\phi^o_{ij,N}(t) = 0$ for $T < t \le L$ and for any $N \ge N^*$.

Proof. (Note: To simplify the notation we drop subindices i, j and superindex 'o'.) Given some $L > T$, pick $\varepsilon > 0$ such that

$$\min_{T < t \le L} (\gamma_0 - |g(t)|) = \varepsilon. \tag{12.29}$$

By Lemma 12.5.1, for every $L > T$ there exists N^* such that

$$\|P_L(g_N - g)\|_\infty < \frac{\varepsilon}{2} \tag{12.30}$$

for all $N > N^*$. First we prove (by contradiction) that $|g_N(t)| < \gamma_{0,N}$ for $T < t < L$ and for any $N \geq N^*$. The result then follows from the alignment conditions.

Given $N > N^*$, assume that $|g_N(t_1)| = \gamma_{0,N}$ for $T < t_1 \leq L$. Then, by Equations (12.29) and (12.30),

$$\gamma_{0,N} - \gamma_0 \leq |g_N(t_1)| - |g(t_1)| - \varepsilon < \frac{\varepsilon}{2} - \varepsilon.$$

Therefore,

$$\gamma_0 - \gamma_{0,N} > \frac{\varepsilon}{2}. \tag{12.31}$$

Next, consider the point $t = T$. From Equation (12.30) and the fact that $|g_N(t)| \leq \gamma_{0,N}$ in general, we have

$$\gamma_0 - \gamma_{0,N} \leq |g(T)| - |g_N(T)| < \frac{\varepsilon}{2}$$

which contradicts Equation (12.31). This implies that $\phi_N(t) = 0$ for $T < t \leq L$ and $N > N^*$, which is the desired result. ∎

In other words, given the conditions of the theorem above, and for N large enough, there is a "gap" of zeros (between T and L) in $\phi_{ij,N}^o(t)$ that gets wider as N increases, i.e., as L increases. However, T does not change, giving precise information on the length of the finitely supported entries of Φ^o. The difficulty is that we do not have an a priori estimate of how large N has to be to capture T.

It is worth pointing out that Theorem 12.5.1 can be applied to the FMV sequence of suboptimal solutions too, since the corresponding duals also have a $weak^*$ convergent subsequence. However, an important difference exists in the way the DA and FMV sequence of solutions behave, which was pointed out in the previous section. Indeed, while the FMV solutions are consistently supported for $t > L$, the DA solutions are not. This observation was crucial in constructing low-order suboptimal controllers. We expand these ideas in the following section.

12.5.1 Observations

This section includes a few observations based on a fair amount of computational experience using the Delay Augmentation method and on some intuitive ideas on the problem of ℓ_1 optimization in general. It is by no means a formal or precise presentation. However, it is intended to give more insight into the DA approximation method.

Consider the way the DA method works. It transforms a general multiblock problem into a square one, thereby generating polynomial superoptimal solutions, Φ_N^o. Without changing the order of inputs and outputs, the sequence Φ_N^o will increase its length as N increases. However, it was noted in previous examples that not every entry of Φ_N^o increases its length in the same way. In fact, a closer look at the sequence Φ_N^o suggests that the support of some of its entries stops changing after some N. This is exactly what happened in Example 12.4.1, case (a) and case (b), where the support of one of the entries of Φ_N^o remained the same after some N regardless of the ordering. In Example 12.4.2, the pattern also occurs but for $N > 12$ (not shown in Table 12.5).

Next, note that $\Phi_{11,N}^o = \bar{\Phi}_{11,N}$ since that block of the problem is not affected by the extra free parameters. Therefore, for each N, $\bar{\Phi}_{11,N}$ is polynomial. If those entries of Φ_N^o that have constant support after some N are collected (by reordering) in $\Phi_{11,N}^o$, then $\bar{\Phi}_{11,N}$ will have constant support. Interestingly, those entries of constant support seem to be always enough to define a one-block partition and therefore fill the necessary entries of $\bar{\Phi}_{11,N}$. Furthermore, most multiblock problems seem to have this property.

A multiblock problem in this class can be viewed as dominated by a one-block partition. In other words, there is an embedded one-block problem that is further constrained by the rank interpolation conditions. Such constraints, however, are not enough to change the polynomial nature of the optimal solution corresponding to that partition, although, in general, they have the effect of increasing its order. With this we extend the notion of TD one-block partitions where the added constraints due to the rank interpolation conditions are redundant.

Definition 12.5.1. Given a multiblock problem, a one-block partition is *Partially Dominant* (PD) if all ℓ_1 optimal solutions are polynomial in the entries corresponding to such partition. ∎

Clearly, a TD one-block partition is also PD but not vice versa. Therefore, if a given multiblock problem has a PD one-block partition, a sequence of DA solutions will eventually capture the support of such partition, independently of the ordering. This observation is supported by a fair amount of numerical experiments covering most combinations (i.e., two-block row and column problems and four-block problems with different input-output dimensions). At the same time, it is consistent with Theorem 12.5.1 but stronger. Indeed, the observation indicates that the superoptimal solution will not be supported for $t \geq L$. If proven correct, this characteristic has interesting consequences. To illustrate some of the ideas involved, consider the following simple two-block column problem:

$$\begin{pmatrix} \phi_1 \\ \phi_2 \end{pmatrix} = \begin{pmatrix} h_1 \\ h_2 \end{pmatrix} - \begin{pmatrix} u_1 \\ u_2 \end{pmatrix} q$$

and assume, without loss of generality, that $\hat{u}_1$ and $\hat{u}_2$ are polynomials (this can always be obtained by polynomial factorization of $\hat{U}$). Further, assume that $(h_1 \quad h_2)^T$ is a polynomial feasible solution and that the outputs are ordered such that ϕ_1 is PD. (These assumptions seem to be quite restrictive but in fact are often satisfied.) Then we have the following equality due to the rank interpolation conditions:

$$\hat{u}_2 \hat{\phi}_1^o - \hat{u}_1 \hat{\phi}_2^o = \hat{u}_2 \hat{h}_1 - \hat{u}_1 \hat{h}_2. \tag{12.32}$$

Assume that all zeros common to $\hat{u}_1$ and $\hat{u}_2$ have been canceled out of the above equation. Clearly, the right-hand side of Equation (12.32) is polynomial, and, furthermore, the first term on the left-hand side is polynomial since we assumed that ϕ_1 is PD. Therefore, the second term on the left-hand side must be polynomial. This implies that two situations are possible: Either $\hat{\phi}_2^o$ is polynomial or it has stable poles that are canceled by stable zeros of $\hat{u}_1$.

This observation has interesting implications. On one hand, there is a class of multiblock problems with polynomial optimal solutions that is characterized by the absence of stable zeros in $\hat{u}_1$. Such solutions can then be computed exactly by either the FMV or the DA method. An example illustrating this situation is presented in the last section of this chapter. On the other hand, if $\hat{u}_1$ has stable zeros and ϕ_2^o is infinitely supported, the rate at which ϕ_2^o decays is given by a subset of the stable zeros of $\hat{u}_1$. This information could be used to transform the original problem into a finite-dimensional one for which exact solutions are computable. Finally, it should be noted that the above ideas can be easily extended to the general multiblock problem.

12.6 REDUNDANCY

In the previous chapter we noted that the feasibility constraints can be defined in a minimal fashion for the exact primal and dual problems. It is interesting to point out that the minimality of constraints is preserved in the DA method (for being a one-block problem) and in the FME method (for taking a subset of linearly independent constraints). The FMV method, however, may generate redundant constraints depending on the problem data. A trivial example is one where the order of approximation N is smaller than the algebraic multiplicity of some zero in Λ_{UV}. In such case, the matrix A_{zero} will not be full rank.

From a numerical point of view, even though the constraints are guaranteed to be independent (for the DA and FME methods), the resulting constraint matrix can be badly conditioned (i.e., numerically rank deficient). Therefore, in practice, it is necessary to take care of "numerically redundant" constraints.

12.7 THE FULL-STATE FEEDBACK PROBLEM

In this section we study the full-state feedback problem in the context of ℓ_1 optimization. More precisely, the conditions under which there exists a static linear controller that achieves optimality. To simplify the analysis, we will study systems with scalar control and scalar disturbance. In particular, two different types of problems within this class of systems will be considered:

1. One-block problems with a scalar regulated output, and
2. Two-block column problems with two regulated outputs, where one of the outputs is the scalar control signal.

For systems in (a), we will show that there exists a static controller that is ℓ_1-optimal if the nonminimum phase zeros of the transfer function from the control input to the regulated output satisfy a simple algebraic condition. Violating such condition, however, may result in a dynamic ℓ_1-optimal controller of possibly high order (generally when the nonminimum phase zeros are "close" to the unit circle). For systems in (b), we show by means of an example that optimal controllers are dynamic in a broad class of cases that are common in control design.

12.7.1 Problem Formulation

Consider the following state-space minimal realization of a full-state feedback system of order n, with scalar input disturbance, scalar control, and scalar regulated output:

$$G = \left[\begin{array}{c|cc} A & b_1 & b_2 \\ \hline c_1 & 0 & d_{12} \\ I & 0 & 0 \end{array} \right],$$

where $A \in \mathbb{R}^{n \times n}$, b_1 and $b_2 \in \mathbb{R}^{n \times 1}$, $c_1 \in \mathbb{R}^{1 \times n}$, and $d_{12} \in \mathbb{R}$. The ℓ_1 problem can be stated as follows:

$$\inf_{q \in \ell_1^{1 \times n}} \|h - uqv\|_1 = \inf_{q \in \ell_1^{1 \times n}} \|\phi\|_1. \tag{12.33}$$

Thus, h and $u \in \ell_1$, and $v \in \ell_1^{n \times 1}$. State-space realizations for h, u, and v can be found by using the state-space formulae given in Chapter 5 with the observer gain matrix equal to $-A$. For this specific choice, the realizations are

$$\hat{h}(\lambda) = \lambda(A_f, A b_1, c_1 + d_{12}f, c_1 b_1), \tag{12.34}$$

$$\hat{u}(\lambda) = (A_f, b_2, c_1 + d_{12}f, d_{12}),$$

$$\hat{v}(\lambda) = (0, b_1, I, 0) = \lambda b_1,$$

where $A_f := A + b_2 f$, and $f \in \mathbb{R}^{1 \times n}$ is chosen such that all the eigenvalues of A_f are inside the unit disc.

12.7.2 One-Block Problems with Minimum Phase Plants

Here we consider the case where the transfer function from the control input to the regulated output is minimum phase except for an integer number of unit delays (i.e., zeros at the origin in the λ-plane). It will be assumed throughout that (A, b_2) is reachable.

Theorem 12.7.1. For such a system, the static state feedback gain, f^*, which places the eigenvalues of $(A + b_2 f^*)$ at the exact location of the minimum phase zeros of (A, b_2, c_1, d_{12}) and the rest at the origin, is ℓ_1-optimal.

Proof. Consider using f^* as the state feedback gain in the parameterization described above. Then, after carrying out all stable pole-zero cancellations,

$$\hat{u}(\lambda) = \gamma_r \lambda^r,$$

where r is the number of unit delays in (A, b_2, c_1, d_{12}) and γ_r is a scalar depending on r. In what follows, the cases where $r = 0$ and $r > 0$ will be treated separately.

1. If $r = 0$, then $d_{12} \neq 0$, $c_1 + d_{12} f^* = 0$, and $\hat{u}(\lambda) = d_{12}$. Also, from Equation (12.34),

$$\hat{h}(\lambda) = c_1 b_1 \lambda = (c_1 b_1 + d_{12} f^* b_1 - d_{12} f^* b_1)\lambda = -d_{12} f^* b_1$$

$$\implies \hat{\phi}(\lambda) = -d_{12} f^* b_1 \lambda - d_{12} \hat{q}(\lambda) b_1 \lambda.$$

Thus, the ℓ_1-optimal solution is given by $\hat{q}^o(\lambda) = -f^*$, and $\hat{\phi}^o(\lambda) = 0$. Furthermore, using the state-space equations for computing the optimal controller, it can be shown after a little algebra that $\hat{k}^o(\lambda) = f^*$.

2. If $r > 0$, then $d_{12} = 0$, $c_1 A_{f*}^r = 0$ by construction since (A, b_2) is reachable. Also $\hat{u}(\lambda) = c_1 A_{f*}^{r-1} b_2 \lambda^r$. Again, from Equation (12.34),

$$\hat{h}(\lambda) = c_1 b_1 \lambda + c_1 A b_1 \lambda^2 + c_1 A_{f*} A b_1 \lambda^3 + \cdots + c_1 A_{f*}^{r-1} A b_1 \lambda^{r+1}.$$

Therefore, the closed-loop pulse response is given by

$$\hat{\phi}(\lambda) = c_1 b_1 \lambda + c_1 A b_1 \lambda^2 + c_1 A_{f*} A b_1 \lambda^3 + \cdots$$
$$+ c_1 A_{f*}^{r-2} b_1 \lambda^r + c_1 A_{f*}^{r-1} (A - b_2 \hat{q}(\lambda)) b_1 \lambda^{r+1}.$$

Clearly, q does not affect the first $r + 1$ elements of ϕ (i.e., $\phi(i), i = 0, 1, \ldots, r$). It follows that the best possible choice of q, in the sense of minimizing the ℓ_1-norm of ϕ, is the one that makes $\phi(i) = 0$ for $i = r + 1, r + 2, \ldots$, and is achieved by letting $\hat{q}^o(\lambda) = -f^*$, since in that case $\phi(r + 1) = c_1 A_{f*}^r b_1 = 0$. Again, the corresponding ℓ_1-optimal controller is f^*. ∎

Corollary 12.7.1. The ℓ_1-optimal closed-loop transfer function of the system considered in Theorem 12.7.1 (with $r > 0$) is given by

$$\hat{\phi}^o(\lambda) = c_1 \sum_{i=1}^{r} A^{i-1} \lambda^i b_1.$$

Proof. It follows directly from the fact that $c_1 A_{f*}^i b_2 = 0$ for $i = 0, 1, \ldots, r - 2$. ∎

Put in words, Theorem 12.7.1 says that there is nothing the controller can do to invert the delays in the system. It can, however, cancel the rest of the dynamics of the system due to the absence of nonminimum phase zeros in the transfer function from the control input to the regulated output. This results in an optimal closed-loop pulse response that is equal to the open-loop pulse response in its first $(r + 1)$ elements and zero thereafter.

12.7.3 One-Block Problems with Nonminimum Phase Plants

This subsection considers those cases where (A, b_2, c_1, d_{12}) has r nonminimum phase zeros not necessarily at the origin (i.e., $\lambda = 0$).

Again, we use the same parameterization as in the previous case. That is, we choose f^* to place $(n - r)$ eigenvalues of A_{f^*} at the exact location of the minimum phase zeros of (A, b_2, c_1, d_{12}) and the rest (r) at the origin. Then, from the discussion in the previous subsection, $\hat{h}(\lambda)$ is polynomial in λ and of order $(r + 1)$, $\hat{u}(\lambda)$ is polynomial too, but of order r, and $\hat{v}(\lambda)$ is simply λb_1. Therefore, the closed-loop transfer function can be written as follows:

$$\hat{\phi}(\lambda) = \left(g_1 \prod_{i=1}^{r} (\lambda - \xi_i) - g_2 \prod_{j=1}^{r} (\lambda - \lambda_j) \hat{\tilde{q}}(\lambda) \right) \lambda =: \hat{\tilde{\phi}}(\lambda) \, \lambda, \qquad (12.35)$$

where $g_1, g_2 \in \mathbb{R}$, the ξ_i's are the zeros of $\hat{h}$, the λ_j's are the zeros of $\hat{u}$ (i.e., unstable zeros of (A, b_2, c_1, d_{12})), and $\hat{\tilde{q}}(\lambda) := \hat{q}(\lambda) b_1 \in \ell_1$. Note that $\|\phi\|_1 \equiv \|\tilde{\phi}\|_1$. Also, by Corollary 12.1.1, $\hat{\tilde{\phi}}^o(\lambda)$ is polynomial in λ, which implies that $\hat{\tilde{q}}^o(\lambda)$ is polynomial in λ. Thus, the optimization problem is equivalent to the following linear programming (primal) problem: for a sufficiently large but finite positive integer N,

$$\min_{\tilde{\phi}} \sum_{i=0}^{N} \left| \tilde{\phi}(i) \right|, \qquad (12.36)$$

subject to

$$\sum_{i=0}^{N} \tilde{\phi}(i) \lambda_j^i = g_1 \prod_{i=1}^{r} (\lambda_j - \xi_i) \, , \quad j = 1, 2, \ldots, r.$$

In the above we have assumed that the λ_j's are simple zeros (i.e., zeros with geometric multiplicity equal to one) to simplify the formulation of the zero interpolation conditions. The following results, however, carry over to the more general case.

The following theorem will prove to be useful (see Exercise 10.7).

Theorem 12.7.2. The order of $\hat{\tilde{\phi}}^o$ in (12.36) equals the number of constraints minus one, $r - 1$, if

$$\sum_{i=0}^{r-1} |a_i| < 1, \qquad (12.37)$$

where $\prod_{j=1}^{r} (\lambda - \lambda_j) = \lambda^r + a_{r-1} \lambda^{r-1} + \cdots + a_1 \lambda + a_0$. ∎

Now we are ready to present the next result.

Theorem 12.7.3. Let (A, b_2, c_1, d_{12}) have r nonminimum phase zeros, then if condition (12.37) is satisfied, f^* is ℓ_1-optimal.

Proof. By theorem 12.7.2, $\hat{\tilde{\phi}}^o(\lambda)$ is of order $(r - 1)$. Then, considering the order of each term in (12.35), it is clear that $\hat{\tilde{q}}^o(\lambda)$ has to be constant and such that $\tilde{\phi}(r) = 0$. Using the state-space formulae (12.34),

$$0 = \tilde{\phi}(r) = (c_1 + d_{12} f^*) A_{f^*}^{r-1} (A b_1 - b_2 \tilde{q}^o(0))$$

$$= (c_1 + d_{12} f^*) A_{f^*}^{r-1} (A - b_2 q^o(0)) b_1.$$

But, by construction, $(c_1 + d_{12}f^*)A^r_{f*} = 0$ due to the stable pole-zero cancellations and the fact that the rest of the poles are placed at the origin. Therefore, $\hat{q}^o = -f^*$ is the required value, and $\hat{k}^o = f^*$. ∎

Observation: It remains to consider those cases where the nonminimum phase zeros of (A, b_2, c_1, d_{12}) are such that they violate condition (12.37). Theorem 12.7.2 established a sufficient condition to determine the order of the optimal response. If condition (12.37) is violated, the optimal closed-loop response may be of higher order, possibly greater than n, but still polynomial. If that is the case, then the ℓ_1-optimal controller is necessarily dynamic, since the highest-order polynomial response that a static controller can generate is n, by placing all closed-loop poles of the plant at the origin. Any polynomial response of order greater than n, say N, requires a dynamic compensator of at least order $N - n$. Thus, $\sum_{i=0}^{r-1} |a_i| \geq 1$ can be viewed as a necessary condition for the optimal controller to be dynamic.

The following example shows that a large class of state feedback one-block problems are dynamic.

Example 12.7.1

Consider the following parameterized family of plants (with parameter κ),

$$P_\kappa(\lambda) = \frac{\lambda(\kappa\lambda^2 - 2.5\lambda + 1)}{(1 - 0.2\lambda)(23\lambda^2 - 2.5\lambda + 1)}.$$

Assume that the controller has access to the state vector and that the disturbance acts at the plant input, i.e.,

$$z = P_\kappa(u + w), \quad y = x.$$

The nonminimum phase zeros relevant to this problem are given by the roots of $\kappa\lambda^2 - 2.5\lambda + 1$, as a function of κ. It is easy to see that for $\kappa > 3.5$ condition (12.37) is satisfied and the optimal controller is f^*. Further, it can be shown that for $\kappa = 3.5$ the optimal solution is not unique. Actually, two possible solutions with $\|\phi^o\|_1 = 7$ are

$$\hat{\phi}^o_{\kappa=3.5} = \begin{cases} \lambda - 2.5\lambda^2 + 3.5\lambda^3 \\ \lambda - 1.1\lambda^2 + 4.9\lambda^4 \end{cases}.$$

The first is achieved with f^* whereas the second requires a first-order controller. Note that for this value of κ, the left-hand side of (12.37) is equal to one.

For $1.5 < \kappa < 3.5$ condition (12.37) is violated and the optimal solution has the following general form:

$$\hat{\phi}^o_{1.5 < \kappa < 3.5} = \lambda + \phi_\kappa(2)\lambda^2 + \phi_\kappa(N_\kappa)\lambda^{N_\kappa}.$$

As $\kappa \searrow 1.5$, one of the nonminimum phase zeros approaches the boundary of the unit disc while $\phi_\kappa(2) \longrightarrow -1.5$, $\phi_\kappa(N_\kappa) \longrightarrow 0.5$, and, most important, $N_\kappa \nearrow \infty$. This implies that the optimal controller can have arbitrarily large order. For instance, if $\kappa = 1.51$, then

$$\hat{\phi}^o_{\kappa=1.51} \simeq \lambda - 1.4907\lambda^2 + 0.5776\lambda^{12}$$

and the optimal compensator is of order 9. It is also interesting to point out that for $\kappa < 1.5$ one of the nonminimum phase zeros leaves the unit disc and condition (12.37) is again satisfied. In this case, $\hat{\phi}^o_{\kappa<1.5} = \lambda - 1.5\lambda^2$ and $\hat{k}^o = f^*$. With regard to the optimal norm, it drops from a value arbitrarily close but greater than 3 to a value of 2.5 in the transition.

We have seen a similar behavior for the case of sensitivity minimization through output feedback. The above example shows that the nature of such solutions have comparable characteristics even under full-state feedback. There is one difference, however, that reflects the structure added to the problem. This setup requires at least a second-order plant with two nonminimum phase zeros away from the origin. Otherwise, condition (12.37) is automatically satisfied.

12.7.4 A Two-Block Column Problem

So far we have considered problems with a scalar-regulated output. We could argue that sensitivity minimization problems, such as the one in the above example, where a measure of the control effort is not included in the cost functional, may have peculiar solutions that could hide the structure of the more general nonsingular case. To clarify this point, we will consider a variation of the above example by including the control effort in the cost functional. That is,

$$v^o = \inf_{k-stab.} \left\| \begin{array}{c} \phi_1 \\ \rho\,\phi_2 \end{array} \right\|_1 \tag{12.38}$$

where ϕ_1 represents the closed-loop map from the disturbance to the output of the plant, ϕ_2 represents the closed-loop map from the disturbance to the control input, and ρ is a positive scalar weight.

Example 12.7.2

Consider Problem (12.38) for the parameterized family of plants of Example 12.7.1. By expanding each term, Equation (12.38) can be rewritten as

$$v^o = \inf_{q \in \ell_1^{1 \times n}} \left\| \left(\begin{array}{c} h_1 \\ \rho h_2 \end{array} \right) - \left(\begin{array}{c} u_1 \\ \rho u_2 \end{array} \right) q v \right\|_1 \tag{12.39}$$

where, according to the previous parameterization (and using the same notation),

$$\begin{aligned} \hat{h}_1(\lambda) &= \lambda[A_{f^*}, Ab_1, c_1 + d_{12}f^*, c_1b_1], \\ \hat{h}_2(\lambda) &= \lambda[A_{f^*}, Ab_1, f^*, 0], \\ \hat{u}_1(\lambda) &= [A_{f^*}, b_2, c_1 + d_{12}f^*, d_{12}], \\ \hat{u}_2(\lambda) &= [A_{f^*}, b_2, f^*, 1], \\ \hat{v}(\lambda) &= \lambda b_1. \end{aligned} \tag{12.40}$$

With the particular problem data and $\kappa = 2$, $\hat{v}(\lambda)$ has a zero at the origin and $(\hat{u}_1(\lambda) \quad \rho\hat{u}_2(\lambda))^T$ has no unstable zeros, i.e., $\hat{u}_1(\lambda)$ and $\hat{u}_2(\lambda)$ are coprime. Furthermore, neither $\hat{u}_1(\lambda)$ nor $\hat{u}_2(\lambda)$ has stable zeros. According to the discussion at the end of Section 12.5, this implies that if any one of the two rows has a polynomial optimal solution, then the other row must also be polynomial, and the exact optimal solution is computable. Indeed, this is the case at hand. By applying the DA method to the problem, for $N = 5$ we obtain:

$$\hat{\phi}_1(\lambda) = \lambda - \frac{887}{558}\lambda^2 + \frac{631}{558}\lambda^4 + \frac{308}{558}\lambda^5, \tag{12.41}$$

$$\hat{\phi}_2(\lambda) = -\frac{998.6}{558}\lambda + \frac{11895.4}{558}\lambda^2 + \frac{8955.4}{558}\lambda^3 + \frac{1282.2}{558}\lambda^4 - \frac{708.4}{558}\lambda^5, \qquad (12.42)$$

for which $\underline{\eta}_5 = \bar{\eta}_5 = 1192/279 \simeq 4.2724$. Thus, Equation (12.41) is the exact optimal solution for the two-block problem, and it achieves a norm $\nu^o = 1192/279$. Therefore, since such solution has finite support and is of fifth order, the optimal controller is necessarily dynamic, specifically, of second order. Also note that the optimal closed-loop response is such that $\|\phi_1\|_1 = \rho\|\phi_2\| = \nu^o$.

It is also interesting to consider the singular problem corresponding to this example (i.e., $\kappa = 2$ and $\rho = 0$). The optimal solution (which is obtained by eliminating the second row and solving the resulting SISO problem) is given by

$$\hat{\phi}_1(\lambda) = \lambda - \frac{90}{68}\lambda^2 + \frac{128}{68}\lambda^5,$$

$$\hat{\phi}_2(\lambda) = -\frac{103.6}{68}\lambda + \frac{1446}{68}\lambda^2 + \frac{1394.4}{68}\lambda^3 + \frac{1136}{68}\lambda^4 - \frac{294.4}{68}\lambda^5,$$

where $\|\phi_1\|_1 = 286/68 \simeq 4.2059$ while $\|\phi_2\|_1 = 4374.4/68 \simeq 64.3294$ is clearly larger since it was left out of the optimization. In fact, the above solution is valid for $\rho \in \left[0, 286/4374.4\right]$ since for any ρ in such interval $\|\phi_1\|_1 \geq \rho\|\phi_2\|_1$. In other words, for any such ρ, ϕ_1 is TD and the ℓ_1-optimal controller is dynamic and of second order. This alone constitutes a family of problems, parameterized by $\rho \in [0, 286/4374.4]$, requiring dynamic optimal controllers.

The above indicates that given a nonsingular (two-block) problem, the optimal controller may very well be dynamic, whether or not there is a TD row in the problem. Further, it can be shown through a numerical example that even when the corresponding one-block problem has a static optimal controller, the two-block problem (which includes the control effort in the cost function) may require a dynamic one.

A last question remains to be answered: Given a full-state feedback problem with a dynamic ℓ_1-optimal controller, is it possible to find a static controller that achieves an ℓ_1-norm arbitrarily close to the optimal? Again, it is easy to show via a numerical counter example that this is not the case. In fact, a simple second-order problem can show that the gap between the norms achieved by the optimal and the static-optimal controller can be significant.

12.8 SUMMARY

In this chapter we have developed ways of solving MIMO ℓ_1 model matching problems, in both the one-block and multiblock case. While one-block problems can be solved exactly, multiblock problems require some form of approximation and only suboptimal solutions can be obtained in general. We have seen essentially two iterative methods of finding suboptimal solutions with a prespecified accuracy: the FMV-FME combination and the Delay Augmentation method. In the first approach, two finite linear programs are solved at each iteration, and the order of solution typically grows with the order of the approximation. In the DA approach, one finite linear program is solved at each iteration, and, by reordering inputs and outputs, it is possible to avoid order inflation. Finally, we discussed the special problem of full-state feedback, and we showed that solutions may be of high order requiring dynamic controllers.

EXERCISES

12.1. In the FMV method, show how one can approximate the problem by one that will always have a feasible FIR solution.

12.2. Derive the support characteristics of the FMV method that parallel the results of the DA method.

12.3. Interpret the DA method directly from the 2-input 2-output description. Show that it corresponds to adding fictitious delayed control inputs and measured outputs.

12.4. Prove Corollary 12.2.4.

12.5. Consider the DA method. Instead of augmenting the matrices U and V with delays, suppose you augment them with constants such that the resulting matrices are square. Now consider solving the square problem and taking the limit as the constants approach zeros. Will this method work? Explain.

12.6. In Chapter 7, we have shown that the robust performance problem in ℓ_1 corresponds to solving

$$\inf_{\text{K stabilizing}} \rho(\widehat{\Phi}) = \inf_{\text{K stabilizing}} \inf_{D \in \mathbf{D}} ||D^{-1}\Phi D||_1.$$

(a) Show that the spectral radius is not a convex function of $\widehat{\Phi}$.

(b) We propose the following iterative scheme: First fix D and minimize over K which is a standard ℓ_1 problem. Then fix the controller that results from this minimization and minimize over D. The latter is an eigenvalue problem. Explain how one can adapt the methods FMV and DA to solve such problems.

(c) Argue that this iteration stops if for some D, the optimal ℓ_1 solution has all row norms equal to the minimum norm. In particular, find the relation between the norm of such a solution and its spectral radius. Is this a serious limitation?

(d) Solve the above problem exactly if $\Phi = H - UQV$ with $\hat{U} = (\lambda - a)I$, $-1 < a < 1$ and $V = I$.

NOTES AND REFERENCES

The use of FIR approximations for designing filters by exploiting linear programming has been extensively discussed in the signal-processing literature; see [CRR69, OS70, Rab72].

Obtaining approximate solutions to the ℓ_1 multiblock problem using (FMV) approximation was proposed in [DP88a] and further developed in [MP91, Sta90]. Although convergence was guaranteed, the method did not produce lower-bound approximations to test for optimality. The use of duality to obtain lower-bound approximations was proposed in [Dah92, Sta91] in which the (FME) method was introduced; see also [DBD93, Sta90]. In [Sta91], an example was computed to show that such approximate methods always yield unnecessarily high-order solutions, and that the optimal may be of much lower order. The Delay Augmentation algorithm was proposed in [DBD93] to obtain further information about the structure of the optimal controller as well as make the computations systematic. A discussion on the equivalence between the duality theory of distance problems and duality of general convex problems using Lagrange multipliers in the context of ℓ_1 minimization was reported in [DK93].

The interest in the state-feedback problem is historical. The parallel $\mathcal{H}_2$ and $\mathcal{H}_\infty$ problems in the case of state feedback (or full information) have been analyzed in [Kuc79, DGKF89, ZK88, IG91] and will be discussed in the next chapter. The result shown in this chapter was reported in [DBD92]. If the feedback is allowed to be nonlinear, then the optimal state feedback can be static [Sha93].

Solutions to the ℓ_1 problem were extended to deal with multirate system [DVV92], sampled data systems [BDP93, DF92], and slowly varying systems (as in [WZ91, ZW91, DD91]) [DD90, VDVb]. Estimation and filtering problems based on ℓ_∞ to ℓ_∞ have been studied in [MB82, Vou93, NP92].

CHAPTER 13 ———————————————

State-Space Solution to the $\mathcal{H}_2$ and $\mathcal{H}_\infty$ Problems

In the last two chapters we have concentrated on the study of the general ℓ_1 problem in order to highlight the various computational techniques that can be utilized for problems with linear objectives and an infinite number of linear constraints. These techniques will be quite useful when we consider mixed objectives as will be seen in Chapter 14. At this point, however, it is interesting to make a digression from the general theme and present the solutions for the general $\mathcal{H}_2$ and $\mathcal{H}_\infty$ problems. The corresponding SISO model matching problems were presented in Chapter 10.

The presentation in this chapter serves several purposes. The solution of the $\mathcal{H}_2$ problem is a very nice application of the ideas presented in Section 5.3.4 in which the algebraic separation extends to a norm separation between the full information problem and the output estimation problem. A similar, but not as transparent, separation appears in the $\mathcal{H}_\infty$ problem. Solutions to both these problems are quite elegant in the sense that formulae for the optimal (or suboptimal) controller are given explicitly. The order of the controller is no higher than the order of the original general 2-input 2-output plant.

In the sequel, we will provide complete proofs for the $\mathcal{H}_2$ problem primarily to highlight the separation ideas. As for the $\mathcal{H}_\infty$ problem, we will be content with the final results with intuitive (not very rigorous) explanations. A complete treatment of the $\mathcal{H}_\infty$ problem will take us too far from the main theme of this book.

13.1 PROBLEM DEFINITION

Let G be given by

$$G = \left[\begin{array}{c|cc} A & B_1 & B_2 \\ \hline C_1 & D_{11} & D_{12} \\ C_2 & D_{21} & 0 \end{array} \right].$$

The following assumptions are made:

1. (A, B_2) is stabilizable and (C_2, A) is detectable.
2. D_{12} has full column rank and there exists a matrix $D_\perp$ such that $(\begin{array}{cc} D_{12} & D_\perp \end{array})$ is unitary.

3. D_{21} has full row rank and there exists a matrix $\tilde{D}_\perp$ such that $\left(\begin{array}{c} D_{21} \\ \tilde{D}_\perp \end{array} \right)$ is unitary.

4. The matrix

$$\left(\begin{array}{cc} A - e^{i\theta} I & B_2 \\ C_1 & D_{12} \end{array} \right)$$

   has full column rank for all $\theta \in [0, 2\pi)$.
5. The matrix

$$\left(\begin{array}{cc} A - e^{i\theta} I & B_1 \\ C_2 & D_{21} \end{array} \right)$$

   has full row rank for all $\theta \in [0, 2\pi)$.
6. D_{22} is equal to zero. This implies that the plant is strictly proper.

Assumption 1 is necessary for the existence of stabilizing controllers. Assumption 2 indicates that all the control signals are penalized. Assumption 3 is the dual of assumption 2. Assumptions 4 and 5 are technical to guarantee the existence of solutions to certain Riccati equations. In the case where $D_{12}^T C_1 = 0$, we can replace condition 4 by assuming that (C_1, A) has no unobservable modes on the unit circle. Similarly for assumption 5. The assumption $D_{22} = 0$ is not necessary but is introduced for simplicity.

13.2 THE $\mathcal{H}_2$ PROBLEM

In this section we present a solution to the $\mathcal{H}_2$ r .mization problem for nonsquare plants. We will show that the separation structure .i stabilizing controllers depicted in Chapter 5 extends to a strong separation in the $\mathcal{H}_2$ minimization problem.

Simplifying assumptions.   To make the presentation more lucid, we make the following extra assumptions:

1. $D_{12}^T (C_1 \quad D_{12}) = (0 \quad I)$.
2. $\begin{pmatrix} B_1 \\ D_{21} \end{pmatrix} D_{21}^T = \begin{pmatrix} 0 \\ I \end{pmatrix}$
3. $D_{11} = 0$.

It follows from these assumptions that assumption (4) reduces to the condition that (C_1, A) has no unobservable modes on the unit circle, and assumption (5) reduces to the condition that (A, B_1) has no unreachable modes on the unit circle.

Performance objective.   The objective is to minimize the $\mathcal{H}_2$ norm of the closed-loop map T_{zw}. In time-domain, this norm is equivalent to computing the energy of the output for an impulse input of unit length. To illustrate this, notice that for any input w, the energy in the regulated output is given by

$$\|z\|_2^2 = \sum_{t=0}^{\infty} z(t)^T z(t)$$

$$= \sum_{t=0}^{\infty} (x(t)^T \quad u(t)^T) \begin{pmatrix} C_1^T \\ D_{12}^T \end{pmatrix} (C_1 \quad D_{12}) \begin{pmatrix} x(t) \\ u(t) \end{pmatrix}$$

$$= \sum_{t=0}^{\infty} x(t)^T C_1^T C_1 x(t) + u(t)^T u(t).$$

To compute the $\mathcal{H}_2$-norm of $\hat{T}_{zw}$, let w_i be the input given by

$$w_i(t) = \begin{pmatrix} 0 \\ \vdots \\ \delta(t+1) \\ 0 \\ \vdots \end{pmatrix}$$

where $\delta(t+1)$ is the Kronecker δ applied at $t = -1$. The regulated output due to this input is denoted by z^i. It follows that

$$\|\hat{T}_{zw}\|_2^2 = \sum_{i=1}^{m} \|z^i\|_2^2.$$

The $\mathcal{H}_2$ problem is defined as

$$\min_{K\text{-stabilizing}} \|\hat{T}_{zw}\|_2.$$

In fact, we will solve the more general problem of characterizing all suboptimal solutions of level γ, i.e., all stabilizing controllers such that

$$\|\hat{T}_{zw}\|_2 < \gamma.$$

Recall from Section 5.3.4 the definitions of the various special problems. In the sequel, we first obtain the solution to the FI (full information) special problem. Using duality and equivalence, we can obtain solutions to the rest of the special problems, including the OE (output estimation) problem. Later on, we demonstrate that the optimal $\mathcal{H}_2$ solution decomposes into the solution of these two special problems, FI and OE.

13.2.1 Solution to the *FI* Problem

The full information (FI) problem is constructed by assuming that both the states and the disturbance input are measured. In this case, the system is given by

$$G_{FI} = \left[\begin{array}{c|cc} A & B_1 & B_2 \\ \hline C_1 & 0 & D_{12} \\ I & 0 & 0 \\ 0 & I & 0 \end{array} \right].$$

We have shown in Theorem 5.3.1 that all stabilizing controllers that parametrize all possible achievable maps are given by

$$K_{FI} = (F \quad Q) \quad \text{for all } Q \in \mathcal{RH}_\infty \text{ and } A + B_2 F \text{ is stable.}$$

We will choose the control gain matrix in a special way. Define the symplectic pair of matrices, H_2, as

$$H_2 = \left[\begin{pmatrix} A & 0 \\ -C_1^T C_1 & I \end{pmatrix}, \begin{pmatrix} I & B_2 B_2^T \\ 0 & A^T \end{pmatrix} \right].$$

Since (A, B_2) is stabilizable and (C_1, A) has no unobservable modes on the unit circle, it follows that $H_2 \in \text{dom}(\text{Ric})$. Let $X_2 = \text{Ric}(H_2)$, then X_2 satisfies the control algebraic Riccati equation (CARE)

$$X_2 = A^T X_2 (I + B_2 B_2^T X_2)^{-1} A + C_1^T C_1.$$

It follows from Appendix B that X_2 satisfies $X_2 = X_2^T$, $X_2 > 0$, and $(I + B_2 B_2^T X_2)^{-1} A$ is stable. Define

$$R_c = I + B_2^T X_2 B_2,$$
$$F_2 = -R_c^{-1} B_2^T X_2 A.$$

The inverse of R_c exists since it is positive definite. We refer to F_2 as the control gain matrix. It follows that

$$A_F = A + B_2 F_2 = (I + B_2 B_2^T X_2)^{-1} A$$

is stable. Using this gain matrix, the class of controllers given by

$$K_{FI} = (F_2 \quad Q) \quad \text{for all } Q \in \mathcal{RH}_\infty$$

parametrizes all achievable closed loop maps for the FI problem, given by

$$T_{zw} = G_c B_1 + U Q \quad \text{for all } Q \in \mathcal{RH}_\infty,$$

where

$$G_c = \left[\begin{array}{c|c} A + B_2 F_2 & I \\ \hline C_1 + D_{12} F_2 & 0 \end{array}\right], \quad U = \left[\begin{array}{c|c} A + B_2 F_2 & B_2 \\ \hline C_1 + D_{12} F_2 & D_{12} \end{array}\right].$$

To obtain a complete solution to the FI, we start with the following lemma.

Lemma 13.2.1. Let G_c and U be defined as before. It follows that

1. $\hat{U}^\sim \hat{U} = R_c$, which is a constant for all frequencies.
2. $\hat{U}^\sim \hat{G}_c \in \mathcal{H}_2^\perp$, or equivalently $\hat{G}_c^\sim \hat{U} \in \mathcal{H}_2$. The latter has a state-space realization given by

$$\hat{G}_c^\sim \hat{U} = \left[\begin{array}{c|c} A_F & B_2 \\ \hline X_2 C_F & X_2 B_2 \end{array}\right],$$

where $A_F = A + B_2 F_2$ and $C_F = C_1 + D_{12} F_2$.

Proof. By direct manipulation of the Riccati equation, we get

$$X_2 = A_F^T X_2 A_F + C_F^T C_F \tag{13.1}$$

and

$$F_2 = -B_2^T X_2 A_F. \tag{13.2}$$

Let $\hat{\phi} = (1/\lambda I - A_F)^{-1}$. To prove the first part, we expand the product:

$$\begin{aligned} \hat{U}^\sim \hat{U} &= D_{12}^T D_{12} + D_{12}^T C_F \hat{\phi} B_2 + B_2^T \hat{\phi}^\sim C_F^T D_{12} + B_2^T \hat{\phi}^\sim C_F^T C_F \hat{\phi} B_2 \\ &= I + F_2 \hat{\phi} B_2 + B_2^T \hat{\phi}^\sim F_2^T + B_2^T \hat{\phi}^\sim C_F^T C_F \hat{\phi} B_2. \end{aligned} \tag{13.3}$$

From Equation (13.1), we get

$$0 = -A_F^T X_2 \left(\frac{1}{\lambda} I - A_F\right) - \frac{1}{\lambda}(\lambda I - A_F^T) X_2 + C_F^T C_F.$$

Multiplying both sides by $\hat{\phi} B_2$ from the right and $B_2^T \hat{\phi}^\sim$ from the left, we get

$$\begin{aligned} 0 &= -B_2^T \hat{\phi}^\sim A_F^T X_2 B_2 - \frac{1}{\lambda} B_2^T X_2 \hat{\phi} B_2 + B_2^T \hat{\phi}^\sim C_F^T C_F \hat{\phi} B_2 \\ &= B_2^T \hat{\phi}^\sim F_2^T - \frac{1}{\lambda} B_2^T X_2 \hat{\phi} B_2 + B_2^T \hat{\phi}^\sim C_F^T C_F \hat{\phi} B_2. \end{aligned} \tag{13.4}$$

However,

$$\begin{aligned} \frac{1}{\lambda} B_2^T X_2 \hat{\phi} B_2 &= B_2^T X_2 B_2 + B_2^T X_2 A_F \hat{\phi} B_2 \\ &= B_2^T X_2 B_2 - F_2 \hat{\phi} B_2. \end{aligned} \tag{13.5}$$

Comparing Equation (13.3) with Equations (13.4, 13.5) we get

$$\hat{U}^{\sim}\hat{U} = R_c.$$

For the second part, by direct expansion we get

$$\begin{aligned}
\hat{G}_c^{\sim}\hat{U} &= \hat{\phi}^{\sim}C_F^T(C_F\hat{\phi}B_2 + D_{12}) \\
&= \hat{\phi}^{\sim}(C_F^TC_F - A_F^TX_2\hat{\phi}^{-1})\hat{\phi}B_2 \\
&= \hat{\phi}^{\sim}\left(X_2 - \frac{1}{\lambda}A_F^TX_2\right)\hat{\phi}B_2 \quad \text{From Equation (13.1)} \\
&= \frac{1}{\lambda}X_2\hat{\phi}B_2 \\
&= X_2B_2 + X_2A_F\hat{\phi}B_2 \in \mathcal{H}_2
\end{aligned}$$

which completes the proof.   ∎

The next theorem presents a complete solution to the FI problem.

Theorem 13.2.1.   The minimum value of the $\mathcal{H}_2$ optimal solution for the FI problem is given by

$$\min_{K\text{-stabilizing}} \|\hat{T}_{zw}\|_2 = \|\hat{G}_cB_1\|_2.$$

All suboptimal solutions satisfying $\|\hat{T}_{zw}\|_2 < \gamma$ are given by

$$\hat{K}_{FI} = (\, F_2 \quad \hat{Q}\,) \qquad \text{for all } \hat{Q} \text{ satisfying} \qquad \|R_c^{1/2}\hat{Q}\|_2^2 < \gamma^2 - \|\hat{G}_cB_1\|_2^2.$$

Proof. This theorem follows from Lemma 13.2.1:

$$\begin{aligned}
\|\hat{T}_{zw}\|_2^2 &= \langle \hat{G}_cB_1 + \hat{U}\hat{Q}, \hat{G}_cB_1 + \hat{U}\hat{Q}\rangle \\
&= \langle \hat{G}_cB_1, \hat{G}_cB_1\rangle + \langle \hat{G}_cB_1, \hat{U}\hat{Q}\rangle + \langle \hat{U}\hat{Q}, \hat{G}_cB_1\rangle + \langle \hat{U}\hat{Q}, \hat{U}\hat{Q}\rangle \\
&= \|\hat{G}_cB_1\|_2^2 + \langle \hat{U}^{\sim}\hat{G}_cB_1, \hat{Q}\rangle + \langle \hat{Q}, \hat{U}^{\sim}\hat{G}_cB_1\rangle + \|R_c^{1/2}\hat{Q}\|_2^2 \\
&= \|\hat{G}_cB_1\|_2^2 + \|R_c^{1/2}\hat{Q}\|_2^2.
\end{aligned}$$

The cross terms are equal to zero since $\hat{U}^{\sim}\hat{G}_cB_1 \in \mathcal{H}_2^{\perp}$ and $\hat{Q} \in \mathcal{H}_2$. The desired results follow from the last equation.   ∎

The solution to all other special problems, and in particular the OE problem, are obtained by duality and equivalence.

Example 13.2.1

Let $\hat{P}$ be given by:

$$\hat{P} = \frac{1}{1-\lambda}.$$

Consider the problem of designing a state feedback controller that minimizes the objective function:

$$\sum_{t=0}^{\infty} \rho z_1(t)^2 + u(t)^2,$$

where

$$z_1 = P(u + w).$$

Assume that $w(t) = \delta(t + 1)$. The 2-input 2-output description is given by

$$G = \left[\begin{array}{c|cc} 1 & 1 & 1 \\ \hline \begin{pmatrix} \rho \\ 0 \end{pmatrix} & \begin{pmatrix} 0 \\ 0 \end{pmatrix} & \begin{pmatrix} 0 \\ 1 \end{pmatrix} \\ 1 & 0 & 0 \end{array} \right].$$

This satisfies the assumptions needed for the FI problem since $D_{12}^T(C_1 \ D_{12}) = (0 \ I)$, and (C_1, A) is observable. The symplectic pair of matrices H_2 is given by:

$$H_2 = \left[\begin{pmatrix} 1 & 0 \\ -\rho^2 & 1 \end{pmatrix}, \begin{pmatrix} 1 & 1 \\ 0 & 1 \end{pmatrix} \right].$$

It follows that X_2 is a positive solution of the CARE

$$X_2 = X_2(1 + X_2)^{-1} + \rho^2.$$

and is given by:

$$X_2 = \frac{\rho^2 + \rho\sqrt{\rho^2 + 4}}{2}.$$

The state feedback matrix is given by

$$F_2 = \frac{-X_2}{1 + X_2},$$

and the closed-loop dynamic matrix is

$$A_F = \frac{1}{1 + X_2} = \frac{2}{2 + \rho^2 + \rho\sqrt{\rho^2 + 4}}.$$

Notice that if $\rho \to 0$, then $A_F \to A = 1$, and if $\rho \to \infty$ then $A_F \to 0$. These situations correspond (indirectly) to the expensive control and cheap control problems, respectively.

13.2.2 Separation Structure of the $\mathcal{H}_2$ Optimal Solution

In this section we follow exactly the ideas of Section 5.3.4. Consider the change of variables

$$u = F_2 x + v.$$

Define G_{temp} to be the map from w and u to v and y. It has the realization

$$G_{temp} = \left[\begin{array}{c|cc} A & B_1 & B_2 \\ \hline -F_2 & 0 & I \\ C_2 & D_{21} & 0 \end{array} \right].$$

Since $A + B_2 F_2$ is stable, G_{temp} is an OE problem. It can be seen that K stabilizes G if and only if K stabilizes G_{temp}, since both have the same G_{22} matrix, and the same A matrix. As shown in Figure 5.12, all closed-loop maps are given by

$$T_{zw} = G_c B_1 + U T_{vw}. \tag{13.6}$$

It follows from the previous section that

$$\|\hat{T}_{zw}\|_2^2 = \|\hat{G}_c B_1\|_2^2 + \|R_c^{1/2} \hat{T}_{vw}\|_2^2,$$

which implies that the minimizing controller for T_{zw} is the same controller that minimizes $\|R_c^{1/2} \hat{T}_{vw}\|_2^2$, i.e.,

$$\min_{K\text{-stabilizing}} \|\hat{T}_{zw}\|_2^2 = \|\hat{G}_c B_1\|_2^2 + \min_{K\text{-stabilizing}} \|R_c^{1/2} \hat{T}_{vw}\|_2^2.$$

The later problem is an OE problem and is solved using the results from the FI problem. To show the solution, define the symplectic pair

$$J_2 = \left[\left(\begin{array}{cc} A^T & 0 \\ -B_1 B_1^T & I \end{array} \right), \ \left(\begin{array}{cc} I & C_2^T C_2 \\ 0 & A \end{array} \right) \right].$$

From Appendix B, $J_2 \in \text{dom(Ric)}$. If $Y_2 = \text{Ric}(J_2)$, then Y_2 satisfies the filter algebraic Riccati equation (FARE)

$$Y_2 = AY_2(I + C_2^T C_2 Y_2)^{-1} A^T + B_1 B_1^T$$

with the properties $Y_2 = Y_2^T$, $Y_2 > 0$, and $A(I + Y_2 C_2^T C_2)^{-1}$ is stable. Define

$$R_f = I + C_2 Y_2 C_2^T,$$
$$L_2 = -AY_2 C_2^T R_f^{-1}.$$

We refer to L_2 as the filter gain matrix. It follows that

$$A_L = A + L_2 C_2 = A(I + Y_2 C_2^T C_2)^{-1}$$

is stable. Finally, define the system

$$G_f = \left[\begin{array}{c|c} A + L_2 C_2 & B_1 + L_2 D_{21} \\ \hline I & 0 \end{array} \right].$$

The next theorem gives a complete solution to the $\mathcal{H}_2$ problem.

Theorem 13.2.2. The minimum value of the $\mathcal{H}_2$ optimal solution for G is given by

$$\min_{K\text{-stabilizing}} \|\hat{T}_{zw}\|_2^2 = \|\hat{G}_c B_1\|_2^2 + \|R_c^{1/2} F_2 \hat{G}_f\|_2^2.$$

All stabilizing controllers that guarantee that

$$\|\hat{T}_{zw}\|_2 < \gamma$$

are given by

$$K(Q) = F_\ell(J, Q) \quad \text{for all } Q \text{ satisfying } \|\hat{Q}\|_2^2 < \gamma^2 - \|\hat{G}_c B_1\|_2^2 - \|R_c^{1/2} F_2 \hat{G}_f\|_2^2$$

and J is given by

$$J = \left[\begin{array}{c|cc} A + B_2 F_2 + L_2 C_2 & L_2 & -B_2 R_c^{-1/2} \\ \hline -F_2 & 0 & R_c^{-1/2} \\ R_f^{-1/2} C_2 & R_f^{-1/2} & 0 \end{array} \right].$$

Proof. We have already shown that the controller has to minimize $\|R_c^{1/2} \hat{T}_{vw}\|_2$ associated with G_{temp}, which is an OE problem. By dualizing this problem to a DF problem, and using the equivalence between FI and DF, we find that all closed-loop maps are given by

$$T_{vw} = F_2 G_f + QV$$

where V is given by

$$V = \left[\begin{array}{c|c} A + L_2 C_2 & B_1 + L_2 D_{21} \\ \hline C_2 & D_{21} \end{array} \right].$$

The systems G_f and V satisfy $\hat{G}_f \hat{V}^\sim \in \mathcal{H}_2^\perp$ and $\hat{V}\hat{V}^\sim = R_f$. All closed-loop maps T_{zw} are given by

$$T_{zw} = G_c B_1 + U F_2 G_f + U Q V.$$

The norm of T_{zw} decomposes as follows:

$$\|\hat{T}_{zw}\|_2^2 = \|\hat{G}_c B_1\|_2^2 + \|R_c^{1/2} F_2 \hat{G}_f\|_2^2 + \|R_c^{1/2} \hat{Q} R_f^{1/2}\|_2^2,$$

from which the characterization of all suboptimal solutions and the optimal solution follow. From Section 5.3.4, we know that the class of controllers achieving these closed-loop maps is given by

$$K(Q) = F_\ell(J, Q)$$

where J is given by the above realization after normalizing Q. This completes the proof.

■

13.2.3 The General Problem

If we only assume that $D_{11} = 0$ the formulae presented in the previous sections have to be adjusted as follows:

$$H_2 = \left[\begin{pmatrix} A - B_2 D_{12}^T C_1 & 0 \\ -C_1^T D_\perp D_\perp^T C_1 & I \end{pmatrix}, \begin{pmatrix} I & B_2 B_2^T \\ 0 & (A - B_2 D_{12}^T C_1)^T \end{pmatrix} \right],$$

$$J_2 = \left[\begin{pmatrix} (A - B_1 D_{21}^T C_2)^T & 0 \\ -B_1 \tilde{D}_\perp^T \tilde{D}_\perp B_1^T & I \end{pmatrix}, \begin{pmatrix} I & C_2^T C_2 \\ 0 & A - B_1 D_{21}^T C_2 \end{pmatrix} \right],$$

$$X_2 = \text{Ric}(H_2),$$

$$Y_2 = \text{Ric}(J_2),$$

$$R_c = I + B_2^T X_2 B_2,$$

$$R_f = I + C_2 Y_2 C_2^T,$$

$$F_2 = -R_c^{-1}(B_2^T X_2 A + D_{12}^T C_1),$$

$$L_2 = -(A Y_2 C_2^T + B_1 D_{21}^T) R_f^{-1}.$$

All the formulae involving the parametrization of suboptimal controllers follow exactly as before, using the above variables.

13.3 THE $\mathcal{H}_\infty$ PROBLEM

In this section we present the solution to the $\mathcal{H}_\infty$ general nonsquare problem. It turns out that the suboptimal controllers have a separation structure analogous to that of the $\mathcal{H}_2$ problem; however, it is less transparent. The clarity of the separation of the $\mathcal{H}_2$ solution stems from the geometry of Hilbert spaces, which $\mathcal{H}_\infty$ lacks. The proofs for the subsequent results tend to be quite messy (certainly much messier than their continuous-time counterparts) and thus will be omitted. Instead, the general idea of how one arrives to such results will be presented.

Performance objective.   The $\mathcal{H}_\infty$ problem is defined as follows: Given a performance level γ, find all possible controllers such that

$$\|\hat{T}_{zw}\|_\infty < \gamma.$$

This performance objective has the following interpretation: Find a control law $u = K y$ such that

$$\sup_{w \in \ell_2} \left(\|z\|_2^2 - \gamma^2 \|w\|_2^2 \right) < 0.$$

With this interpretation, we view the $\mathcal{H}_\infty$ problem as a game with w as the adversary and u as the minimizing input.

Contrary to the $\mathcal{H}_2$ problem, we will not try to find the structure of the controller at the optimum γ; however, we will be content with suboptimal parametrizations.

$J_{m,n}$-factorization.   In order to define the parameters in the parametrization of suboptimal controllers we need to perform a specific kind of factorization of matrices, known as $J_{m,n}$-factorization.

Definition 13.3.1.   A matrix $R \in \mathbb{C}^{(m+n)\times(m+n)}$ satisfying $R^* = R$ has $J_{m,n}$-factorization if there exists a matrix $T \in \mathbb{C}^{(m+n)\times(m+n)}$ such that

$$R = T^* \begin{pmatrix} -I_m & 0 \\ 0 & I_n \end{pmatrix}.$$

The following proposition gives conditions on a matrix so that it has $J_{m,n}$-factorization.

Proposition 13.3.1.   Suppose

$$R = \begin{pmatrix} A & B \\ B^T & D \end{pmatrix}$$

and assume R^{-1} exists and D is a symmetric positive definite matrix. Then R has a $J_{m,n}$-factorization, with $T_{12} = 0$, if and only if

$$\Delta := A - BD^{-1}B^T < 0.$$

Proof. Let $D = D_1 D_1^T$ and $\Delta = -\Delta_1 \Delta_1^T$. The following identity can be verified by direct computation:

$$R = \begin{pmatrix} A & B \\ B^T & D \end{pmatrix} = \begin{pmatrix} \Delta_1 & BD^{-1}D_1 \\ 0 & D_1 \end{pmatrix} \begin{pmatrix} -I_m & 0 \\ 0 & I_n \end{pmatrix} \begin{pmatrix} \Delta_1 & BD^{-1}D_1 \\ 0 & D_1 \end{pmatrix}^T$$

which establishes the sufficiency. On the other hand, if R has a $J_{m,n}$-factorization, then it follows that

$$T_{22}^T T_{22} = D,$$
$$T_{21}^T T_{22} = B,$$
$$-T_{11}^T T_{11} + T_{21}^T T_{21} = A.$$

By backward substitution, it follows that

$$\Delta = A - BD^{-1}B^T = -T_{11}^T T_{11} \le 0.$$

However, since R has an inverse, it follows that $\det(\Delta) \ne 0$, which establishes the necessity.   ∎

13.3.1 Solution to the *FI* Problem

Here we present a parametrization of all controllers that guarantee a norm less than γ when both the states and the disturbance input are available for feedback.

Definitions.   Define the matrix R as follows:

$$R := \begin{pmatrix} D_{11}^T \\ D_{12}^T \end{pmatrix} (D_{11} \quad D_{12}) + \begin{pmatrix} -\gamma^2 I_{n_w} & 0 \\ 0 & 0 \end{pmatrix}.$$

Let H_∞ denote the symplectic pair of matrices:

$$H_\infty := \left[\begin{pmatrix} A - BR^{-1}(D_{11} \quad D_{12})^T C_1 & 0 \\ -C_1^T (I - (D_{11} \quad D_{12}) R^{-1}(D_{11} \quad D_{12})^T) C_1 & I \end{pmatrix}, \right.$$

$$\left. \begin{pmatrix} I & BR^{-1}B^T \\ 0 & (A - BR^{-1}(D_{11} \quad D_{12})^T C_1)^T \end{pmatrix} \right].$$

where $B = (B_1 \quad B_2)$. If $H_\infty \in \text{dom(Ric)}$ and $X_\infty = \text{Ric}(H_\infty)$, then we can define the gain matrix F as

$$F := \begin{pmatrix} F_1 \\ F_2 \end{pmatrix} := -(R + B^T X_\infty B)^{-1} \left[\begin{pmatrix} D_{11}^T \\ D_{12}^T \end{pmatrix} C_1 + B^T X_\infty A \right].$$

Finally, suppose that $R + B^T X_\infty B$ has a J_{n_w, n_u}-factorization given by

$$R + B^T X_\infty B = \begin{pmatrix} T_{11}^T & T_{21}^T \\ 0 & T_{22}^T \end{pmatrix} \begin{pmatrix} -I_{n_w} & 0 \\ 0 & I_{n_u} \end{pmatrix} \begin{pmatrix} T_{11} & 0 \\ T_{21} & T_{22} \end{pmatrix},$$

then it follows that the following identities hold:

$$T_{22}^T T_{22} = I + B_2^T X_\infty B_2$$

$$T_{22}^T T_{21} = (D_{12}^T D_{11} + B_2^T X_\infty B_1)$$

$$T_{11}^T T_{11} = T_{21}^T T_{21} + \gamma^2 I - B_1^T X_\infty B_1 - D_{11}^T D_{11}.$$

Main result for the *FI* problem.   Let G be given by

$$G = \left[\begin{array}{c|cc} A & B_1 & B_2 \\ \hline C_1 & D_{11} & D_{12} \\ I & 0 & 0 \\ 0 & I & 0 \end{array} \right].$$

Assume also that assumptions 1, 2, and 4 are satisfied.

Theorem 13.3.1.   Let G be given as above. Then

1. There exists a stabilizing controller satisfying $\|T_{zw}\|_\infty < \gamma$ if and only if $H_\infty \in \text{dom(Ric)}$, $X_\infty = \text{Ric}(H_\infty) \geq 0$, and $R + B^T X_\infty B$ has J_{n_w, n_u}-factorization.

2. All admissible controllers are given by

$$K = T_{22}^{-1} (-Q \quad I) \begin{pmatrix} T_{11} & 0 \\ T_{21} & T_{22} \end{pmatrix} \begin{pmatrix} F_1 & -I \\ F_2 & 0 \end{pmatrix}$$

for any $Q \in \mathcal{RH}_\infty$ with $\|\hat{Q}\|_\infty < \gamma$.   ∎

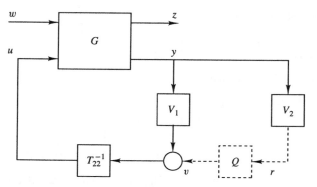

Figure 13.1 Decomposition of the FI controller.

Sufficiency: Basic idea.   After appropriate normalization we can assume that $\gamma = 1$. The block diagram in Figure 13.1 gives a decomposition of the controllers shown in part 2 of the above theorem. The variables are defined as follows:

$$v = T_{22}u - (T_{21} \quad T_{22})\,Fx + T_{21}w,$$

$$r = T_{11}(w - F_1 x),$$

from which V_1 and V_2 can be defined. Using these variables we get the following lemma.

Lemma 13.3.1.   For all u and w, the following identity holds:

$$\|z\|_2^2 - \|w\|_2^2 = \|v\|_2^2 - \|r\|_2^2.$$

Proof. First notice that X_∞ satisfies the equation

$$X_\infty = A^T X_\infty A + C_1^T C_1 - F^T (R + B^T X_\infty B)F.$$

Consider the following completion of squares:

$$\left[\begin{pmatrix} w(t) \\ u(t) \end{pmatrix} - Fx(t)\right]^T (R + B^T X_\infty B)\left[\begin{pmatrix} w(t) \\ u(t) \end{pmatrix} - Fx(t)\right]$$

$$= -x(t)^T (C_1^T C_1 - F^T (R + B^T X_\infty B)F)x(t) - w(t)^T w(t)$$

$$+ (w(t)^T \quad u(t)^T)\,B^T X_\infty B\,(w(t) \quad u(t))$$

$$+ z(t)^T z(t) + (w(t)^T \quad u(t)^T)\,B^T X_\infty Ax(t) + x(t)^T A^T X_\infty B\,(w(t) \quad u(t))$$

$$= z(t)^T z(t) - w(t)^T w(t) - x(t)^T X_\infty x(t) + x(t+1)^T X_\infty x(t+1).$$

If the closed-loop system is stable, it follows that $x(\infty) = 0$. Assuming zero initial condition, and summing the last equation from zero to infinity, we get

$$\|z\|_2^2 - \|w\|_2^2 = \|T_{22}u - (T_{21} \quad T_{22})\,Fx + T_{21}w\|_2^2 - \|T_{11}(w - F_1 x)\|_2^2,$$

which verifies the result.   ∎

Observe that for $\|z\|_2^2 - \|w\|_2^2 < 0$ for all w, it is essential that

$$\|T_{22}u - (T_{21} \quad T_{22})\,Fx + T_{21}w\|_2^2 < \|T_{11}(w - F_1 x)\|_2^2 \quad \text{for all } w.$$

The worst-case disturbance is clearly given by

$$w = F_1 x.$$

An admissible control input is given by

$$u = T_{22}^{-1} \left[(T_{21} \quad T_{22}) \, Fx - T_{21} w \right].$$

After simple manipulations of the diagram in Figure 13.1, we get the system P in Figure 13.2. It is expected from the above lemma that P is stable and satisfies $\hat{P}^\sim \hat{P} = I$. From that it follows that $\|\hat{T}_{zw}\|_\infty < 1$.

Finally, we note that the nominal control law u $(Q = 0)$ depends on both the states and the disturbance (i.e., requires all the available information). If T_{21} is equal to zero, then the nominal control law is a state feedback. From the definition of T_{21} it follows that

$$T_{21} = 0 \iff D_{12}^T D_{11} + B_2^T X_\infty B_1 = 0.$$

There are no obvious prior assumptions on the system that guarantee that this condition will be satisfied.

13.3.2 Output Feedback

The solution to the output feedback case is presented next. Unfortunately, the formulae that define the parameters of the admissible controllers are quite messy, and thus verifying the results is quite a cumbersome task. We choose not to do that here and we will be content with the final results.

Definitions.   Define the matrix $\tilde{R}$ as follows:

$$\tilde{R} := \begin{pmatrix} D_{11} \\ D_{21} \end{pmatrix} (D_{11}^T \quad D_{21}^T) + \begin{pmatrix} -\gamma^2 I_{n_z} & 0 \\ 0 & 0 \end{pmatrix}.$$

Let J_∞ denote the symplectic pair of matrices:

$$J_\infty := \left[\begin{pmatrix} (A - B_1(D_{11}^T \quad D_{21}^T)\tilde{R}^{-1}C)^T & 0 \\ -B_1(I - (D_{11}^T \quad D_{21}^T)\tilde{R}^{-1}(D_{11}^T \quad D_{21}^T)^T)B_1^T & I \end{pmatrix}, \right.$$
$$\left. \begin{pmatrix} I & C^T \tilde{R}^{-1} C \\ 0 & A - B_1(D_{11}^T \quad D_{21}^T)\tilde{R}^{-1}C \end{pmatrix} \right].$$

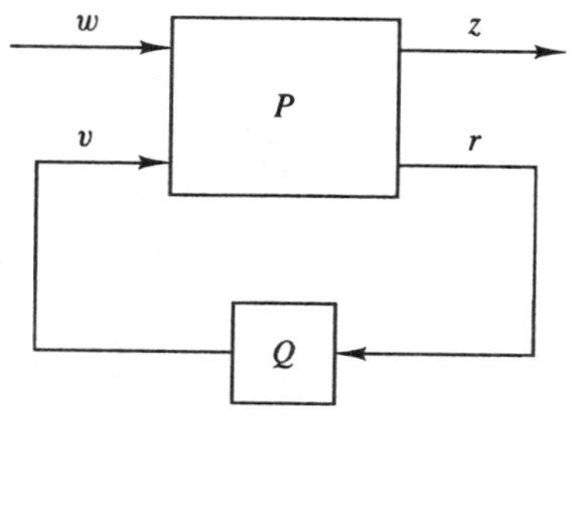

Figure 13.2   General setup.

If $J_\infty \in \text{dom(Ric)}$ and $Y_\infty = \text{Ric}(J_\infty)$, then we can define the gain matrix L as

$$L := (L_1 \quad L_2) := -\left[B_1 \left(D_{11}^T \quad D_{21}^T\right) + AY_\infty C^T\right] (\tilde{R} + CY_\infty C^T)^{-1}.$$

Also, define $Z_\infty = I - \gamma^{-2} Y_\infty X_\infty$.

Next, we define a set of auxiliary variables. We will assume that the problem is normalized so that $\gamma = 1$. Let E be any matrix that satisfies

$$E^T E = I - T_{11}^T T_{11}.$$

Define:

$$N := \begin{pmatrix} N_{11} & N_{12} \\ N_{21} & N_{22} \end{pmatrix} := \begin{pmatrix} E\tilde{D}_\perp^T & ED_{21}^T \\ T_{21}\tilde{D}_\perp^T & T_{21}D_{21}^T \end{pmatrix},$$

$$M_2 := C_2 + D_{21}F_1,$$

$$U_\infty := I + M_2 Y_\infty Z_\infty^{T^{-1}} M_2^T,$$

$$V_\infty := -N_{12}U_\infty^{-1}M_2 Y_\infty Z_\infty^{T^{-1}}[F_2^T T_{22}^T + M_2^T N_{22}^T] - N_{11}N_{21}^T,$$

$$W_\infty := I - N_{11}N_{11}^T - N_{12}M_2 Y_\infty Z_\infty^{T^{-1}} M_2^T U_\infty^{-1} N_{12}^T.$$

Solution to the _OF_ problem and the separation structure. The next theorem gives a complete characterization of all suboptimal controllers satisfying $\|\hat{T}_{zw}\|_\infty < 1$.

Theorem 13.3.2. Given the system G as before. Then
a. There exists a stabilizing controller K such that $\|\hat{T}_{zw}\|_\infty < 1$ if and only if the following hold:

1. $\max(\sigma_{max}[D_\perp^T D_{11}], \sigma_{max}[\tilde{D}_\perp^T D_{11}^T]) < 1$.
2. $H_\infty \in \text{dom(Ric)}$, $X_\infty = \text{Ric}(H_\infty) \geq 0$, and $R + B^T X_\infty B$ has J_{n_w,n_u}-factorization.
3. $J_\infty \in \text{dom(Ric)}$, $Y_\infty = \text{Ric}(J_\infty) \geq 0$, and $\tilde{R} + CY_\infty C^T$ has J_{n_z,n_y}-factorization.
4. $\rho(Y_\infty X_\infty) < 1$.

b. Given the above conditions, all rational admissible controllers are given by $F_\ell(K_o, Q)$ with $Q \in \mathcal{RH}_\infty$, $\|\hat{Q}\|_\infty < 1$ and K_o equals

$$K_o = \left[\begin{array}{c|cc} \bar{A} & \bar{B}_1 & \bar{B}_2 \\ \hline \bar{C}_1 & \bar{D}_{11} & \bar{D}_{12} \\ \bar{C}_2 & \bar{D}_{21} & 0 \end{array}\right]$$

where

$$\bar{D}_{11} = T_{22}^{-1}(V_\infty^T W_\infty^{-1} N_{12} - N_{22} + T_{22}F_2 Y_\infty Z_\infty^{T^{-1}} M_2^T)U_\infty^{-1},$$

$\bar{D}_{12}$ is any matrix satisfying

$$
\begin{aligned}
\bar{D}_{12}\bar{D}_{12}^T :=\ & T_{22}^{-1}(I - N_{21}N_{21}^T - N_{22}N_{22}^T)T_{22}^{T^{-1}} \\
& - F_2 Y_\infty Z_\infty^{T\,-1} F_2^T - T_{22}^{-1} V_\infty^T W_\infty^{-1} V_\infty T_{22}^{T^{-1}} \\
& + (T_{22}^{-1} N_{22} - F_2 Y_\infty Z_\infty^{T\,-1} M_2^T) U_\infty^{-1}(T_{22}^{-1} N_{22} - F_2 Y_\infty Z_\infty^{T\,-1} M_2^T)^T,
\end{aligned}
$$

$\bar{D}_{21}$ is any matrix that satisfies

$$
\bar{D}_{21}^T \bar{D}_{21} := U_\infty^{-1} - U_\infty^{-1} N_{12}^T W_\infty^{-1} N_{12} U_\infty^{-1}.
$$

The rest of the matrices defining K_o are given by:

$$
\begin{aligned}
\bar{B}_2 &:= Z_\infty^{-1}(B_2 + L_1 D_{12})\bar{D}_{12}, \\
\bar{B}_1 &:= -Z_\infty^{-1} L_2 + \bar{B}_2 \bar{D}_{12}^{-1} \bar{D}_{11}, \\
\bar{C}_2 &:= -\bar{D}_{21}(C_2 + D_{21} F_1), \\
\bar{C}_1 &:= F_2 + \bar{D}_{11} \bar{D}_{21}^{-1} \bar{C}_2, \\
\bar{A} &:= A + BF + \bar{B}_1 \bar{D}_{21}^{-1} \bar{C}_2,
\end{aligned}
$$
∎

The main idea in the proof of the above theorem is the conversion of the OF problem to an OE problem. From the discussion in the previous sections, we have shown that

$$
\|z\|_2^2 < \|w\|_2^2 \iff \|r\|_2^2 < \|v\|_2^2.
$$

It turns out, after some nontrivial manipulations, that the OF problem with r as the exogenous input and v as the regulated output is an output estimation (OE) problem. The solution for the latter is then given by the solution of the full information problem (FI) via duality and equivalence. In this sense, the $\mathcal{H}_\infty$ problem has a separation structure. Notice, however, that the OE problem depends on the solution of the FI problem, which is not the case for the $\mathcal{H}_2$ problem.

13.4 SUMMARY

In this chapter we have presented a complete treatment of the state-space solutions of the $\mathcal{H}_2$ problem as well as the main state-space results regarding the $\mathcal{H}_\infty$ problem. This material is presented primarily for two purposes: The first is to show an example where the algebraic separation structure of stabilizing controllers extends to a norm separation, and the second purpose is to summarize these results in conjunction with the previous results on the ℓ_1 problem. A detailed treatment of the $\mathcal{H}_\infty$ problem will take us too far from the main theme of this book, and hence is omitted.

EXERCISES

13.1. Given the dynamic system

$$x(k+1) = Ax(k) + B_1 w(k) + B_2 u(k)$$

$$z(k) = C_1 x(k) + D_{12} u(k)$$

with all the assumptions for the FI $\mathcal{H}_2$ problem satisfied. Assume that all the states are available for control. If $w(k) \in \ell_2(\mathbf{Z}_+)$, show that the optimal input u that minimizes $||z||_2$ is given by

$$u = F_2 x + R_c^{-1} \Pi_{\ell_2(\mathbf{Z}_+)} U^\sim G_c B_1 w.$$

Give a time domain interpretation of this control input and argue that u depends in a noncausal fashion on w. Specialize to the case where w is a delta function.

13.2. Consider the full state feedback problem

$$\left[\begin{array}{c|cc} A & b_1 & b_2 \\ \hline c_1 & 0 & d_{21} \\ I & 0 & 0 \end{array} \right]$$

with scalar control and scalar disturbance. Assume that the transfer function from the control input to the regulated output is minimum phase, except for r zeros at the origin (corresponding to delays). Find the $\mathcal{H}_2$ optimal state feedback controller. Compare to the optimal ℓ_1 state feedback controller solved in the previous chapter.

13.3. Suppose $||\hat{G}||_\infty < 1$. For simplicity, assume that $D = 0$. Show that for any $x(0) = x_0$,

$$\sup_{w \in \ell_2(\mathbf{Z}_+)} \{||z||_2^2 - ||w||_2^2\} = -x_0^T X x_0,$$

where X is the solution to the Riccati equation with the symplectic matrix P as in Equation (4.30). (Hint: Show that X satisfies:

$$X = A^T X A - C^T C - (A^T X B)(I + B^T X B)^{-1}(A^T X B)^T.$$

Consider the difference $x(k+1)^T X x(k+1) - x(k)^T X x(k)$ and complete the squares.)

13.4. In Theorem 13.3.1, show that if $D_{11} = 0$ then $R + B^T X_\infty B$ has $J_{n_w \times n_u}$-factorization if and only if

$$I - \gamma^{-2} B_1^T X_\infty (I + B_2 B_2^T X_\infty)^{-1} B_1 > 0.$$

13.5. Let $\hat{R}, \hat{V} \in \mathcal{RL}_\infty$, both scalars. Suppose we want to find a $Q \in \mathcal{H}_\infty$ such that

$$\left\| \begin{pmatrix} \hat{R} - \hat{Q} \\ \hat{V} \end{pmatrix} \right\|_\infty < \gamma.$$

This is a generalization of the standard Nehari problem discussed in Chapter 4. It is necessary that $\gamma > ||\hat{V}||_\infty$. Let $\hat{Y}$ satisfy:

$$\hat{Y}^\sim \hat{Y} = \gamma^2 - \hat{V}^\sim \hat{V}$$

with the property that $\hat{Y}$ is stable minimum phase. Argue that such a Y always exists. Show that for any $Q \in \mathcal{H}_\infty$

$$\left\| \begin{pmatrix} \hat{R} - \hat{Q} \\ \hat{V} \end{pmatrix} \right\|_\infty < \gamma \iff ||\hat{Y}^{-1}(\hat{R} - \hat{Q})||_\infty < 1.$$

The latter is a standard Nehari problem. Suggest a bisection algorithm for solving the first problem.

13.6. Given the 2×1 matrix in $\mathcal{RH}_\infty$

$$\hat{U} = \begin{pmatrix} \hat{u}_1 \\ \hat{u}_2 \end{pmatrix}.$$

We want to show that we can factor $\hat{U}$ as $\hat{U} = \hat{U}_i \hat{u}_0$ with $\hat{U}_i^\sim \hat{U}_i = 1$, $\hat{u}_0$ has no unstable zeros, and both factors are stable. Show that this is always possible and describe a procedure for doing it.

13.7. We are interested in the following nonsquare $\mathcal{H}_\infty$ problem: Find $\hat{Q} \in \mathcal{H}_\infty$ such that

$$\left\| \begin{pmatrix} \hat{h}_1 \\ \hat{h}_2 \end{pmatrix} - \begin{pmatrix} \hat{u}_1 \\ \hat{u}_2 \end{pmatrix} \hat{Q} \right\|_\infty < \gamma.$$

All the above are scalar quantities. Show that this problem can be mapped to a generalized Nehari problem as in Exercise 13.5 if we can perform the following stable inner-outer factorization:

$$\begin{pmatrix} \hat{u}_1 \\ \hat{u}_2 \end{pmatrix} = \begin{pmatrix} \hat{U}_i & \hat{U}_\perp \end{pmatrix} \begin{pmatrix} \hat{u}_0 \\ 0 \end{pmatrix},$$

with

$$\begin{pmatrix} \hat{U}_i & \hat{U}_\perp \end{pmatrix}^\sim \begin{pmatrix} \hat{U}_i & \hat{U}_\perp \end{pmatrix} = I$$

and $\hat{u}_0$ has no unstable zeros. Show that you really don't have to compute $\hat{U}_\perp$ to find $\hat{Q}$.

NOTES AND REFERENCES

The theory of linear optimal quadratic control has been extensively analyzed in several books; see, for instance [AM90, AF66, BH75, KS72, Kuc79]. The solution presented here is the discrete-time analog of the solution presented in [DGKF89].

State-space solutions of the $\mathcal{H}_\infty$ problem were first derived in [DGKF89, GD88] for continuous-time plants. The discrete-time analogs can be found in [IG91, LGW89, Sto92]. The solution presented here follows [IG91].

CHAPTER 14 —————————————————

Special Problems, Design and Examples

The previous chapters offered a complete account to the solution of the ℓ_1 optimal multiblock problem. Duality theory was utilized to provide methods for generating accurate approximations of the solution. This, in turn, motivated the formulation of the Delay Augmentation problem as a way of approximation, making the computations more efficient and systematic.

In here we apply the theory developed in previous chapters to specific control problems. This will help illustrate the use of the Delay Augmentation method in more realistic cases. To give some perspective to this presentation, we compare the characteristics of ℓ_1-based designs to that of $\mathcal{H}_\infty$-based designs in some of the examples.

Although the ℓ_1 optimal control problem captures a wide variety of design requirements, it does not directly incorporate specific time-domain and frequency-domain specifications such as overshoot of a response to a fixed input, or a bandwidth constraint on the controller. In the first part of this chapter we present solutions to such problems that rely on the mathematical machinery developed for the standard ℓ_1 problem.

14.1 SPECIAL PROBLEMS

We have seen in Chapter 3 that those constraints representing time-domain and frequency-domain specifications can be augmented into the ℓ_1 minimization problem to form an infinite-dimensional linear programming problem. Here we discuss methods for solving such problems, which are basically inherited from the methods for solving the standard

ℓ_1 problem. With the nonstandard inequality constraints due to the added performance specifications, the infinite-dimensional linear programs will have both (generally infinite) equality and inequality constraints, making it more difficult to obtain exact duality results. In particular, the resulting dual problem may not have the same value as the primal problem. We will not attempt to derive a general theory for such problems. However, we will continue to exploit duality to derive upper and lower approximations of the solution.

14.1.1 Weak Duality Revisited

The general standard problem for mixed objectives can be written as

$$v^p = \inf \|x\|_1$$
$$\text{subject to} \qquad\qquad\qquad\qquad (P)$$
$$\mathcal{A}x \le b,$$

where $\mathcal{A} : X \to Z$, $b \in Z$ and Z is a normed linear space, and the inequality is with respect to a positive cone in Z denoted by $\mathcal{P}_Z$. Define the dual problem:

$$v^d = \max \langle b, \gamma \rangle$$
$$\text{subject to}$$
$$\|\mathcal{A}^*\gamma\|_{X^*} \le 1, \qquad\qquad (DP)$$
$$\gamma \in Z^*, \ \gamma \le 0.$$

The inequality is with respect to the conjugate cone. The following characterization is known as weak duality.

Theorem 14.1.1.   Let v^p and v^d be the solutions to (P) and (DP), respectively. Then

$$v^p \ge v^d.$$

Proof. Consider any $\gamma \in Z^*$ such that $\|\mathcal{A}^*\gamma\|_\infty \le 1$ and $\gamma \le 0$. Then, for all $x \in X$ such that $\mathcal{A}x \le b$ we have

$$\|x\|_1 \ge \|x\| \|\mathcal{A}^*\gamma\|$$
$$\ge \langle x, \mathcal{A}^*\gamma \rangle$$
$$\ge \langle \mathcal{A}x, \gamma \rangle$$
$$\ge \langle b, \gamma \rangle.$$

Since the left-hand side is independent of γ and the right-hand side is independent of x, the inequality holds for the infimum on the left-hand side and the supremum on the right-hand side. From this the result follows. ∎

The theorem states that the minimization problem is always lower bounded by the dual problem. Whether equality holds or not depends very much on the specific problem. In the case of overshoot and undershoot constraints, we will present an intuitive argument showing that equality holds. When $v^p > v^d$, we say that there is a duality gap for that particular problem.

Finally, if a problem has both equality and inequality constraints, we can replace the equality constraints by two inequality constraints:

$$\mathcal{A}x = b \quad \Longleftrightarrow \quad \begin{pmatrix} \mathcal{A} \\ -\mathcal{A} \end{pmatrix} x \le \begin{pmatrix} b \\ -b \end{pmatrix}.$$

14.1.2 Bounded Overshoot and Undershoot

We discuss in some detail the problem of designing a controller that results in steady-state tracking of a step, and has bounded overshoot and undershoot. Consider the system shown in Figure 14.1. As was discussed in Section 5.1, the controller is forced to have a pole at $\lambda = 1$ to guarantee a zero steady-state error (assuming the plant does not have a pure integrator). Let z_i, $i = 1, \ldots, m$ and p_i, $i = 1, \ldots, n$ be, respectively, the zeros and poles of the plant inside the unit disc (for simplicity, we assume they are distinct). A given error signal $z = r - y$ is feasible if and only if

$$\begin{aligned} \langle z, \underline{z_i} \rangle &= \frac{1}{1 - z_i}, \quad i = 1, \ldots m, \\ \langle z, \underline{p_i} \rangle &= 0, \quad i = 1, \ldots n. \end{aligned}$$

These conditions are represented in the operator

$$V_\infty^T z = b.$$

The matrix V_∞^T is the standard Vandermonde matrix, which has an infinite number of columns, and a finite number of rows.

The overshoot OS and undershoot US of the response y are defined as

$$OS = \sup_t y(t) - 1, \qquad US = \sup_t -y(t).$$

We first present the general problem of bounded overshoot and undershoot. Let g_u, $g_l \in \ell_1$ be any two signals. The problem is to find any feasible z that lies in between

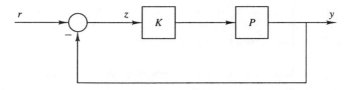

Figure 14.1 Tracking.

these functions. In particular, we will find a z with the minimum ℓ_1 norm (the choice of the norm on z is arbitrary). The general problem is given by

$$v^p = \inf \|z(t)\|_1$$

subject to

$$V_\infty^T z = b,$$ (14.1)

$$g_l \leq z \leq g_u.$$

The problem is an infinite-dimensional linear program. Its dual problem is given by

$$v^d = \max_{\gamma_1, \gamma_2, \gamma_3} \langle \gamma_1, b \rangle + \langle \gamma_2, g_u \rangle + \langle \gamma_3, g_l \rangle$$

subject to

$$\|V_\infty \gamma_1 + \gamma_2 - \gamma_3\|_\infty \leq 1,$$ (14.2)

$$\gamma_1 \in \mathbb{R}^{c_z}, \ \gamma_2, \ \gamma_3 \in \ell_\infty,$$

$$\gamma_2, \gamma_3 \leq 0.$$

It follows that both the primal and dual problems are infinite dimensional. Thus, approximate solutions can be computed by truncating both problems. Of course, depending on the choice of g_l, g_u, it may not be possible to obtain feasible suboptimal solutions. We will not analyze this problem in the general case. Instead, we will study the no-overshoot problem only.

14.1.3 No-Overshoot Problem

For this problem:

$$g_l = 0 \qquad g_u = \infty$$

where ∞ means that the constraint does not exist. In this case, the dual problem is given by

$$v^d = \max_{\gamma_1, \gamma_3} \langle \gamma_1, b \rangle$$

subject to

$$\|V_\infty \gamma_1 - \gamma_3\|_\infty \leq 1,$$ (14.3)

$$\gamma_1 \in \mathbb{R}^{c_z}, \ \gamma_3 \in \ell_\infty,$$

$$\gamma_3 \leq 0.$$

Notice that

$$\|V_\infty \gamma_1 - \gamma_3\|_\infty \leq 1 \quad \text{for some } \gamma_3 \leq 0 \iff V_\infty \gamma_1 \leq 1.$$

This follows from the fact that γ_3 is an arbitrary negative element, and that it does not enter the objective function. Thus, the dual problem is equal to

$$v^d = \max_{\gamma} \langle \gamma, b \rangle$$

subject to

$$V_\infty \gamma \leq 1, \tag{14.4}$$

$$\gamma \in \mathbb{R}^{c_z}.$$

Problem (14.4) has a finite number of variables and an infinite number of constraints. This problem resembles the standard dual problem in ℓ_1 with the difference that the constraints are not norm constraints.

Duality gap. It turns out that the no-overshoot problem (as well as Problem (14.1)) has no duality gap. We will not give a complete proof of this statement; however, we will give an intuitive justification based on approximation. First, notice that the collection of all FIR z (z with finite length) approximates feasible solutions arbitrarily closely. It follows that whenever the no-overshoot problem has a solution, there exists an FIR solution z of length N such that $V_\infty^T z = b$ and $z \geq -\epsilon$ for any prespecified positive ϵ. Define the following optimization problem

$$\nu_N(\epsilon) = \inf \|z(t)\|_1$$

subject to

$$V_\infty^T z = b, \tag{14.5}$$

$$z \geq -\epsilon,$$

$$z \in \mathbb{R}^N. \tag{14.6}$$

Because this is a finite-dimensional problem, the dual problem will have the same value as the primal and is given by

$$\nu_N(\epsilon) = \max_{\gamma_1, \gamma_3} \langle \gamma_1, b \rangle + \epsilon \langle 1, \gamma_3 \rangle$$

subject to

$$\|V_\infty \gamma_1 - \gamma_3\|_\infty \leq 1, \tag{14.7}$$

$$\gamma_1 \in \mathbb{R}^{c_z}, \ \gamma_3 \in \mathbb{R}^N,$$

$$\gamma_3 \leq 0,$$

where 1 is the sequence $(1, 1, \ldots)$. It can be verified by inspection that as ϵ goes to zero, and N goes to infinity, $\nu_N(\epsilon)$ converges to v^d. This implies that there is no duality gap for the overshoot problem.

Existence of solutions. The following results parallel the standard results from the ℓ_1 model matching problem.

Theorem 14.1.2.   Consider the "no-overshoot problem" and its dual.

1. If there is no feasible solution to the primal problem, the dual will have a value of infinity.
2. If the feasible set is not empty, then a solution to the primal problem exists if the system has no poles or zeros on the unit circle.
3. If the feasible set is not empty, then the dual problem has a solution.

Proof

1. Follows immediately from duality theory.
2. To show that a solution is guaranteed to exist, we show that the set of all feasible solutions is *weak** closed.

Consider the set

$$\{z \in \ell_1 \mid V_\infty^T z = b\}.$$

We have shown in Chapter 10 that this set is *weak** closed if all the poles and zeros are in the open unit disc. On the other hand, the set

$$\{z \in \ell_1 \mid z \geq 0\}$$

is *weak** closed. If $z_k \to z^o$ (*weak**), and $z^o(l) < 0$ for some l, then $z_k(l)$ will not converge to $z^o(l)$, contradicting *weak** convergence. Thus the feasibility set

$$\{z \in \ell_1 \mid V_\infty^T z = b\} \cap \{z \in \ell_1 \mid z \geq 0\}$$

is *weak** closed. Following the discussion in Section 9.5, an optimal solution exists.

∎

Structure of optimal solutions.   The optimal solution of the primal problem is deadbeat (i.e., z has a finite impulse response). This claim is verified by noting that for any maximizing solution of the dual problem, only finitely many constraints will be supported owing to the fact that the coefficients of the matrix V_∞ decay to zero. From the alignment conditions, the optimal solution to the primal problem is FIR.

Computation.   Even though the dual problem is finite dimensional with decaying constraints, it is difficult to derive an a priori bound on the number of active constraints, since such constraints are not norm bounds. One can, however, solve the problem by iterations. This means that only a finite number of constraints are retained in each iteration. This corresponds to solving the primal problem with z truncated to a finite number of variables. The solution of each truncated dual problem is checked to verify whether it satisfies the rest of the constraints. By increasing the number of constraints (in the dual) or variables (in the primal), we are guaranteed to obtain the exact solution in a finite number of steps.

Example 14.1.1

Consider a tracking problem as in Figure 14.1 where

$$\hat{P}(\lambda) = \frac{\lambda(\lambda + 0.8)}{(\lambda - 1)(\lambda - 2/3)}.$$

The plant is unstable and nonminimum phase. Hence, given any stabilizing controller, the output will generally undershoot and overshoot when tracking a step. We want to analyze how the closed-loop system performance changes when we include an overshoot bound in the design.

Assume that the design specifications call for good tracking (in the ℓ_∞ sense) and an overshoot no greater than γ when the system is driven by a unit step input. The problem can be posed as follows:

$$\nu^o = \inf \| S \|_1$$

subject to

$$S \text{ feasible},$$

$$-z \leq \gamma,$$

where S is the sensitivity. The exact solution to the above problem for different values of γ is shown in Table 14.1. This family of solutions shows explicitly the design trade-offs among overshoot, undershoot, tracking performance, and controller structure, as well as the fundamental limitations of a linear design. For instance, it shows that there is no stabilizing linear controller that can achieve an overshoot smaller than 0.5. Figure 14.2 depicts how the tracking performance deteriorates as we demand more stringent bounds on the overshoot. Similar trends can be observed in the level of undershoot.

TABLE 14.1 TRADE-OFFS IN THE BOUNDED OVERSHOOT EXAMPLE

γ	ν^o	Undershoot	ord(K)	Comments
∞	3.672	0	11	$OS = 0.836$
0.8	4.430	0	11	
0.7	6.549	0	11	
0.65	7.609	0.504	11	
0.6	9.850	1.708	13	
0.58	12.035	1.541	15	
0.56	15.403	1.778	17	
0.54	22.796	3.799	19	
0.53	30.268	8.556	21	
0.52	47.567	12.149	22	
0.51	104.294	37.739	23	
0.50	–	–	–	Unfeasible

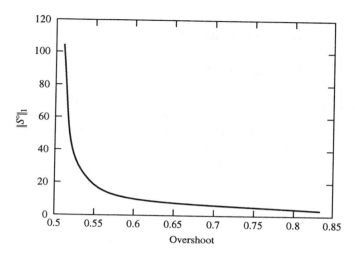

Figure 14.2   Tracking performance vs. overshoot.

14.1.4 Bounded Undershoot

For the bounded undershoot problem, $g_l = -\infty$ (i.e., the constraint does not exist). In this case Problem 14.2 simplifies to

$$\max_{\gamma_1, \gamma_2} \langle \gamma_1, b \rangle + \langle \gamma_2, g_u \rangle$$

subject to

$$\|V_\infty \gamma_1 + \gamma_2\|_\infty \leq 1, \tag{14.8}$$

$$\gamma_1 \in \mathbb{R}^{c_z}, \ \gamma_2, \in \ell_\infty,$$

$$\gamma_2, \leq 0.$$

The above problem does not simplify any further, and, clearly, both primal and dual problems are infinite-dimensional. By truncating both primal and dual solutions, we can get lower and upper bound approximations of the objective function.

14.1.5 General Mixed Objectives

In Chapter 3 we presented general performance specifications in both time and frequency domains. A design problem will in general have a combination of such specifications. For instance, on top of some template constraints on a fixed command, the controller may be required to have good disturbance rejection properties as well as bandwidth constraints. Many of these specifications can be either exactly represented or approximated by linear constraints as was shown in Chapter 3. In this section we address prototypes of such problems, and we show that weak duality can be used to obtain upper and lower approximations of the optimal solutions.

14.1.6 ℓ_1 Minimization with $\mathcal{H}_\infty$ Constraints

Recall from Chapter 3 that $\mathcal{H}_\infty$ constraints can be approximated by a finite set of linear constraints, represented in the operator $\mathcal{A}_{\mathcal{H}_\infty} : \ell_1^{n_z \times n_w} \to \mathbb{R}^s$. Of course, the feasibility conditions can be represented in terms of two operators: $\mathcal{A}_{\text{zero}} : \ell_1^{n_z \times n_w} \to \mathbb{R}^{c_z}$ for the zeros interpolation conditions and $\mathcal{A}_{\text{rank}} : \ell_1^{n_z \times n_w} \to \ell_1^{(n_z - n_u) \times n_w}$ for the rank interpolation conditions. Formally, the problem we want to solve is:

$$v^p = \inf_{\Phi \in \ell_1^{n_z \times n_w}} \|\Phi\|_1$$

subject to

$$\mathcal{A}_{\text{feas}} \Phi = b_{\text{feas}}, \tag{14.9}$$
$$\mathcal{A}_{H_\infty} \Phi \le b_{H_\infty},$$

where $\mathcal{A}_{\text{feas}}$ is the combined feasibility operator. To derive lower bounds for v^p, we consider the dual problem.

Dual problem.   The dual problem is defined as follows:

$$v^d = \max_{\gamma_1, \gamma_2} \langle \gamma_1, b_{\text{feas}} \rangle + \langle \gamma_2, b_{\mathcal{H}_\infty} \rangle$$

subject to

$$\|\mathcal{A}_{\text{feas}}^* \gamma_1 + \mathcal{A}_{\mathcal{H}_\infty}^* \gamma_2\|_\infty \le 1, \tag{14.10}$$
$$\gamma_1 \in \ell_\infty, \ \gamma_2 \in \mathbb{R}^s,$$
$$\gamma_2 \le 0.$$

Computations.   The Finitely Many Variables (FMV) problem is given by

$$\overline{v}_N = \inf \|\Phi\|_1$$

subject to

$$\mathcal{A}_{\text{feas}} \Phi = b_{\text{feas}},$$
$$\mathcal{A}_{\mathcal{H}_\infty} \Phi \le b_{\mathcal{H}_\infty},$$
$$\Phi(k) = 0 \quad \text{for all } k > N.$$

As in the standard ℓ_1 problem, this is a finite-dimensional linear program with the property that $\overline{v}_N \ge v^p$. If the above problem has feasible solutions, then it follows that

$$\lim_{N \to \infty} \overline{v}_N = v.$$

The Finitely Many Equations (FME) problem is given by

$$\underline{v}_N = \max_{\gamma_1, \gamma_2} \langle \gamma_1, b_{\text{feas}} \rangle + \langle \gamma_2, b_{\mathcal{H}_\infty} \rangle$$

subject to

$$\|\mathcal{A}_{\text{feas}}^* \gamma_1 + \mathcal{A}_{\mathcal{H}_\infty}^* \gamma_2\|_\infty \le 1, \tag{14.11}$$

$$\gamma_1 \in \mathbb{R}^N, \; \gamma_2 \in \mathbb{R}^s,$$

$$\gamma_2 \le 0.$$

It is straightforward to verify that $\underline{v}_N \le v^d$ and that

$$\lim_{N \to \infty} \underline{v}_N = v^d.$$

In here, we have not explicitly shown the two types of constraint operators reflecting the zero conditions and the rank conditions. With that in mind, the above problem resembles the standard ℓ_1 FME problem with one difference, namely the constraints attributable to $\mathcal{A}_{\mathcal{H}_\infty}$ are infinite and do not decay. Thus, the above problem has finitely many variables and infinitely many constraints that do not decay. To understand these constraints, consider the case of constraining the magnitude at one frequency point. The exact set of constraints is given by (SISO system):

$$|\hat{\Phi}(e^{iw_0})| \le 1 \Longleftrightarrow \sum_{k=0}^{\infty} \Phi(k) \cos(kw_0 - \theta) \le 1 \quad \text{for all } \theta \in [0, 2\pi).$$

These constraints can be approximated by:

$$|\hat{\Phi}(e^{iw_0})| \le 1 \Longrightarrow \sum_{k=0}^{\infty} \Phi(k) \cos(kw_0 - \theta_m) \le 1 \quad m = 0, 1, \ldots s - 1,$$

where $\theta_m = 2\pi m/s \in [0, 2\pi)$. The operator $\mathcal{A}_{\mathcal{H}_\infty}^*$ has the representation

$$(\mathcal{A}_{\mathcal{H}_\infty}^* \gamma_2)(k) = \sum_{m=0}^{s-1} \gamma_2(m) \cos(kw_0 - \theta_m).$$

When $\gamma_1 \in \mathbb{R}^N$, the constraints due to $\mathcal{A}_{\text{feas}}$, decay to zero. The above constraints due to $\mathcal{A}_{\mathcal{H}_\infty}$ remain for all k. However, if w_0 is chosen to be a multiple of the frequency $2\pi/rs$ for any integer r and for s as in the approximation above, then the constraints are periodic and only one period should be retained. This shows that the FME problem can be well approximated by a finite linear program.

The Delay Augmentation can also be applied to this case. The new augmented problem will have a feasible operator $\mathcal{A}_{\text{feas},N}$ with a finite-dimensional range for every set of delays of order N. The convergence results of that scheme are more subtle than those of the FMV or FME and will not be discussed.

14.1.7 Mixed Time and Frequency Constraints

A more general problem can be formulated in which both time and frequency domain constraints are added to a standard ℓ_1 minimization problem. Formally, the problem is

$$v^p = \inf_{\Phi \in \ell_1^{n_z \times n_w}} \|\Phi\|_1$$

subject to

$$\mathcal{A}_{\text{feas}} \Phi = b_{\text{feas}},$$
$$\mathcal{A}_{\mathcal{H}_\infty} \Phi \le b_{\mathcal{H}_\infty}, \tag{14.12}$$
$$\mathcal{A}_{\text{temp}} \Phi \le b_{\text{temp}},$$

where $\mathcal{A}_{\text{feas}}$ is the feasibility operator, and $\mathcal{A}_{\text{temp}}$ reflects the time domain constraints due to a fixed input. For example, typical performance specifications involve constraints on the step response of the system in terms of overshoot, undershoot, settling time, and so forth. For such a problem, the operator $\mathcal{A}_{\text{temp}}$ is explicitly given by

$$\begin{bmatrix} \mathcal{A}_{\text{step}} \\ -\mathcal{A}_{\text{step}} \end{bmatrix} \Phi \le \begin{bmatrix} \overline{b}_{\text{step}} \\ \underline{b}_{\text{step}} \end{bmatrix}, \tag{14.13}$$

where $\overline{b}_{\text{step}}$, $\underline{b}_{\text{step}}$ determine a template for the regulated response, and $\mathcal{A}_{\text{step}}$ is the convolution operator associated with a step input, i.e.,

$$\mathcal{A}_{\text{step}}(\Phi) = w_f * \Phi.$$

For example, in the SISO case, $\mathcal{A}_{\text{step}} : \ell_1 \to \ell_\infty$ is the following Toeplitz matrix:

$$\mathcal{A}_{\text{step}} = \begin{pmatrix} 1 & 0 & 0 & 0 & 0 & \cdots \\ 1 & 1 & 0 & 0 & 0 & \ddots \\ 1 & 1 & 1 & 0 & 0 & \ddots \\ 1 & 1 & 1 & 1 & 0 & \ddots \\ \vdots & & & & \ddots & \ddots \end{pmatrix}.$$

Computations.　We first note that the operator $\mathcal{A}_{\text{step}}$ does not have fading memory and continues to act on the whole time sequence. This makes the approximation of the primal problem a difficult task. For instance, the FMV approach will yield a linear programming problem with finitely many variables but with infinitely many constraints that do not decay. One way of getting around this problem is to impose the template on the step response for only a finite duration, i.e.,

$$\underline{b}_{\text{step}}(t) \le w_f * \Phi(t) \le \overline{b}_{\text{step}}(t) \quad t = 0, \ldots M - 1.$$

Explicitly, the problem becomes

$$\nu_M^p = \inf \| \Phi \|_1$$

subject to

$$A_{\text{feas}} \Phi = b_{\text{feas}},$$

$$A_{\mathcal{H}_\infty} \Phi \leq b_{\mathcal{H}_\infty}, \qquad\qquad (14.14)$$

$$A_M \Phi \leq \overline{b}_M,$$

$$-A_M \Phi \leq \underline{b}_M,$$

with

$$A_M = \left. \begin{pmatrix} 1\;0\;0\;0\;0\;\cdots \\ 1\;1\;0\;0\;0\;\cdots \\ 1\;1\;1\;0\;0\;\cdots \\ 1\;1\;1\;1\;0\;\cdots \end{pmatrix} \right\} M, \quad \overline{b}_M = \begin{pmatrix} \overline{b}_{\text{step}}(0) \\ \vdots \\ \overline{b}_{\text{step}}(M-1) \end{pmatrix}, \quad \underline{b}_M = \begin{pmatrix} \underline{b}_{\text{step}}(0) \\ \vdots \\ \underline{b}_{\text{step}}(M-1) \end{pmatrix}.$$

A large class of problems is such that, if M is chosen large enough, then the template constraints for $t \geq M$ become inactive. This is a result of the underlying ℓ_1 optimization, which will produce a Φ that decays to zero, possibly at a rate faster than $\underline{b}_{\text{step}}$ and $\overline{b}_{\text{step}}$. Later in this chapter we will show examples of such behavior, in particular, problems with template constraints that do not decay.

Notice that if Problem (14.14) does not have a feasible solution then neither does Problem (14.12) with Equation (14.13). Moreover, the sequence $\{\nu_M^p\}$ is monotonically nondecreasing as M increases. Finally, we can obtain lower bound approximations for both Problems (14.14) and (14.12) by finding the dual problems.

An alternative computational approach is the Delay Augmentation method. It is particularly well suited for solving problems with truncated template constraints as in Problem (14.14). Recall from the previous chapter that the DA solution is such that $\Phi_N^o(t) = \overline{\Phi}_N(t)$ for $0 \leq t \leq N-1$, where N is the order of the delays. This implies that if N is greater than M, the truncated template constraints that are satisfied by Φ_N^o will also be satisfied by $\overline{\Phi}_N$. In other words, the augmentation does not distort the time response of the solution up to time index $N-1$. Note that this is not the case with frequency-domain constraints, since the response of a system at a given frequency is affected by all the elements in the impulse response. Consequently, a Φ_N^o that satisfies certain frequency-domain constraints will result, in general, in a $\overline{\Phi}_N$ that does not satisfy such constraints (i.e., that is not feasible). A possible way to circumvent this problem is to leave those entries of Φ that are frequency-constrained unaugmented (e.g., order them into Φ_{11}).

14.2 DESIGN AND EXAMPLES

As explained in Chapter 3, we are interested in solving several analysis and design examples where the specifications are given mostly in time domain, although frequency-domain constraints are considered too. All designs consider various control specifications on two different processes or plant models: the pitch axis dynamics of the X29 aircraft and a simple flexible beam. (See Chapter 3 for a discussion on these models and the general specifications of interest.)

Through the rest of this chapter, the reader should get a fairly broad exposure to the different classes of problems that can be posed and solved by the theory presented, from questions of fundamental limitations of constrained linear designs, to problems of stability and performance robustness. However, this presentation is not intended to be complete, but rather, to demonstrate the potential behind this approach.

Note: This section contains several numerical results. A reader trying to reproduce these results should bear in mind that the algorithms used, such as the Linear Programming solver, require the setting of a few numerical tolerances (for instance, to define the numerical zero). These settings may have, in some cases, a noticeable effect on the final answer. This is particularly so when the value in question is of a discrete nature, such as the order of the solution. However, the most important digits of real variables (such as the value of an optimization) should be reproducible with any set of reasonable tolerances.

14.2.1 Pitch Axis Control of the X29 Aircraft

We refer to the simplified plant model presented in Chapter 3.

ℓ_1 **performance objective.**   Consider the following formal synthesis problem: Find a stabilizing discrete-time controller such that the ℓ_1-norm of the transfer function from the disturbance, w, to the weighted control sequence, z_1, and the weighted output, z_2, is minimized. That is,

$$\inf_{K \text{ stab.}} \left\| \begin{matrix} W_1 K S \\ W_2 S \end{matrix} \right\|_1$$

where $S := (I - PK)^{-1}$ is the sensitivity function.

With this problem setup we are ready to apply the Delay Augmentation algorithm as described in Chapter 12. Table 14.2 shows the sequence of results obtained in this case, starting with $N = 4$.

Note how the length of the response corresponding to the weighted sensitivity stops increasing after $N = 7$, suggesting that such row is PD. For $N = 80$ the achieved ℓ_1 norm is within 1% of the optimal so we stop the iteration process. It is interesting to note that the upper bound converges rather slowly to the optimal value. This behavior is consistent with the observations made in Chapter 12 regarding the rate of decay of Φ when one row is PD. Indeed, if the row corresponding to the weighted sensitivity is PD, then the rate of decay of the second row ($W_1 K S$) is dictated by the stable zeros of $\hat{U}_1(\lambda)$. It is easy to check that such a transfer function contains two stable zeros that are close to the unit circle. Then, if the optimal $K S$ decays slowly, the extra free parameter (Q_{21}) corresponding to the DA solution will be significant even for large values of N.

The 6th order suboptimal controller corresponding to $N = 80$ is given below.

$$\hat{K}(\lambda) = \frac{-41.763\lambda^6 + 218.478\lambda^5 - 293.168\lambda^4 - 121.311\lambda^3 + 343.000\lambda^2 + 80.680\lambda - 186.872}{\lambda^6 + 1.9728\lambda^5 - 1.7703\lambda^4 - 3.0297\lambda^3 - 0.8446\lambda^2 + 0.1489\lambda + 2.5231}$$

TABLE 14.2 X29 SYNTHESIS PROBLEM: DELAY
AUGMENTATION ALGORITHM

N	$\underline{\eta}_N$	$\bar{\eta}_N$	$\text{len}(\Phi_N^o)^T$	$\text{ord}(K)$	Comments
4	3.254	1256.4	(10 5)	11	Reorder outputs
5	4.024	7.619	(5 5)	6	Keep order
6	4.045	5.059	(5 6)	6	,,
7	4.048	4.652	(6 7)	6	,,
8	4.051	4.318	(6 8)	6	,,
9	4.051	4.317	(6 9)	6	,,
10	4.052	4.224	(6 10)	6	,,
$\vdots$					
20	4.053	4.196	(6 19)	6	,,
$\vdots$					
40	4.053	4.160	(6 35)	6	,,
$\vdots$					
80	4.054	4.098	(6 65)	6	,,

In the next section we compare the time and frequency domain characteristics of the ℓ_1 suboptimal design corresponding to $N = 80$ with an $\mathcal{H}_\infty$ (sub-)optimal design.

The ℓ_1 and $\mathcal{H}_\infty$ designs.   Here we include an $\mathcal{H}_\infty$ solution as a reference or base design since its characteristics are well understood. The comparison is based on three different aspects of the solutions: (1) operator norms, (2) frequency-response characteristics, and (3) time-response characteristics.

Table 14.3 shows how the ℓ_1 and $\mathcal{H}_\infty$ norms of the two solutions compare. As expected, the $\mathcal{H}_\infty$ design achieves better $\mathcal{H}_\infty$ norms while the ℓ_1 design achieves better ℓ_1 norms. A cross-examination shows that both solutions are fairly good in terms of both measures. In view of the norm inequalities presented in Chapters 2 and 4 (valid for

TABLE 14.3 OPERATOR NORM
COMPARISON ($T_S = 1/30$)

		$\|\cdot\|_{\mathcal{H}_\infty}$	$\|\cdot\|_1$	$\text{ord}(K)$
$\mathcal{H}_\infty$ design	Φ	2.3	5.2	5
	$W_1 K S$	2.3	4.3	
	$W_2 S$	2.3	5.2	
ℓ_1 design	Φ	3.8	4.1	6
	$W_1 K S$	2.8	4.1	
	$W_2 S$	3.6	4.1	

any finite-dimensional system H):

$$\|H\|_{\mathcal{H}_\infty} \leq \sqrt{r}\|H\|_1 \leq \sqrt{rm}(2n+1)\|H\|_{\mathcal{H}_\infty}$$

where $\hat{H}$ is $r \times m$ and n is its McMillan degree, minimizing any of the two norms will also upper-bound the other one. These bounds are particularly effective in a low-order problem with few zero interpolations as the one under consideration. Hence, the similarity in both solutions. However, as pointed out in Section 10.10, solutions may be significantly different in general.

Next, let us examine the frequency domain features. Both designs have fairly similar frequency domain characteristics as shown in Figures 14.3–14.6. While the ℓ_1 design has better disturbance rejection at low and medium frequencies, it is worse at high frequencies where the $\mathcal{H}_\infty$ norm is achieved (see Figures 14.3 and 14.6). In fact, Figure 14.7 shows that both controllers have essentially the same response, the only significant difference being at frequencies close to π/T_s. An interesting difference, though, is that the ℓ_1 design results in an unstable controller while the $\mathcal{H}_\infty$ design does not for this example.

Finally, we compare the weighted and unweighted step response of both designs (Figures 14.8, 14.9, and 14.10). Note how the output of the plant y converges to zero faster in the ℓ_1 design than in the $\mathcal{H}_\infty$ design (Figure 14.10). This is a direct result of the smaller weighted steady-state error in the ℓ_1 design (see Figure 14.8) and the pole of $\hat{W}_2$ at 0.9999 (almost a pure integrator).

To conclude, there is an interesting question in the context of this design example. How is the ℓ_1 design affected by an increase in sampling frequency? Consider doubling the sampling frequency, that is, letting $T_s = 1/60$, and repeating the design process starting with the discretization of Equation (3.18). Formally, the overhead factor in Equation (3.18) is adjusted to reflect the digital implementation with a higher sampling frequency. In this case, however, we will not modify the continuous model of the plant to

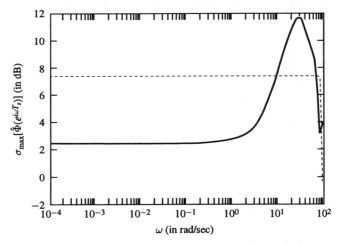

Figure 14.3 X29: Frequency response of $\hat{\Phi}$ for ℓ_1 (solid line) and $\mathcal{H}_\infty$ (dashed) designs.

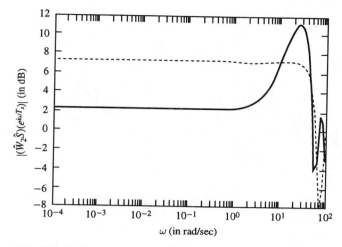

Figure 14.4   X29: Frequency response of $\hat{W}_2\hat{S}$ for ℓ_1 (solid) and $\mathcal{H}_\infty$ (dashed) designs.

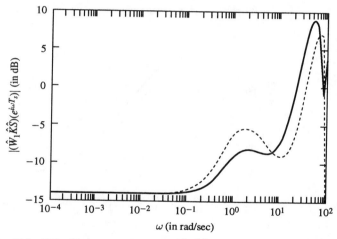

Figure 14.5   X29: Frequency response of $\hat{W}_1\hat{K}\hat{S}$ for ℓ_1 (solid) and $\mathcal{H}_\infty$ (dashed) designs.

simplify the comparison. Table 14.4 shows that the optimal norms are slightly improved with both design methodologies (an average 10%). The most noticeable change is the increase in the order of the ℓ_1 suboptimal controller (from 7th to 15th order). This observation is consistent with the fact that the zero interpolation locations approach the unit circle when the sampling frequency increases. This, in turn, implies that the zero interpolation functionals will have a slower rate of decay; thus the optimal response will be, in general, of higher order. In the limit, as $T_s \rightarrow 0$, the problem can be thought of as a continuous-time $\mathcal{L}_1$ optimal design, which is known to have infinite-dimensional optimal controllers (i.e., controllers with pure delays).

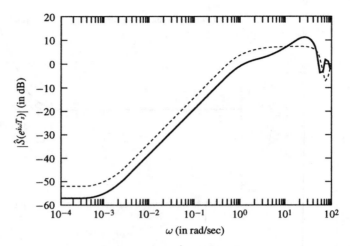

Figure 14.6   X29: Frequency response of $\hat{S}$ for ℓ_1 (solid) and $\mathcal{H}_\infty$ (dashed) designs.

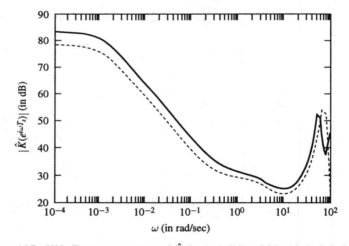

Figure 14.7   X29: Frequency response of $\hat{K}$ for ℓ_1 (solid) and $\mathcal{H}_\infty$ (dashed) designs.

The ℓ_1 performance objective with frequency-domain constraints.   Let us go back to the first model obtained with $T_s = 1/30$ seconds. From Table 14.3 we see that $\|\hat{W}_2\hat{S}\|_\infty = 3.6$ with a peak at high frequency. Assume that the design specifications require that the frequency response never exceeds $\gamma_{\mathcal{H}_\infty}$. Following the approach presented in Chapter 3, we augment the basic ℓ_1 design with frequency domain constraints as given in Equation (3.13). In particular, we will consider two different bounds: $\gamma_{\mathcal{H}_\infty} = 2.8$ and $\gamma_{\mathcal{H}_\infty} = 2.6$.

From Figure 14.4 we see that the weighted sensitivity violates the frequency constraints in the high frequency region. Hence, we choose five frequency points, ω_i for $i = 1, \ldots, 5$, between 10 and 40 radians per second, evenly spaced in a log scale. Fur-

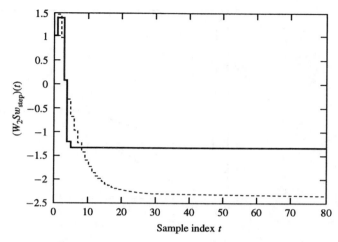

Figure 14.8   X29: Step response of $W_2 S$ for ℓ_1 (solid) and $\mathcal{H}_\infty$ (dashed) designs.

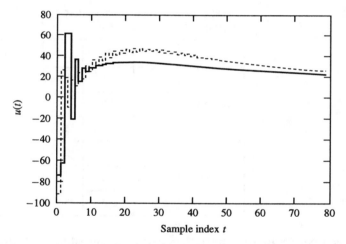

Figure 14.9   X29: Step response of KS for ℓ_1 (solid) and $\mathcal{H}_\infty$ (dashed) designs.

ther, for each frequency point we discretize the unit circle in the complex plane with 16 points (i.e., $\theta_i = 0, \pi/8, \pi/4, \ldots, 2\pi$). This specifies completely the operator $A_{\mathcal{H}_\infty}$ and we can proceed with the solution. The results are presented in Table 14.5.

The corresponding frequency responses are given in Figures 14.11 and 14.12 where the original response is included for comparison. Note that the response may overshoot in-between the chosen frequency points, as shown in Figure 14.11. Nevertheless, the number of points need not be too dense in order to capture the required $\mathcal{H}_\infty$ constraint, particularly if the solution is of low order. Another interesting observation is that the weighted sensitivity remains PD in spite of the new constraints. However, the added constraints demand more degrees of freedom from the controller, which increases its

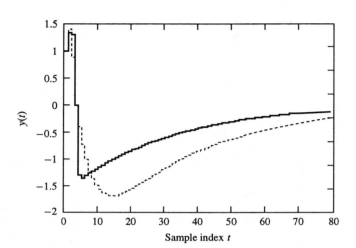

Figure 14.10   X29: Step response of S for ℓ_1 (solid) and $\mathcal{H}_\infty$ (dashed) designs.

TABLE 14.4   OPERATOR NORM COMPARISON ($T_S = 1/60$)

		$\lVert \cdot \rVert_{\mathcal{H}_\infty}$	$\lVert \cdot \rVert_1$	ord(K)
	Φ	2.2	5.0	
$\mathcal{H}_\infty$ design	$W_1 K S$	2.0	3.4	5
	$W_2 S$	2.2	5.0	
	Φ	3.5	3.9	
ℓ_1 design	$W_1 K S$	3.2	3.9	15
	$W_2 S$	3.4	3.9	

order from 6th to 10th and 13th, respectively, revealing explicitly the trade-offs in the design. Finally, the fact that the frequency-domain constraints are applied to the PD block (i.e., the weighted sensitivity) allows us to make an efficient use of the DA method.

The ℓ_1 performance objective with fixed input constraints. Consider the case where the specifications given are such that the control signal resulting from a step input must be constrained uniformly in time (e.g., to avoid actuator saturation). That is, we want to bound the controller response to a step input and at the same time minimize the ℓ_1 norm of Φ. In such a case we augment the basic ℓ_1 problem in the following way:

$$\inf_{K \text{ stab.}} \left\lVert \begin{array}{c} W_1 K S \\ W_2 S \end{array} \right\rVert_1$$

TABLE 14.5 X29: DESIGN WITH
FREQUENCY-DOMAIN CONSTRAINTS

		$\|\cdot\|_{\mathcal{H}_\infty}$	$\|\cdot\|_1$	ord(K)
$\gamma\mathcal{H}_\infty = 2.8$	Φ	3.7	4.3	
	W_1KS	3.4	4.3	10
	W_2S	3.0	4.3	
$\gamma\mathcal{H}_\infty = 2.6$	Φ	3.8	4.5	
	W_1KS	3.5	4.5	13
	W_2S	2.6	4.4	

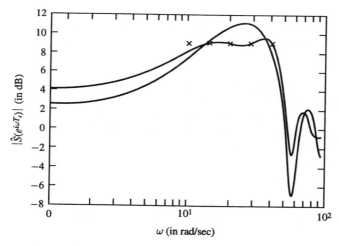

Figure 14.11 X29: Frequency response of S with and without $\mathcal{H}_\infty$ constraints ($\gamma\mathcal{H}_\infty = 2.8$).

subject to

$$\|KSw_f\|_\infty \le U_{max}$$

where w_f is a unit step input disturbance and U_{max} is the specified bound. Clearly, the above results in adding infinitely many constraints on the sequence KS (i.e., convolution of a unit step with KS). However, since the peak is typically achieved in early samples, only a finite number of constraints need to be included in general (the rest being inactive). This is a particular case of nondecaying template constraints, which arise frequently in control system design.

Trade-offs in design.   We take these specifications a step further by asking the following questions: What are the trade-offs in the design? How does the bound on the

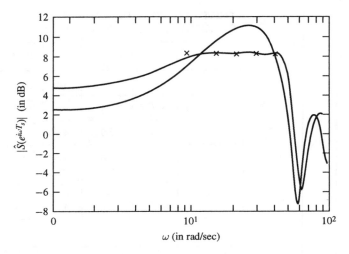

Figure 14.12 X29: Frequency response of S with and without $\mathcal{H}_\infty$ constraints ($\gamma_{\mathcal{H}_\infty} = 2.6$).

control signal step response affect the overall performance? And, how does it affect the structure of the optimal solution?

These questions can be readily answered with the ℓ_1 machinery. It amounts to solving a family of mixed ℓ_1 problems parametrized in $U_{\max}$. Solutions for a range of values of $U_{\max}$ are presented in Figures 14.13 and 14.14 by showing the performance degradation and the controller order growth as $U_{\max}$ decreases. The following conclusions can be drawn from this analysis:

1. For $U_{\max} \geq 74$ the solution is unconstrained (i.e., the step input constraints are all inactive).

2. For $U_{\max} \leq 35.57$ the plant can no longer be stabilized (i.e., the feasible set is empty).

3. As $U_{\max}$ approaches 35.57, the value of the optimization and (possibly) the order of the controller grow unboundedly.

4. The performance suffers very little in the range $40 \leq U_{\max} \leq 74$. For instance, with $U_{\max} = 40$ there is a 46% reduction in the control peak with only a 4.4% increase in the value of the norm. At the same time, the controller only requires two extra states.

The solutions to these problems were computed with the DA method, such that the suboptimal norms were always within 0.1% of the optimal. Note that we were able to exploit the structure of the optimal solution in constructing low order controllers, even with the extra template constraints. For instance, with $U_{\max} = 40$, the number of delays

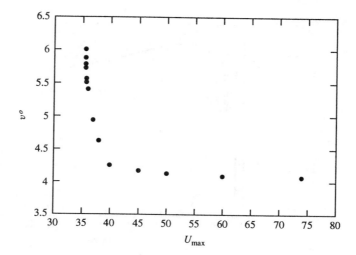

Figure 14.13   X29: Trade-offs in performance vs. control signal bound.

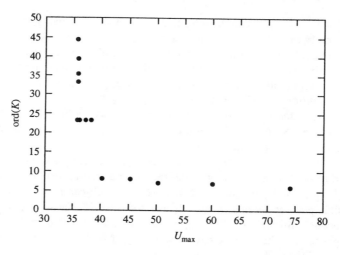

Figure 14.14   X29: Trade-offs in controller order vs. control signal bound.

$N = 30$, and the number of step input constraints $M = 30$, the results are the following:

$$\begin{cases} \underline{\eta}_N = 4.2514 \\ \bar{\eta}_N = 4.2514 \\ \text{supp}(\Phi^o_N)^T = (7 \quad 29) \\ \text{ord}(K) = 8 \end{cases}$$

Figure 14.15 shows that the design goal is achieved. Note that the weighted sensitivity remains PD in spite of the extra constraints. In fact, a simple computation shows that

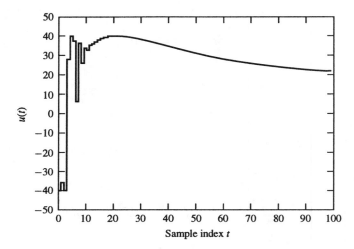

Figure 14.15   X29: KS step response with $U_{\text{max}} = 40$.

$\|W_1 K S\|_1 = 2.3767$, indicating that such row is inactive in the ℓ_1 objective. This does not imply that $W_2 S$ is TD and that we can reduce the problem to a one-block one, since we are dealing with a mixed problem (not purely ℓ_1). The above remark implies that the constraint $\|W_1 K S\|_1 \leq \nu$ can be eliminated without changing the solution. However, the rank interpolation conditions must remain in place to account for the interplay between the conditions $\|W_2 S\|_1 \leq \nu$ and $\|K S w_{\text{step}}\|_\infty \leq U_{\text{max}}$.

The step response of a more extreme case, with $U_{\text{max}} = 35.58$, is shown in Figure 14.16. Note how the high-order controller is able to maintain the response at the boundary for approximately forty samples.

It should be pointed out that these types of template specifications are usually very hard to deal with indirectly, often requiring a trial-and-error procedure that iterates, say, on frequency-dependent weights. Such approach can result in a long and nonsystematic design process, which is not always successful. As an example, let $U_{\text{max}} = 40$ and consider minimizing $\|W_1 K S\|_1$ as an indirect way of bounding the control signal step response. For the X29 we have that $\inf_{K \ \text{stab.}} \|W_1 K S\| = 0.8813$ and $\|K S w_f\|_\infty = 47.33$, which violates the design constraints.

14.2.2 Flexible Beam

The flexible beam, as modeled in Chapter 3, has an interesting feature: It has a pure integrator in the plant (i.e., an interpolation at $\lambda = 1$). Next, we show how we can generate reasonable controllers using the ℓ_1 methodology, even with boundary interpolations.

We start with a few nominal performance designs in order to demonstrate the flexibility of the approach. Later, we will solve a performance robustness problem with a reduced model of the beam. The discrete-time model is obtained by sampling at $T_s = \frac{1}{3}$ seconds.

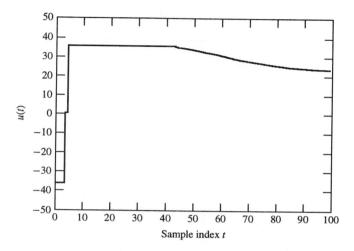

Figure 14.16   X29: KS step response with $U_{\max} = 35.58$.

The ℓ_1 performance objective.
The design objective is the following:

$$\inf_{K \text{ stab.}} \left\| \begin{matrix} W_1 S \\ W_2 K S \end{matrix} \right\|_1 .$$

Due to the large low-frequency gain of the plant, it is possible to obtain reasonable designs without frequency weights. Hence, assuming that the exogenous disturbance has no particular frequency content, we choose $W_1 = W_2 = 1$ for simplicity. Note that this would not be possible in an $\mathcal{H}_\infty$-based design.

The results of applying the Delay Augmentation algorithm are, for $N = 40$:

$$\begin{cases} \underline{\eta}_N = 2.00000 \\ \bar{\eta}_N = 2.00000 \\ \text{supp}(\Phi_N^o)^T = (20 \quad 39) \\ \text{ord}(K) = 18 \end{cases}$$

The support structure indicates that the sensitivity is PD. The time response for a step input is shown in Figure 14.17. Clearly, the response is not satisfactory since the controller seems to delay its action for approximately ten sample periods. Actually, there is a very small control signal during this initial interval, which is not revealed in Figure 14.17. The reason for such interesting behavior is that we have not included frequency weights in the design. Indeed, consider that the system has a pure integrator in the forward loop. Hence, it is clear that for any stabilizing controller, $\|S\|_1 \geq 2$. That is, the closed-loop system cannot track an arbitrary ℓ_∞ command with a tracking error smaller than 2. (Command the system with a positive unit step and let it settle at 1. Then switch the command to a negative unit step. This signal will produce a tracking error of at least 2 at the switch point due to the inertia effect of the pure integrator, for any stabilizing controller.) Furthermore, since the plant is nonminimum phase, there will always be some level of undershoot. Therefore, we can only have $\|S\|_1 > 2$.

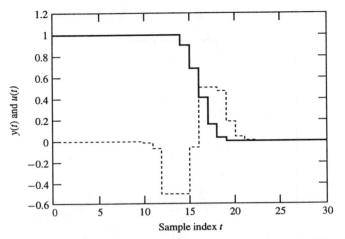

Figure 14.17 Flexible Beam: Step response of unweighted ℓ_1 design, y (solid) and u (dashed).

Why is the controller in the above design almost idle during the first ten sample periods? Clearly, it is trading speed of response for level of undershoot. In fact, pure delays are not penalized by the ℓ_1 norm (i.e., $\|KS\|_1 = \|\lambda^n KS\|_1$), so the design objective reduces to minimizing the undershoot. Note that the cost function corresponding to this design is very close to 2 (within 10^{-6}), which is the theoretical lower bound.

The most obvious way to avoid this kind of behavior is to add a frequency weight to the sensitivity function, so that the low frequency range is penalized. However, a low-pass weighting function will affect the time response of the solution in other ways that are hard to anticipate. It may, for instance, increase the level of overshoot or undershoot. At the same time, note that the time response of the previous design, apart from the delayed action, seems quite desirable.

A simple way to retain the general features of the time response and at the same time eliminate the "almost pure delays" in the controller is to use a weighted ℓ_1 norm as in Equation (2.46) (which we rewrite for convenience):

$$\|\Phi\|_C = \max_{1 \le i \le n_z} \sum_{j=1}^{n_w} \sum_{t=0}^{\infty} |c_{ij}(t)\phi_{ij}(t)|. \tag{14.15}$$

Note that, from the point of view of the DA algorithm, this only changes the construction of the norm constraints. That is,

$$\|\phi_{ij}\|_C = \sum_{k=0}^{\infty} |c_{ij}(k)|(\phi_{ij}^+(k) + \phi_{ij}^-(k)).$$

The rest of the machinery remains intact. A sensible selection of a weight sequence could be an exponential one, where $c_{ij} = a^t$ and $a > 1$. This choice has the desired effect of penalizing late errors over early ones. Also, it bounds the exponential decay of the solution by a^{-t} (see Exercise 10.10).

Let $a = 1.01$; then the DA suboptimal solution with $N = 40$ is as follows:

$$\begin{cases} \|\Phi\|_C = 2.1258 \\ \|S\|_1 = 2.00578 \\ \|KS\|_1 = 1.9369 \\ \text{supp}(\Phi_N^o)^T = (10 \quad 39) \\ \text{ord}(K) = 11 \end{cases}$$

which is within 10^{-7} of the optimal. Figure 14.18 shows that the new design retains a similar response but without delaying the controller action.

Yet another way of posing the problem is by perturbing the plant. Indeed, consider perturbing the pure integrator pole slightly into the unstable region, that is, let $\lambda = 1 - \epsilon$ where $0 < \epsilon << 1$. With this new plant, the controller no longer can take a long time to avoid a small undershoot since there is a growing error that needs to be stabilized. In this setting, the results are as follows: for $N = 30$ and $\epsilon = 0.001$,

$$\begin{cases} \underline{\eta}_N = 2.00693 \\ \overline{\eta}_N = 2.00693 \\ \text{supp}(\Phi_N^o)^T = (10 \quad 23) \\ \text{ord}(K) = 9 \end{cases}$$

and the suboptimal controller is $\hat{K}(\lambda) = \hat{N}(\lambda)/\hat{D}(\lambda)$ where

$$\begin{aligned} \hat{N}(\lambda) = {} & 21.278\lambda^9 + 88.434\lambda^8 + 95.022\lambda^7 + 119.490\lambda^6 + 6.5082\lambda^5 \\ & -97.832\lambda^4 - 97.734\lambda^3 - 116.856\lambda^2 - 15.761\lambda - 2.8573 \end{aligned}$$

and

$$\begin{aligned} \hat{D}(\lambda) = {} & \lambda^9 + 1.3367\lambda^8 - 18.591\lambda^7 - 65.031\lambda^6 - 119.857\lambda^5 \\ & -170.175\lambda^4 - 192.853\lambda^3 - 192.660\lambda^2 - 192.486\lambda - 235.993 \end{aligned}$$

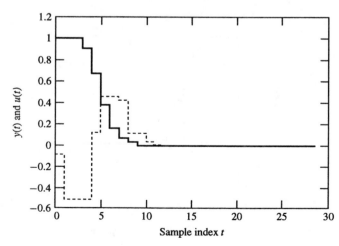

Figure 14.18 Flexible Beam: Step response of weighted ℓ_1 design, y (solid) and u (dashed).

Again, the support structure indicates that the sensitivity is PD. Notice that recomputing the norm with $\epsilon = 0$ gives 2.0087, only a slight degradation of the original value. Figures 14.19 and 14.20 show the sensitivity frequency response and the step response of the closed-loop system. As expected, the time-domain features of the response are very similar to the previous design. In the sequel, we will use this approach to compute designs with different specifications.

The ℓ_1 performance objective with ℓ_∞ constraints.   Consider the perturbed plant. Suppose that we want to bound with one the response of the control signal

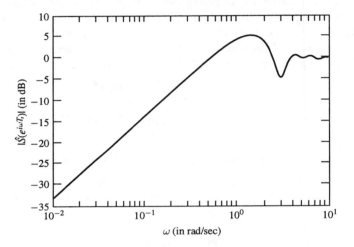

Figure 14.19   Flexible Beam: Sensitivity frequency response.

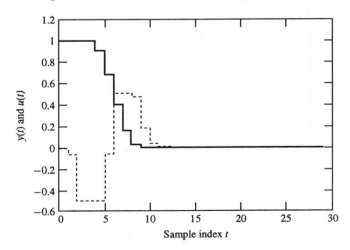

Figure 14.20   Flexible Beam: Step response of "perturbed" design, y (solid) and u (dashed).

for all possible bounded disturbances of magnitude less than one, and that we require the tracking performance to be better than, say, $\gamma = 2.2$ (clearly, it cannot be smaller than 2.00693). That is,

$$\inf_{K \text{ stab.}} \left\| \begin{array}{c} \frac{1}{\gamma} S \\ \frac{1}{U_{max}} K S \end{array} \right\|_1$$

where $U_{max} = 1$. The above is a standard ℓ_1 problem. If the resulting norm is less than one, then the design objectives are achieved. After applying the Delay Augmentation method, we get (for $N = 30$):

$$\begin{cases} \bar{\eta}_N = 0.9113 \\ \underline{\eta}_N = 0.9128 \\ \text{supp}(\Phi_N^o)^T = (19 \quad 28) \\ \text{ord}(K) = 13 \end{cases}$$

This implies that a performance of 2.2 is guaranteed along with the bound on $\|KS\|_1$. To find out what is the best achievable performance without iterating on γ we solve the following problem:

$$\inf_{K \text{ stab.}} \|S\|_1$$
$$\text{subject to} \tag{14.16}$$
$$\|KS\|_1 \leq U_{max}.$$

Again, the above is a standard ℓ_1 problem with an added norm constraint on KS. The DA results are, with $N = 30$ and $\epsilon = 0.001$,

$$\begin{cases} \underline{\eta}_N = 2.0079 \\ \bar{\eta}_N = 2.0079 \\ \text{supp}(\Phi_N^o)^T = (16 \quad 29) \\ \text{ord}(K) = 14 \end{cases}$$

which are almost as good as without the constraint on $\|KS\|_1$. Recomputing the norms for $\epsilon = 0$ results in $\|S\|_1 = 2.0052$ and $\|KS\|_1 = 1.001$. (Note that we could start the design with a bound slightly smaller than one in order to achieve a bound of one with the unperturbed plant.) Figure 14.21 shows the step response for this last design.

The ℓ_1 performance with fixed input constraints.   Here we consider an ℓ_1 performance objective further constrained by a uniform bound on the control signal step response. In other words, we want to obtain the best tracking performance in the ℓ_1 sense, subject to the magnitude of the control signal not exceeding a given value, say U_{max}, when the input disturbance is a unit step. That is,

$$\inf_{K \text{ stab.}} \left\| \begin{array}{c} S \\ K S \end{array} \right\|_1$$
$$\text{subject to}$$
$$\|K S w_f\|_\infty \leq U_{max}$$

where w_f is a unit step.

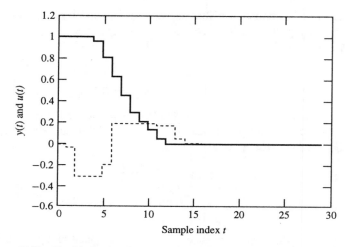

Figure 14.21 Flexible Beam: Step response of design with ℓ_1 constraints, y (solid) and u (dashed).

Given that the step response of the pure ℓ_1 design results in $\|u\|_\infty = 0.5$ (see Figure 14.20), let us say that the control signal should not exceed $U_{max} = 0.3$. Under these conditions the DA solution is: for $N = 40$ and $\epsilon = 0.001$,

$$\begin{cases} \underline{\eta}_N = 2.0078 \\ \bar{\eta}_N = 2.0365 \\ \operatorname{supp}(\Phi_N^o)^T = (13 \quad 39) \\ \operatorname{ord}(K) = 14 \end{cases}$$

and, with $\epsilon = 0$, $\|S\|_1 = 2.0078$ while $\|KS\|_1 = 2.0365$. Notice how the more constrained step-response does not result in a smaller $\|KS\|_1$. Figure 14.22 shows the step response of this design. Clearly, the constraint on the control signal is satisfied.

An ℓ_∞ robust performance problem. Continuing with the flexible beam, we now consider a robust performance specification. Assume that only the rigid body dynamics of the flexible beam were accurately modeled (see Subsection 3.8.2). Then the nominal plant consists of a second-order system with a pure integrator and a slightly damped real pole:

$$P_0(s) = \frac{-0.0032s + 1.26}{s(s + 0.0007)}.$$

We want to design a controller that achieves a certain level of ℓ_∞ performance robustness. To account for the high frequency unmodeled dynamics, we include an additive perturbation of the form

$$P = P_0 + W_1 \Delta_1, \qquad \|\Delta_1\|_{\ell_\infty - \text{ind}} < 1$$

where

$$W_1(s) = \frac{s + 0.5}{s + 5}.$$

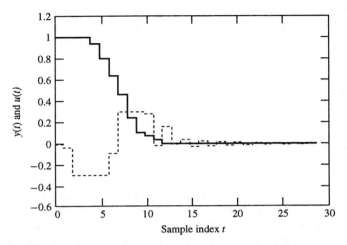

Figure 14.22   Flexible Beam: Step response of design with fixed input constraints, y (solid) and u (dashed).

reflects the accuracy of the nominal model at different frequencies. We would like to solve the following problem: Given $\gamma > 0$, find a robustly stabilizing controller K, such that $\|(I - PK)^{-1}\|_{\ell_\infty-\text{ind}} \leq \gamma$ for all P's in the class described above.

Consider the case where Δ is in Δ_{LTV} or Δ_{NL}. This will allow us to exploit the equivalence between ℓ_∞ stability robustness and ℓ_∞ performance robustness as shown in Theorem 7.7.1 parts (3) and (4). In other words, the problem is equivalent to the ℓ_∞ stability robustness of the system in Figure 14.23.

Hence, we can restate the specifications as follows: Given $\gamma > 0$, find a controller K such that $(I - M\Delta)^{-1}$ is ℓ_∞-stable for all $\Delta \in \{\text{diag}(\Delta_1, \Delta_2) \mid \|\Delta_i\|_{\ell_\infty-\text{ind}} < 1\}$, where

$$M = \begin{pmatrix} 1/\gamma\, S_0 & 1/\gamma\, S_0 \\ W K S_0 & W K S_0 \end{pmatrix}$$

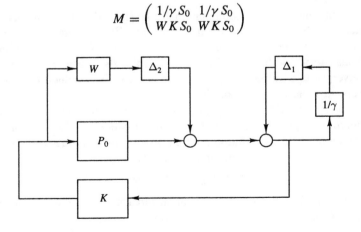

Figure 14.23   Equivalent robust stability problem.

is the transfer function seen by the diagonal perturbation. Theorem 7.3.2 gives noncon-servative conditions for the ℓ_∞ stability robustness of the system in Figure 14.23. That is, the closed-loop system is robustly stable if and only if

$$SN_{\Delta_{LTV}, \infty}(M) = \min_{D \in \mathbf{D}} \|D^{-1}MD\|_1 = \rho(\widehat{M}) \leq 1$$

where

$$\widehat{M} = \begin{pmatrix} 1/\gamma \|S_0\|_1 & 1/\gamma \|S_0\|_1 \\ \|WKS_0\|_1 & \|WKS_0\|_1 \end{pmatrix}.$$

Therefore, the synthesis problem that we are interested in can be posed as

$$\inf_{K \text{ stab.}} \rho(\widehat{M}) = \inf_{K \text{ stab.}} \min_{D \in \mathbf{D}} \|D^{-1}MD\|_1. \tag{14.17}$$

If the value of the above optimization problem is less than one, then there exists a controller that satisfies the design specifications. Finding a solution to Problem 14.17, however, is not an easy task since $\rho(\widehat{M})$ is not convex in $\widehat{M}$.

A first approach to finding suboptimal solutions of Problem 14.17 is based on the following observation: For a fixed D, the problem $\inf_K \|D^{-1}MD\|_1$ is convex (i.e., it is a standard ℓ_1 optimization). Furthermore, for a fixed K, the problem of finding D that minimizes $\|D^{-1}MD\|_1$ is equivalent to an eigenvalue problem. This gives rise to the method known as *D–K iteration*, where Problem 14.17 is approximated by a sequence of two independent optimization problems. It can be summarized as follows:

1. Pick a diagonal scale D.
2. Solve the standard ℓ_1 optimization problem $\inf_K \|D^{-1}MD\|_1$.
3. Compute $\rho(\widehat{M})$. If $\rho(\widehat{M}) \leq 1$ or, if it does not change from the previous iteration (within some tolerance), stop. Otherwise,
4. Assign the eigenvector associated with $\rho(\widehat{M})$ as the diagonal of the new D scaling and go to step 2.

In spite of being computationally feasible, the above method is not without draw-backs. To begin with, the iterative process will converge to a local minimum in general since the problem is not jointly convex in K and D. That is, different initial choices of D may result in different final values of the algorithm. Hence, if a particular iteration converges to a value greater than one, we cannot conclude that the robust performance problem at hand has no solution.

The next difficulty arises as a consequence of the structure of MIMO ℓ_1 optimal solutions. Consider the case of SISO perturbation blocks for simplicity. Assume that at a given point in the iteration problem, the solution to $\inf_K \|D^{-1}MD\|_1$ in step 2 is such that all the rows have equal norm. That is,

$$\|(D^{-1}MD)_i\|_1 = \|D^{-1}MD\|_1 \quad \text{for} \quad i = 1, \dots, n$$

where n is the number of perturbation blocks. In other words, the solution is such that all rows are active. Then, it is easy to show that

$$\|D^{-1}MD\|_1 = \rho(\widehat{M})$$

and,

$$\widehat{M} \begin{pmatrix} d_1 \\ \vdots \\ d_n \end{pmatrix} = \rho(\widehat{M}) \begin{pmatrix} d_1 \\ \vdots \\ d_n \end{pmatrix}.$$

The above implies that in step 4 of the D–K iteration, the D scale remains unchanged and, hence, the iteration converges (i.e., it stops changing). This behavior would be inconsequential if it were not for the fact that the equal-row-norm property occurs frequently in ℓ_1 optimization.

An alternative approach to D–K iteration is to carry out an exhaustive search by gridding the space of diagonal scales. Being computationally very intensive, this method can only handle problems with a small number of perturbation blocks. Since the flexible beam problem has only two blocks of uncertainty, it is a good candidate to demonstrate this method of solution.

Another approach is based on the principle of Sensitivity Analysis in Linear Programming (how sensitive is the solution of the LP to small changes in D?). The basic idea is as follows: Instead of carrying out a search over a uniform grid in the space of diagonal scales, use sensitivity information from the solution of $\inf_K \|D^{-1}MD\|_1$ to choose the next D scale. In this way the resulting grid will be nonuniform and more efficient. Note that the D scale does not affect the feasibility constraints; it only affects the norm constraints (n of them), making the sensitivity analysis tractable.

In the case of the flexible beam problem, because there are only two perturbation blocks we can write the following:

$$D^{-1}MD = \begin{pmatrix} m_{11} & \alpha\, m_{12} \\ \frac{1}{\alpha} m_{21} & m_{22} \end{pmatrix}$$

where $\text{diag}(D) = (1, \alpha)$ and α is the (one-dimensional) search variable. We solve the problem over a log spaced grid in the search variable α, between the values of 10^{-3} and 1. That is, for each value of α we solve $\inf_K \|D^{-1}MD\|_1$ and compute $\rho(\widehat{M})$. Results for $\gamma = 2.2$, and such that $\bar{\eta} - \underline{\eta} < 10^{-4}$ are shown in Table 14.6 and in Figure 14.24.

TABLE 14.6 FLEXIBLE BEAM:
GRID SEARCH VALUES.

$1/\alpha$	$\|1/\gamma\, S_0\|_1$	$\|WKS_0\|_1$	$\rho(\widehat{M})$
1	0.911	0.911	1.822
2	0.912	0.456	1.368
4	0.912	0.228	1.140
8	0.913	0.114	1.027
16	0.914	0.0572	0.972
32	0.916	0.0286	0.944
64	0.918	0.0143	0.932
128	0.920	0.0072	0.928
256	0.924	0.0036	0.928
512	0.930	0.0018	0.931
1024	0.937	0.0009	0.938

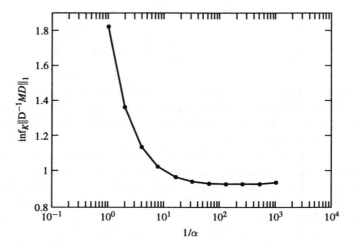

Figure 14.24 Flexible Beam: Grid search visualization.

It is interesting to note that for all values of α in this grid the solution is such that both rows have equal norm. Hence, the D–K iteration method would have failed to converge to the minimum, starting at any point in the grid.

It is clear from Table 14.6 that the 26th order controller that is obtained by solving $\inf_K \|D^{-1}MD\|_1$ with a scale $\alpha = 1/16$, solves the problem at hand. One could choose a controller close to the minimum, say for $\alpha = 1/128$. However, it has the disadvantage of being of higher order ($\simeq 60$) and having worse nominal performance. Finally, we verify that the robustness condition is satisfied with the original nominal plant (i.e., for $\epsilon = 0$). We have that $\rho(\widehat{M}) = 0.9675$ for the controller corresponding to $\alpha = 1/16$.

So far we have looked at approximate methods for solving the ℓ_∞ robust performance problem at hand. In fact, we will show that this particular problem can be solved "exactly" due to the special structure of M. Note that M is rank one. It is easy to verify that

$$\rho(\widehat{M}) = \|1/\gamma\, S_0\|_1 + \|WKS_0\|_1 = \|1/\gamma\, S_0 \qquad WKS_0\|_1.$$

That is, the spectral radius of $\widehat{M}$ is equal to the ℓ_1 norm of the two-block row system $\Phi = (1/\gamma\, S_0 \quad WKS_0)$. Hence,

$$\inf_{K \text{ stab.}} \rho(\widehat{M}) = \inf_{K \text{ stab.}} \|\Phi\|_1$$

and the robustness problem reduces to a standard ℓ_1 problem, the solution of which provides a convenient verification of our previous calculations. The results are, for

$\gamma = 2.2$,

$$\begin{cases} \rho(\widehat{M}) = 0.9273 \ (0.9166) \\ \|S_0\|_1 = 2.029 \ (2.005) \\ \|WKS_0\|_1 = 0.0052 \ (0.0053) \\ \alpha_{optimal} = 1/176.5 \\ \text{ord}(K) = 55 \end{cases}$$

where the numbers in parentheses show the actual values on original plant (with $\epsilon = 0$). It is clear from the above that the grid-search method provided consistent results. Actually, the grid-search solution is richer in the sense that several controllers that satisfy the design requirements are generated. Hence, one has the luxury of choosing one based on other considerations, such as controller order.

Finally, we check that the performance robustness bound is satisfied with the 4th order beam model:

$$\|S\|_1 = 2.02 < 2.2$$

where we have used the 26th order controller obtained with $\alpha = 1/16$. Figure 14.25 shows how the closed-loop system responds to a unit step. Notice how the robustness constraints have effectively reduced the bandwidth of the closed-loop system.

The μ analysis. We end this section by addressing the following question. What is the level of ℓ_2 performance robustness that is achieved by the design presented above, if the plant perturbation is LTI? We know from Chapter 3 that a necessary and sufficient condition for ℓ_2 performance robustness in the presence of LTI perturbations is given by

$$\mu_\Delta[\widehat{M}(e^{i\theta})] \leq 1 \quad \text{for all } \theta \in [0, \pi].$$

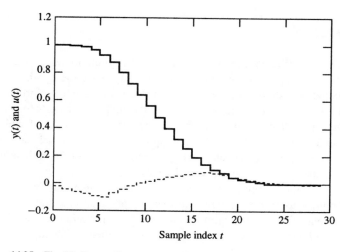

Figure 14.25   Flexible Beam: Step response of robust design, y (solid) and u (dashed).

We also know that the controller in question should satisfy this condition, since it is less stringent than the spectral radius condition required for ℓ_∞ performance robustness. Nevertheless, it is generally of interest to determine what is the actual margin attained by a give design, in terms of $\mathcal{H}_\infty$ performance and LTI plant perturbations.

Figure 14.26 plots the value of $\mu_\Delta[\hat{M}(e^{i\omega T_s})]$ in the frequency interval of interest. Note that μ_Δ attains a maximum value of 0.721 at a frequency of approximately 0.7 radians per second. This indicates that the system can tolerate arbitrary LTI plant perturbations up to 1.387 ℓ_2 gain, and that the performance $\|\hat{S}\|_\infty$ is guaranteed to be not worse than 1.586 in all such cases.

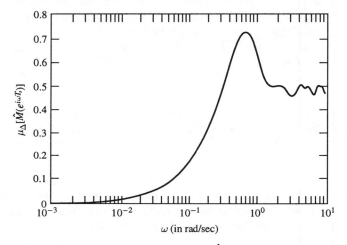

Figure 14.26   Flexible Beam: Plot of $\mu_\Delta[\hat{M}]$ for ℓ_1 robust design.

14.3 SUMMARY

We have presented a broad class of control problems with specifications that arise often in practical applications, either in the analysis or the synthesis phase. These include bounded overshoot and undershoot problems, nominal performance problems, problems with step input, ℓ_∞ and $\mathcal{H}_\infty$ constraints, and robust performance problems. All were solved using the ℓ_1 mathematical machinery, at the heart of which is a linear program. These examples show how the linear programming approach can provide not only feasible controllers but also valuable insights into the trade-offs built into the design as well as the fundamental limitations of a linear approach.

The most important aspect of any design methodology in engineering reduces to the ability to address interesting design specifications under realistic assumptions in a comprehensive and straightforward way. By including this collection of examples, we hope to have demonstrated the power of the theory developed in this book.

APPENDIX A

Elements of Functional Analysis

In this appendix we present the necessary concepts from functional analysis that are needed for the development of this book. In some places we have included complete proofs; in others, we opted to state the final conclusions. The results in this chapter are standard and can be found in many functional analysis textbooks.

A.1 NORMED LINEAR SPACES (NLS)

We begin with the following definition.

Definition A.1.1. Let X be a linear space. A norm on X denoted by $\| \cdot \|$ is a bounded function from X to the positive real numbers satisfying the following axioms:

1. $\|x\| \geq 0$ for all $x \in X$. Also $\|x\| = 0 \iff x = 0$.
2. $\|\alpha x\| = |\alpha| \|x\|$ for all $\alpha \in \mathbb{R}$.
3. $\|x + y\| \leq \|x\| + \|y\|$.

These conditions are known as the positivity, homogeneity, and triangle inequality. A vector space with the above norm is called a normed linear space. With the norm, a distance function between elements of the space can be defined as $\|x_1 - x_2\|$. ∎

There are many examples of normed linear spaces, both finite dimensional and infinite dimensional. In the sequel a few are presented.

Example A.1.1

A familiar normed linear space is the space $\mathbb{R}^n$ with a norm defined as

$$|x|_p = \left(\sum_{i=1}^{n} |x_i|^p \right)^{\frac{1}{p}} \tag{A.1}$$

for any $1 \leq p \leq \infty$. For $p = \infty$, the norm is the pointwise limit of the above definition and is given by

$$|x|_\infty = \max_i |x_i|. \tag{A.2}$$

Example A.1.2

A natural extension of the above finite-dimensional space is the space of infinite sequences. Denote by ℓ_p the space of all bounded sequences with bounded ℓ_p-norm defined as

$$\|x\|_p = \left(\sum_{k=0}^{\infty} |x(k)|^p \right)^{\frac{1}{p}} \tag{A.3}$$

for any $1 \leq p \leq \infty$. For $p = \infty$, the norm is the pointwise limit of the above definition and is given by

$$\|x\|_\infty = \sup_{k \geq 0} |x(k)|. \tag{A.4}$$

The subspace of ℓ_∞ that contains all the sequences converging to zero is denoted by c_0. On its own, c_0 is a normed linear space.

Example A.1.3

Let $\mathcal{L}_p(B)$ denote the space of all Lebesgue integrable functions on a closed set B with bounded $\mathcal{L}_p$-norm defined as

$$\|x\|_p = \left(\int_B |x(t)|^p dt \right)^{\frac{1}{p}} \tag{A.5}$$

for any $1 \leq p \leq \infty$. For $p = \infty$, the norm is defined as

$$\|x\|_\infty = ess \sup_{t \in B} |x(t)| \tag{A.6}$$

$$:= \inf_{y = x \, a.e.} \sup_t |y(t)|, \tag{A.7}$$

where $a.e.$ reads almost everywhere, i.e., except on a set of measure zero.

In the above definition, the set B can be quite general. For instance, B can be the whole real line, the positive real line, a closed interval, the unit circle, the imaginary axis, and so on. These functions can be real valued or complex valued. As this book discusses primarily discrete-time systems, these spaces are used in general to denote spaces of Fourier transforms, and hence B is usually the unit circle. There are some examples, however, in which these spaces are defined over the real numbers.

Example A.1.4

The space of continuous functions on the interval $[a, b]$ is denoted by $C[a, b]$. In general, this space is equipped with the $\mathcal{L}_\infty$-norm. Any other $\mathcal{L}_p$ norm defines a new normed linear space.

A.2 OPEN AND CLOSED SETS

There exists a unique notion of an open set associated with a normed linear space X. The norm induces a topology on the vector space, which is simply a collection of open sets. The following definition provides the basic idea behind the characterization of open sets inside normed linear spaces.

Definition A.2.1.   Let M be a subset of a normed linear space X. A point $x \in M$ is an interior point of M if there exists a ball of radius ϵ defined as

$$B(x, \epsilon) = \{y|\ \|x - y\| < \epsilon,\ y \in X\}$$

which is contained in M. The set M is open if every point is an interior point.   ∎

Clearly, the set $B(x, r)$ is open for any $x \in X$ and $r \in \mathbb{R},\ r > 0$. The whole space and the empty set are both open sets. The following is the definition of a closed set.

Definition A.2.2.   A set $M \subset X$ is closed if and only if its complement M^c is open.   ∎

The set

$$\overline{B(x, r)} = \{y|\ \|x - y\| \leq r,\ y \in X\}$$

is clearly closed.

A.3 CONVERGENCE OF SEQUENCES

Let $x_n,\ n \geq 0$ be a sequence in a normed linear space X. The sequence converges to an element $x \in X$ if

$$\lim_{n \to \infty} \|x - x_n\| = 0.$$

From a topological point of view, x_n converges to x if and only if every open set containing x intersects the tail of the sequence x_n. It is evident that this is equivalent to the above definition. As an extension of this definition, suppose that every open set containing x intersects an infinite number of elements in the tail of x_n, but not necessarily all the elements; then x is said to be a *cluster* point of the sequence x_n. Equivalently, there exists a subsequence of x_n, denoted by x_{nj}, that converges to x.

If x_n converges to x, then it is necessary that its elements get closer and closer to each other. A sequence with such a property is known as a *Cauchy sequence*.

Definition A.3.1.   A sequence x_n is said to be a Cauchy sequence if for every $\epsilon > 0$ there exists an N such that

$$\|x_n - x_m\| < \epsilon \quad \text{for all } n, m \geq N. \qquad \blacksquare$$

The following proposition is immediate from the definition.

Proposition A.3.1.   Every convergent sequence is Cauchy. $\blacksquare$

The converse of this proposition is not true.

Example A.3.1

Consider the space of continuous functions on the interval $[0, 1]$ equipped with the $\mathcal{L}_1$-norm. Let x_n be the sequence of functions defined for all $n \geq 2$ as

$$x_n(t) = \begin{cases} 0 & 0 \leq t \leq .5 - \frac{1}{n} \\ n(t - .5) + 1 & .5 - \frac{1}{n} < t \leq .5 \\ 1 & .5 < t \leq 1 \end{cases}.$$

It follows that

$$\|x_n - x_m\|_1 \leq \min\left\{\frac{1}{n}, \frac{1}{m}\right\} \to 0 \text{ as } n, m \to \infty.$$

On the other hand, the unique limit of this function is

$$x(t) = \begin{cases} 0 & 0 \leq t \leq .5 \\ 1 & .5 \leq t \leq 1 \end{cases},$$

which is not continuous. Hence, the limit of the sequence does not lie inside the space.

Note that if the space is equipped with the $\mathcal{L}_p$-norm for $p \neq \infty$, then the same conclusion holds as in the above example. With the $\mathcal{L}_\infty$-norm, the conclusion does not hold since the sequence x_n is not cauchy:

$$\lim_{m \to \infty} \|x_n - x_m\|_\infty = 1 \quad \text{for all } n.$$

A.4 COMPLETE NORMED LINEAR SPACES

Definition A.4.1.   A *complete NLS* is a *NLS* X in which every Cauchy sequence has a limit in X. A complete *NLS* is called a *Banach space*. $\blacksquare$

From Example A.3.1, the space of continuous functions with the $\mathcal{L}_1$-norm is not complete. The following examples are stated without proofs.

Example A.4.1

The space $(\mathbb{R}^n, |\cdot|_p)$ is a Banach space. In fact, it is not hard to show that every finite-dimensional *NLS* is complete.

Example A.4.2

The space ℓ_p is complete for any $p \geq 1$.

Example A.4.3

It was seen that the space of continuous functions on the interval $[a, b]$ is not complete when equipped with the p-norm, with $p \neq \infty$. The space $C[a, b]$, however, is complete. This is due to the fact that convergence in the ∞-norm is equivalent to uniform convergence of functions on the interval $[a, b]$. Thus, if x_n converges to x in the ∞-norm, then x_n converges to x uniformly on $[a, b]$, which implies that x must be continuous, i.e., $x \in C[a, b]$.

Example A.4.4

The space $\mathcal{L}_p(B)$ is a Banach space for any $p \geq 1$.

A.5 HILBERT SPACES

Definition A.5.1.   An inner product space X is a linear space equipped with a complex-valued bilinear function denoted by $\langle \cdot, \cdot \rangle$ satisfying the following axioms:

1. $\langle x, y \rangle = \overline{\langle y, x \rangle}$.
2. $\langle x, x \rangle \geq 0$ and $\langle x, x \rangle = 0 \iff x = 0$.
3. $\langle x + y, z \rangle = \langle x, z \rangle + \langle y, z \rangle$.
4. $\langle \alpha x, y \rangle = \alpha \langle x, y \rangle$ for all $\alpha \in \mathbf{C}$. ∎

Definition A.5.2.   An inner product space is a linear space equipped with an inner product. An inner product space is immediately a normed linear space with the norm defined as

$$\|x\| = \sqrt{\langle x, x \rangle}.$$ ∎

Definition A.5.3.   The space X is a Hilbert space if it is a complete inner product space. ∎

If the space X is real valued, the bilinear function $\langle \cdot, \cdot \rangle$ is real valued. Whether an inner product space is real valued or complex valued will be understood from the context.

Example A.5.1

The space $(\mathbb{R}^n, |\cdot|_2)$ is a Hilbert space. This is the standard Euclidean space with the standard inner product.

Example A.5.2

The space ℓ_2 is a Hilbert space. The inner product is defined as

$$\langle x, y \rangle = \sum_{k=0}^{\infty} y^*(k) x(k).$$

The resulting norm is precisely the 2-norm we defined earlier on this space.

Example A.5.3

The space $\mathcal{L}_2(B)$ is a Hilbert space. The inner product is given by

$$\langle x, y \rangle = \int_B y^*(t)\, x(t)\, dt.$$

The resulting norm is precisely the 2-norm we defined earlier on this space.

A consequence of the inner product is the Cauchy-Schwartz inequality.

Theorem A.5.1.   For all x, y in a Hilbert space, $|\langle x, y \rangle| \leq \|x\| \|y\|$. Equality holds if and only if $x = cy$ for some scalar $c \in \mathbf{C}$.

Proof. Consider the following inequality for any scalar α:

$$0 \leq \langle x - \alpha y, x - \alpha y \rangle = \|x\|^2 - \alpha\overline{\langle x, y \rangle} - \overline{\alpha}\langle x, y \rangle + |\alpha|^2 \|y\|^2.$$

For the specific choice of

$$\alpha = \frac{\langle x, y \rangle}{\|y\|^2},$$

we get the result. If equality holds, then $\|x - \alpha y\| = 0$ for the above α, which proves the second part.   ∎

This added structure on a normed linear space gives a notion of orthogonality, and orthogonal complements.

Definition A.5.4.   Two elements in a Hilbert space are orthogonal if and only if $\langle x, y \rangle = 0$. This is denoted as $x \perp y$.   ∎

The following two properties are true in any inner product space, and can be verified by direct expansion.

Proposition A.5.1.   Let x, $y \in X$, an inner product space.

1. $\|x + y\|^2 + \|x - y\|^2 = 2\|x\|^2 + 2\|y\|^2$.
2. if $x \perp y$ then $\|x + y\|^2 = \|x\|^2 + \|y\|^2$.   ∎

The projection theorem.   Given a Hilbert space X, a subspace M, and a fixed element $x_0 \in X$, we desire to find the element in M that is the closest to x_0. This can be formulated as the solution to the following optimization problem:

$$d = \inf_{x \in M} \|x_0 - x\|. \tag{A.8}$$

A complete characterization of the solution to this problem is given by the projection theorem.

Theorem A.5.2.   Let M be a subspace of a Hilbert space X, and $x_0 \in X$ is fixed. Define the following minimum distance problem:

$$d = \inf_{x \in M} \|x_0 - x\|.$$

It follows that:

1. If x^* solves the minimization problem, then $x_0 - x^* \perp M$.
2. If x^* exists, then it is unique.
3. If M is closed, then a solution is guaranteed to exist.

Proof. (1) To show this part, assume to the contrary that there exists a $m \in M$ with $\|m\| = 1$, such that $\langle m, x_0 - x^* \rangle = \delta \neq 0$. Define $x_1 = x^* + \delta m$. It follows that

$$\|x_0 - x_1\|^2 = \|x_0 - x^* - \delta m\|^2$$
$$= \|x_0 - x^*\|^2 - |\delta|^2$$
$$< \|x_0 - x^*\|^2,$$

which contradicts optimality of x^*.

(2) Let x_1^*, x_2^* be solutions to the minimum distance problem. Then

$$\|x_0 - x_1^*\|^2 = \|x_0 - x_2^* + x_2^* - x_1^*\|^2 = \|x_0 - x_2^*\|^2 + \|x_2^* - x_1^*\|^2$$

where the last equality follows from Propositions A.5.1. Since both solutions are optimal, this implies that

$$\|x_2^* - x_1^*\| = 0 \implies x_1^* = x_2^*.$$

(3) To show this part, assume that $x_n \in M$ is a sequence satisfying

$$\lim_{n \to \infty} \|x_0 - x_n\| = d.$$

If we show that x_n is a cauchy sequence, then it follows from the completeness of X, and the closedness of M, that x_n has a limit $x^* \in M$. Thus x^* will be a candidate for the optimal solution.

From Proposition A.5.1, it follows that

$$\|(x_j - x_0) + (x_0 - x_i)\|^2 + \|(x_j - x_0) - (x_0 - x_i)\|^2 = 2\|x_j - x_0\|^2 + 2\|x_0 - x_i\|^2.$$

Equivalently

$$\|x_j - x_i\|^2 = 2\|x_j - x_0\|^2 + 2\|x_0 - x_i\|^2 - 4\left\|x_0 - \frac{x_i + x_j}{2}\right\|^2.$$

Since M is a subspace, $\frac{x_i + x_j}{2} \in M$ and $\|x_0 - \frac{x_i + x_j}{2}\| \geq d$. Thus, it follows that

$$\|x_j - x_i\|^2 \leq 2\|x_j - x_0\|^2 + 2\|x_0 - x_i\|^2 - 4d^2.$$

The right-hand side of this inequality goes to zeros as i, j go to infinity. This implies that x_n is Cauchy, and has a limit $x^* \in M$. From the continuity of the norm, we get

$$\|x_0 - x^*\| = \lim_{n \to \infty} \|x_0 - x_n\| = d$$

and the result follows.   ∎

It follows from the projection theorem that given any closed set in a Hilbert space, the whole space can be decomposed uniquely as the sum of this set and its orthogonal complement.

Definition A.5.5.   A linear space X is a direct sum of two subspaces Y, Z if every element $x \in X$ can be written uniquely as the sum of two elements $x = y + z$ where $y \in Y$ and $z \in Z$. This is denoted by

$$X = Y \oplus Z.$$ ∎

From the projection theorem, every closed subspace M of a Hilbert space X can be complemented in the sense described above.

Theorem A.5.3.   Let M be a closed subspace of a Hilbert space X. Then

$$X = M \oplus M^\perp$$

where

$$M^\perp = \{x \in X \,|\, x \perp M\}.$$

Proof. For any $x \in X$, let x_1 be the solution of the distance problem in Equation (A.8), and define $x_2 := x - x_1$. From Theorem A.5.2, $x_2 \perp M$, and thus $x_2 \in M^\perp$. To show uniqueness, assume that $x = y_1 + y_2$ with $y_1 \in M$ and $y_2 \in M^\perp$, then

$$x = x_1 + x_2 = y_1 + y_2 \implies x_1 - y_1 = y_2 - x_2.$$

But $x_1 - y_1 \in M$ and $y_2 - x_2 \in M^\perp$. This implies that $x_1 = y_1$ and $x_2 = y_2$.   ∎

A.6 COMPACT SETS

Definition A.6.1.   A subset M of a normed linear space is compact if every sequence in M has a convergent subsequence inside M.   ∎

Example A.6.1

If X is finite dimensional, then M is compact if and only if it is closed and bounded.

Example A.6.2

The closed unit ball in ℓ_p is not compact. Let the sequence x_n be defined as: $x_n(k) = \delta_n(k)$ where $\delta_n(k)$ is the Kronecker delta. This sequence does not have any convergent subsequences, and thus the set is not compact. Notice that the unit ball is closed and bounded.

The above example shows that there are spaces (infinite dimensional) in which closed and bounded sets are not compact. This is not a coincidence, and the following proposition can be shown. It is stated without proof.

Proposition A.6.1.   If the unit ball in a normed linear space X is compact, then the space is finite dimensional.   ∎

Because compactness is a desirable property to have when dealing with optimization problems, the above proposition indicates a difficulty when optimization is performed over infinite-dimensional spaces. To remedy this problem, it is convenient to define a weaker sense of compactness that allows the unit ball to be compact. A detailed discussion of such an alternative using duality theory is discussed in Chapter 9.

A.7 OPERATORS

Let X, Y be two normed linear spaces. An operator (map) T is a relation that maps each element in a domain of X, say D, to one element in Y. The following definition is an extension of the standard definition of continuity.

Definition A.7.1.   An operator T is continuous at a point x_0 if for every given $\epsilon > 0$ there exists a $\delta > 0$ such that

$$\text{for all } x \text{ such that } \|x - x_0\| < \delta \implies \|T(x) - T(x_0)\| < \epsilon.$$

T is continuous if it is continuous at each $x \in X$.   ∎

A particular class of interesting operators are the linear ones. A linear operator satisfies

$$T(\alpha x_1 + \beta x_2) = \alpha T(x_1) + \beta T(x_2) \text{ for all } \alpha, \beta \in \mathbb{R} \text{ and } x_1, x_2 \in X. \tag{A.9}$$

Definition A.7.2.   Let T be a linear operator from X to Y. The range of T is defined as

$$\mathcal{R}(T) = \{y = Tx \mid x \in X\}.$$

The null space of T is defined as

$$\mathcal{N}(T) = \{x \mid Tx = 0\}.$$ ∎

For linear operators, the notion of continuity is closely tied with a notion of boundedness. This is shown in the following theorem.

Theorem A.7.1.   Let T be a linear operator from the space X to Y; then the following are equivalent:

1. T is continuous at one point, x_0.
2. T is continuous everywhere.
3. There exists a positive number K such that $\|T(x)\| \leq K \|x\|$ for all $x \in X$.

Proof. $1 \Longrightarrow 2$: Pick any $x \in X$. Then

$$\|T(x) - T(y)\| = \|T(x_0) - T(x - y + x_0)\|.$$

The continuity at x follows immediately from the continuity at x_0.

$2 \Longrightarrow 3$: Since T is continuous at 0, then for $\epsilon = 1$ there exists a $\delta > 0$ such that

$$\left\| T \left(\frac{\delta}{\|x\|} x \right) \right\| < 1 \qquad \text{for all } x \neq 0.$$

Using linearity, this implies that

$$\|T(x)\| \leq \frac{1}{\delta} \|x\|,$$

which implies 3. Finally, the implication $3 \Longrightarrow 1$ is straightforward. ∎

The above theorem implies that a linear operator is continuous if the maximum amplification it exerts on bounded inputs is finite. This provides a measure on the space of bounded operators.

Definition A.7.3.   The collection of bounded linear operators from X to Y is a linear space. It can be made a normed linear space by defining the following norm on each element:

$$\sup_{x \neq 0} \frac{\|Tx\|}{\|x\|}.$$

This is known as the induced operator norm. ∎

It is straightforward to verify that the induced norm is given by

$$\|T\| = \sup_{x \neq 0} \frac{\|Tx\|}{\|x\|}$$

$$= \sup_{\|x\| \leq 1} \|Tx\|$$

$$= \inf\{K \in \mathbb{R} \mid \|Tx\| \leq K \|x\|\}.$$

Example A.7.1

Given any matrix $A = (a_{ij}) \in \mathbb{R}^{m \times n}$. The induced norm of A as an operator from $(\mathbb{R}^n, |.|_p)$ to $(\mathbb{R}^m, |.|_p)$ is given by

$$\|A\| = |A|_q = \sup_{|x|_p \leq 1} |Ax|_p, \qquad \frac{1}{p} + \frac{1}{q} = 1. \tag{A.10}$$

The induced norm can be computed exactly for $p = \infty$, 1, 2:

$$
\begin{aligned}
|A|_1 &= \max_i \sum_j |a_{ij}|, \\
|A|_\infty &= \max_j \sum_i |a_{ij}|, \\
|A|_2 &= \sigma_{\max}(A).
\end{aligned}
\tag{A.11}
$$

The relation between $|A|_1$ and $|A|_2$ is given by

$$
\frac{1}{\sqrt{n}}|A|_1 \leq |A|_2 \leq \sqrt{m}|A|_1.
\tag{A.12}
$$

The induced norm of A as an operator from $(\mathbb{R}^n, |.|_\infty)$ to $(\mathbb{R}^m, |.|_1)$ is given by

$$
|A|_{1,\infty} = \sup_{i,j} |a_{ij}|.
$$

Common linear operators are convolution operators. Next we will show some examples of such operators.

Example A.7.2

Consider the operator

$$
T(x)(t) = \sum_{k=0}^{t} x(k).
$$

This is a linear operator from ℓ to ℓ (ℓ is the space of all sequences). It is not a bounded operator on ℓ_∞. To see that, let $x = (1, 1, 1, \ldots)$. Then $T(x)$ is unbounded, and thus the induced norm is infinite.

On the other hand, the operator

$$
T(x)(t) = \sum_{k=0}^{t} a^{t-k} x(k), \qquad |a| < 1
$$

is a bounded operator on ℓ_∞. Its induced norm is given by

$$
\|T\| = \sum_{t=0}^{\infty} |a|^t = \frac{1}{1 - |a|} < \infty.
$$

Convergence of sequences of operators.   We discuss two kinds of convergence of sequences of operators. Consider general bounded operators on NLS, mapping X to Y.

1. *Uniform Operator Convergence.* If T_n is a sequence of operators, then it is uniformly convergent to an operator T if it converges in the induced norm, i.e.,

$$
\lim_{n \to \infty} \|T_n - T\| = 0.
$$

2. *Strong Operator Convergence.* If T_n is a sequence of operators, then it is strongly convergent to an operator T if $T_n x$ converges in Y to Tx for every $x \in X$.

The following result holds.

Theorem A.7.2. Let T_n be a sequence of bounded linear operators from X to Y, such that $T_n x$ converges strongly to Tx in Y for any $x \in X$ (i.e., T_n is strongly operator convergent with limit T), then the following inequality holds:

$$\|T\| \le \liminf_{n \to \infty} \|T_n\|. \qquad \blacksquare$$

A.8 HILBERT SPACE ADJOINTS

Let T be a bounded operator from X to Y (two Hilbert spaces). We can associate with T an operator T^* from Y to X. This operator will be linear, bounded, and will have the same norm as T.

Definition A.8.1. To each bounded linear operator $T : X \to Y$, with X, Y both Hilbert spaces, there corresponds a unique linear bounded operator $T^* : Y \to X$ known as the adjoint operator, such that for all $x \in X$ and all $y \in Y$,

$$\langle Tx, y^* \rangle = \langle x, T^* y^* \rangle. \qquad \blacksquare$$

Theorem A.8.1. Let $T : X \to Y$ be a bounded operator, with X, Y both Hilbert spaces. Then, it follows that

$$\|T\| = \|T^*\|,$$

and

$$\|T\| = \|T T^*\|^{\frac{1}{2}} = \|T^* T\|^{\frac{1}{2}}.$$

Proof. From the definition of induced norms and inner products, we have

$$\begin{aligned}
\|T\| &= \sup_{\|x\| \le 1} \|Tx\| \\
&= \sup_{\|x\| \le 1} \sup_{\|y\| \le 1} \langle Tx, y \rangle \\
&= \sup_{\|x\| \le 1} \sup_{\|y\| \le 1} \langle x, T^* y \rangle \\
&= \sup_{\|y\| \le 1} \sup_{\|x\| \le 1} \langle x, T^* y \rangle \\
&= \sup_{\|y\| \le 1} \|T^* y\| \\
&= \|T^*\|.
\end{aligned}$$

To prove the second equality, it follows from above that

$$\|T^* T\| \le \|T\|^2. \qquad (A.13)$$

On the other hand, we have

$$\|Tx\|^2 = \langle Tx, Tx \rangle$$
$$= \langle T^*Tx, x \rangle$$
$$\leq \|T^*Tx\| \|x\|$$
$$\leq \|T^*T\| \|x\|^2,$$

which implies that

$$\|T\|^2 \leq \|T^*T\|. \tag{A.14}$$

The result follows from Equations (A.13, A.14).   ∎

An operator T on a Hilbert space is self adjoint if $T = T^*$. The following result holds for self-adjoint operators.

Theorem A.8.2.   If T satisfies $T = T^*$, then there exists an operator G such that $T = G^*G$ if and only if $\langle Tx, x \rangle \geq 0$ for all $x \in X$. G is usually denoted by $T^{\frac{1}{2}}$.   ∎

Definition A.8.2.   An operator T on a Hilbert space X is positive semi-definite (denoted by $T \geq 0$) if

$$\langle Tx, x \rangle \geq 0 \quad \forall \quad x \in X.$$

It is positive definite (denoted by $T > 0$) if

$$\langle Tx, x \rangle > 0 \quad \forall \quad x \neq 0, x \in X.$$ ∎

Finally, the following theorem holds.

Theorem A.8.3.   Let G and H be two operators from X to Y, both Hilbert spaces, such that

$$\|Gu\| \leq \|Hu\| \quad \text{for all } u \in X.$$

Then it follows that there exists an operator T from Y to Y such that

$$G = TH \quad \text{and} \quad \|T\| \leq 1.$$ ∎

A.9 OPERATOR DILATIONS

In this section we present a few results on operator dilation problems. In the following A, B, C, X are operators on sequences of appropriate lengths (i.e., subspaces in $\ell_2(\mathbf{Z})$) and the $\| \cdot \|$ indicates the induced norm.

Lemma A.9.1. The following are true:

1.

$$\min_{X} \left\| \begin{pmatrix} X \\ A \end{pmatrix} \right\| = \|A\|.$$

2. $\left\| \begin{pmatrix} X \\ A \end{pmatrix} \right\| \le \gamma$ if and only if there exists an operator Y with $\|Y\| \le 1$ such that $X = Y(\gamma^2 I - A^*A)^{\frac{1}{2}}$.

Proof. The first part follows trivially. To prove the second part, notice that

$$\left\| \begin{pmatrix} X \\ A \end{pmatrix} \right\| \le \gamma \iff X^*X \le \gamma^2 I - A^*A =: G^*G,$$

where $G = (\gamma^2 I - A^*A)^{\frac{1}{2}}$. It also follows that $\|Xu\| \le \|Gu\|$ for every input u, which implies (by Theorem A.8.3) that $X = YG$ for some Y with $\|Y\| \le 1$. ∎

Lemma A.9.2. The following are true:

1.

$$\min_{X} \|(X \quad A)\| = \|A\|.$$

2. $\|(X \quad A)\| \le \gamma$ if and only if there exists an operator Z with $\|Z\| \le 1$ such that $X = (\gamma^2 I - AA^*)^{\frac{1}{2}}Z$. ∎

Using the above two lemmas, we prove the following result on operator dilation.

Theorem A.9.1.

$$\gamma_0 := \min_{X} \left\| \begin{pmatrix} X & B \\ C & A \end{pmatrix} \right\| = \max \left\{ \|(C \quad A)\|, \left\| \begin{pmatrix} B \\ A \end{pmatrix} \right\| \right\}.$$

Proof. Let γ_1 be defined as

$$\gamma_1 = \max \left\{ \|(C \quad A)\|, \left\| \begin{pmatrix} B \\ A \end{pmatrix} \right\| \right\}.$$

It follows immediately that

$$\gamma_0 \ge \gamma_1.$$

It follows from Lemmas A.9.1, A.9.2 that there exists two operators Y and Z with norms less than or equal to one such that

$$B = Y(\gamma_1^2 I - A^*A)^{\frac{1}{2}}, \quad C = (\gamma_1^2 I - AA^*)^{\frac{1}{2}}Z.$$

Define a candidate solution to be $\tilde{X} = -YA^*Z$. Then by direct substitution we have

$$\left\| \begin{pmatrix} \tilde{X} & B \\ C & A \end{pmatrix} \right\| = \left\| \begin{pmatrix} -YA^*Z & Y(\gamma_1^2 I - A^*A)^{\frac{1}{2}} \\ (\gamma_1^2 I - AA^*)^{\frac{1}{2}}Z & A \end{pmatrix} \right\|$$

$$= \left\| \begin{pmatrix} Y & 0 \\ 0 & I \end{pmatrix} \begin{pmatrix} -A^* & (\gamma_1^2 I - A^*A)^{\frac{1}{2}} \\ (\gamma_1^2 I - AA^*)^{\frac{1}{2}} & A \end{pmatrix} \begin{pmatrix} Z & 0 \\ 0 & I \end{pmatrix} \right\|$$

$$\leq \gamma_1$$

where the last inequality follows by direct computation. This implies that

$$\gamma_0 \leq \gamma_1$$

which proves the result.   ■

A.10 CONVEXITY

Convexity plays an important role in optimization theory. In this section we present the basic definitions of convex sets and functions.

Definition A.10.1.   A set C is convex if for every x_1 and x_2 in C, $\alpha x_1 + (1-\alpha)x_2$ is also in C for all $0 < \alpha < 1$.   ■

Example A.10.1

Sets that are characterized by linear constraints are convex. Consider a $m \times n$ real matrix A and a fixed vector $b \in \mathbb{R}^m$. The set

$$C = \{x \in \mathbb{R}^n \,|\, Ax \leq b\}$$

is clearly convex.

Example A.10.2

The unit ball in a any normed linear space is convex.

A special class of convex sets are convex cones.

Definition A.10.2.   A subset $\mathcal{P}$ of a vector space X is a convex cone if it is convex and if x is in $\mathcal{P}$ then αx is in $\mathcal{P}$ for all $\alpha \geq 0$. $\mathcal{P}$ is a convex cone with vertex b if $\mathcal{P} - b$ is a convex cone.   ■

Definition A.10.3.   Suppose f is a map from a vector space X to $\mathbb{R}$ with domain D (such a map is usually referred to as a function, or functional if it is linear). Assume D is convex. We say f is convex if for every x_1 and x_2 in D,

$$f(\alpha x_1 + (1-\alpha)x_2) \leq \alpha f(x_1) + (1-\alpha)f(x_2) \quad \text{for all } 0 < \alpha < 1.$$ ■

Example A.10.3

   If X is a normed linear space, then $\| \cdot \|$ is a convex function.

NOTES AND REFERENCES

The material discussed in this appendix is standard and can be found in many functional-analysis books; for example, see [Con85, Kre89, Rud73b].

APPENDIX B ———————————

Discrete-Time Riccati Equations

In this appendix we will review standard material on Riccati equations for discrete-time systems. Such equations play an instrumental role in the solution of the general $\mathcal{H}_2$ and $\mathcal{H}_\infty$ problems.

B.1 SYMPLECTIC MATRICES

Definition B.1.1. A pair of matrices H_1, $H_2 \in \mathbf{C}^{2n \times 2n}$ is called symplectic if it satisfies

$$H_1 J H_1^* = H_2 J H_2^*,$$

where J is the matrix

$$J = \begin{pmatrix} 0 & I \\ -I & 0 \end{pmatrix}.$$

Let H denote the symplectic pair $[H_1, H_2]$ where

$$H_1 = \begin{pmatrix} A & 0 \\ -Q & I \end{pmatrix} \text{ and } H_2 = \begin{pmatrix} I & R \\ 0 & A^* \end{pmatrix},$$

with $Q = Q^*$ and $R = R^*$. ■

Definition B.1.2. A complex number β is a generalized eigenvalue of H if there exists a vector x such that:

$$H_1 x = \beta H_2 x.$$

Equivalently, β is a root of the polynomial $\det (H_1 - \lambda H_2)$.

If β has multiplicity r, then a set of vectors $\{x_1, \ldots, x_\ell\}$ satisfying

$$H_1 x_1 = \beta H_2 x_1$$

$$(H_1 - \beta H_2) x_k = H_2 x_{k-1}, \quad k = 2, \ldots, \ell \quad \ell \le r$$

defines a chain of generalized principal vectors.   ∎

Lemma B.1.1.   If $\beta \ne 0$ is a generalized eigenvalue, then $1/\bar{\beta}$ is also a generalized eigenvalue with the same multiplicity. If $\beta = 0$ is a generalized eigenvalue of multiplicity r then the symplectic pair has $2n - r$ finite generalized eigenvalues.   ∎

In the case where the pair H has no generalized eigenvalues on the unit circle, then it has n generalized eigenvalues inside the unit disc. Let $\mathcal{X}_-(H)$ denote the n-dimensional spectral subspace associated with the generalized eigenvalues inside $\mathcal{D}$. This subspace can be represented as the span of the principal eigenvectors associated with each generalized eigenvalue. Stacking up these vectors, we can write

$$\mathcal{X}_-(H) = \mathcal{R}\begin{pmatrix} X_1 \\ X_2 \end{pmatrix},$$

where $X_1, X_2 \in \mathbf{C}^{n \times n}$. By definition, it follows that

$$H_1 \begin{pmatrix} X_1 \\ X_2 \end{pmatrix} = H_2 \begin{pmatrix} X_1 \\ X_2 \end{pmatrix} T, \tag{B.1}$$

where $T \in \mathbf{C}^{n \times n}$ satisfying $\rho(T) < 1$. If X_1 is nonsingular, it follows that

$$\mathcal{R}\begin{pmatrix} X_1 \\ X_2 \end{pmatrix} = \mathcal{R}\begin{pmatrix} I \\ X_2 X_1^{-1} \end{pmatrix}.$$

Equivalently

$$\mathcal{X}_-(H) \text{ and } \mathcal{R}\begin{pmatrix} 0 \\ I \end{pmatrix}$$

are complementary subspaces.

B.2 THE RICCATI OPERATOR

If X_1^{-1} exists, then we can define a function $\mathrm{Ric}(\cdot)$ that maps the symplectic pair H to the matrix $X := X_2 X_1^{-1}$. We say H is in the domain of Ric, denoted by $\mathrm{dom}(\mathrm{Ric})$ if

1. H has no generalized eigenvalues on the unit circle.
2. X_1^{-1} exists.
3. $I + RX$ is invertible.

In this case we have $\mathrm{Ric}(H) = X$.

Theorem B.2.1.   Let $H \in \mathrm{dom}(\mathrm{Ric})$ and $X = \mathrm{Ric}(H)$. The following statements hold:

1. $X = X^*$.
2. X satisfies the discrete Riccati equation

$$X = A^* X (I + RX)^{-1} A + Q.$$

3. The matrix $(I + RX)^{-1} A$ is stable.

Proof. To show the first part, we verify that $X_1^* X_2 = X_2^* X_1$. From Equation (B.1), it follows that

$$AX_1 = (X_1 + RX_2)T, \tag{B.2}$$

$$X_2 = QX_1 + A^* X_2 T. \tag{B.3}$$

From Equation (B.3), we compute $X_1^* X_2 - X_2^* X_1$:

$$X_1^* X_2 - X_2^* X_1 = (AX_1)^* X_2 T - T^* X_2^* (AX_1).$$

Using Equation (B.2), it follows that

$$X_1^* X_2 - X_2^* X_1 = T^* (X_1^* X_2 - X_2^* X_1)T.$$

This is a Lyapunov equation with a stable T, which implies that

$$X_1^* X_2 - X_2^* X_1 = 0.$$

This proves the first part of the theorem. Notice that since $H \in \mathrm{dom}(\mathrm{Ric})$, the matrix $(I + RX)$ has an inverse. From Equation (B.2), it follows that

$$T = X_1^{-1}(I + RX)^{-1} AX_1.$$

Since T is stable, so is $(I + RX)^{-1}A$, which proves part 3. Substituting this identity in Equation (B.3), we get the standard Riccati equation in part 2.   ■

One symplectic matrix.   If A^{-1} exists, then the spectral subspace can be computed directly from the matrix $H := H_2^{-1} H_1$. H is given by

$$H = \begin{pmatrix} A + RA^{*-1}Q & -RA^{*-1} \\ -A^{*-1}Q & A^{*-1} \end{pmatrix},$$

and it satisfies

$$H^* J H = J.$$

Since H replaces the symplectic pair, we used the same symbol to represent it. It turns out in this case that condition (3) defining the $\mathrm{dom}(\mathrm{Ric})$ is redundant.

Proposition B.2.1.   Suppose A^{-1} exists. Then if H has no eigenvalues on the unit circle and X_1^{-1} exists, then $(I + RX)$ is invertible.

Proof. This follows by simply taking the determinant of Equation (B.2).   ■

The following result applies for a class of symplectic pairs that arise in the $\mathcal{H}_2$ problem. We will assume that all the parameters (A, R, Q) are real.

Theorem B.2.2.   Suppose that $R = BB^T$ and $Q = C^T C$ such that (A, B) is stabilizable (i.e., the unreachable modes are stable) and (C, A) has no unobservable modes on the unit circle. Then $H \in \text{dom(Ric)}$ and $X = \text{Ric}(H) \geq 0$ (positive semidefinite). X is positive definite if and only if all the eigenvalues in the disc are observable.   ∎

NOTES AND REFERENCES

Details on discrete-time Riccati equations can be found in [Kuc79], [Wim91].

Notation, Symbols and Acronyms

$\mathbf{Z}$	Set of all integers.
$\mathbf{Z}_+$	Set of all nonnegative integers.
$\mathbf{Z}_-$	Set of all negative integers.
$\ell(\mathbf{Z})$	Space of all real sequences supported on the integers.
$\ell_p(\mathbf{Z})$	Space of real sequences supported on the integers such that if $x \in \ell_p(\mathbf{Z})$ then $$\|x\|_p = \sum_{k=-\infty}^{\infty} |x(k)|^p < \infty.$$
$\ell_\infty(\mathbf{Z})$	Space of all bounded sequences of real numbers supported on the integers such that if $x \in \ell_\infty(\mathbf{Z})$ then $\|x\|_\infty := \sup_k |x(k)| < \infty$.
ℓ	Space of all real sequences supported on the nonnegative integers.
ℓ_p	Space of real sequences supported on the nonnegative integers such that if $x \in \ell_p$ then $\|x\|_p = \sum_{k=0}^{\infty} |x(k)|^p < \infty.$
ℓ_∞	Space of all bounded sequences of real numbers supported on the nonnegative integers. If $x \in \ell_\infty$ then $\|x\|_\infty := \sup_k |x(k)| < \infty$.
c_0	Subspace of ℓ_∞ consisting of all elements whose entries decay to zero; i.e., $\lim_{k\to\infty} m(k) = 0$.
$\mathbb{R}$	The real numbers.
$\mathbb{R}_+$	The nonnegative real numbers.
$\mathbf{C}$	The complex numbers.
$\mathcal{D}$	The open unit disc.

$\mathcal{L}_p(B)$ Space of Lebesgue integrable function on $B \subset \mathbb{R}$ such that $\|x\|_p = \int_B |x(t)|_p < \infty$.

$\mathcal{L}_\infty(B)$ Space of Lebesgue integrable function on $B \subset \mathbb{R}$ such that $\|x\|_\infty = \operatorname*{ess\,sup}_{t \in B} |x(t)| < \infty$. (ess $\sup_B$ means the supremum over B except for sets of zero measure.)

$\mathcal{H}_2$ The Hilbert space of square integrable complex valued functions on the unit circle that have analytic continuation in the unit disc.

$\mathcal{H}_2^\perp$ The Hilbert space of square integrable complex valued functions on the unit circle that have analytic continuation outside the unit disc.

$\mathcal{L}_\infty$ Space of complex-valued functions on the unit circle that are bounded.

$\mathcal{H}_\infty$ Space of complex-valued functions analytic in $\mathcal{D}$ bounded on the unit circle.

$\mathcal{RH}_\infty$ Space of real rational functions inside $\mathcal{H}_\infty$.

$X^{p \times q}$ Space of $p \times q$ matrices with entries in X.

λ Complex variable representing the unit delay. Given $M \in \ell^{p \times q}$, define $\hat{M}(\lambda) := \sum_{k=0}^{\infty} M(k)\lambda^k$ as the λ-transform of M.

P_k The truncation operator on sequences. Hence, if $x = \{x(i)\}_{i=0}^{\infty}$ is any sequence, then $P_k x = \{x(0), x(1), \ldots, x(k), 0, \ldots\}$.

Π_M Projection on a subspace M.

S_k Right shift by k positions. If $x = \{x(i)\}_{i=0}^{\infty}$ is any sequence and k is a nonnegative integer, then $S_k x = \{\overbrace{0, \ldots, 0}^{k}, x(0), x(1), \ldots\}$.

S_{-k} Left shift by k positions. This operator is well defined on the space $S_k \ell$.

P   Reachability Gramian.

Q   Observability Gramian.

$\rho(A)$ Spectral radius of a matrix A.

$\lambda_{\max}(A)$ Maximum eigenvalue of a matrix A.

$\sigma_{\max}(A)$ Maximum singular of a matrix A.

$SN_{\Delta, p}(M)$ Structured norm.

$\mu_\Delta[\hat{M}(e^{i\theta})]$ Structured singular value.

A^* Adjoint of an operator A.

$\hat{G}^\sim$ Adjoint of a rational function; $\hat{G}^\sim(\lambda) = \hat{G}^T\left(\frac{1}{\lambda}\right)$.

$\mathcal{N}(A)$ Symbol for null space.

$\mathcal{R}(A)$ Symbol for the range.

$\overline{a}$ Complex conjugate of $a \in \mathbf{C}$.

$\angle a$ The angle of a complex number a.

$\limsup$ The largest cluster point.

$\liminf$ The smallest cluster point.

$\overline{M}$	Closure of a set M.
BM	Elements of M with norm less than one.
X^*	Dual space of X.
$S^\perp$	Annihilator subspace of $S \subset X$ ($\subset X^*$).
$^\perp S$	Annihilator subspace of $S \subset X^*$ ($\subset X$).
$\langle x, y \rangle$	The action of $y \in X^*$ on $x \in X$.
DA	Delay augmentation algorithm.
FDLTI	Finite dimensional linear time invariant.
FIR	Finite impulse response.
ITAE	Integral-time absolute error.
LFT	Linear fractional transformation.
LMI	Linear matrix inequality.
LTI	Linear time invariant.
LTV	Linear time varying.
MIMO	Multi-input multi-output.
NLTI	Nonlinear time invariant.
NLTV	Nonlinear time varying.
NLS	Normed linear space.
PD	Partially dominant.
RMS	Root mean square.
SISO	Single-input single-output.
TD	Totally dominant.

We use different representations of operators. We denote the operator by standard symbols, say M. If time invariant, then the λ-transform is denoted by $\hat{M}$. The time-domain impulse response is denoted by M. So we do not distinguish between the operator and the time-domain representation.

In general, all the operators are denoted by capital letters. Sometimes, to emphasize that something is SISO, we use small letters. This, however, is not our convention. Whether something is SISO or MIMO follows from the context.

Except if necessary, we do not show the dimension of the space; i.e., we use X instead of $X^{p \times q}$. The dimensions follow from the context.

The space $\mathcal{RH}_\infty$ contains all real rational stable functions. It is also thought of as the space of systems that have a finite-dimensional realization and are stable. Thus, we say $\hat{M} \in \mathcal{RH}_\infty$ or $M \in \mathcal{RH}_\infty$.

In block diagrams systems are represented by boxes, and a summation of signals is represented by a circle. If we intend to subtract signals, we explicitly put a minus sign near the circle.

Bibliography

[AAK71]   V.M. Adamjan, D.Z. Arov, and M.G. Krein. Analytic properties of Schmidt pairs for a Hankel operator and the generalized Schur-Takagin problem. *Math USSR Sbornik*, 15, 1971.

[Ack85]   J. Ackermann. *Sampled-Data Control Systems*. Springer-Verlag, 1985.

[AD84]   V. Anantharam and C.A. Desoer. On the stabilization of nonlinear systems. *IEEE Trans. on Automatic Control*, 29, 6, June 1984.

[AF66]   M. Athans and P. Falb. *Optimal Control*. McGraw-Hill, 1966.

[AM90]   B.D.O. Anderson and J.B. Moore. *Optimal Control: Linear Quadratic Methods*. Prentice-Hall, 1990.

[AN87]   E.J. Anderson and P. Nash. *Linear Programming in Infinite-Dimensional Spaces*. John Wiley, 1987.

[AP84]   E.J. Anderson and A.B. Philpott. *Infinite Programming*. Springer-Verlag, September 1984. Lecture Notes in Economic and Mathematical Systems.

[AW90]   K.J. Astrom and B. Wittenmark. *Computer Controller Systems: Theory and Design (2nd ed.)*. Prentice-Hall, 1990.

[BS93]   F. Blanchini and M. Sznaier. "Rational $\mathcal{L}^1$ Subophimal Compensators for Continuous time systems". *Proceedings of the American Control Conference,* June 1993.

[BG84]   A.E. Barabanov and O.N. Granichin. *Optimal Controller for a linear plant with bounded noise*. Translated from Automatika i Telemechanika, No. 5, p.p. 39–46, May 1984.

[Bar89]   B.R. Barmish. A generalization of Kharitonov's four polynomial concept for robust stability problems with linearly dependent coefficient perturbations. *IEEE Trans. on Automatic Control*, 34, February 1989.

[Bar84]   S. Barnett. *Matrices in Control Theory*. Robert E. Krieger Publishing Co., 1984.

[BB91]   S.P. Boyd and C.H. Barratt. *Linear Controller Design: Limits of Performance*. Prentice Hall, 1991.

[BB92]   V. Balakrishnan and S.P. Boyd. On computing the worst-case peak gain of linear systems. *Systems and Control Letters*, 19, October 1992.

[BD87]   S.P. Boyd and J.C. Doyle. Comparison of peak and RMS gains for discrete-time systems. *Systems and Control Letters*, 9, June 1987.

[BD92]   B. Bamieh and M. Dahleh. On robust stability with structured time-invariant perturbations. In *Systems and Control Letters*, 21, August 1993.

[BDP93]   B.A. Bamieh, M.A. Dahleh, and J.B. Pearson. Minimization of the $\mathcal{L}_\infty$-induced norm for sampled-data systems. *IEEE Trans. on Automatic Control*, 38, May 1993.

[BGR88]   J.A. Ball, I. Gohberg, and L. Rodman. Realization and interpolation of rational matrix functions. *Operator Theory: Advances and Applications*, 33, 1988.

[BH75]   A.E. Bryson and Y.C. Ho. *Applied Optimal Control*. Hemisphere Publishing, 1975.

[Bha87]   S.P. Bhattacharyya. *Robust Stabilization Against Structured Perturbations*. Springer-Verlag, 1987.

[BMM90]   J.S. Baras, D.C. MacEnany, and R.L. Munach. Fast error-free algorithms for polynomial matrix computations. *Proceedings of the 29th IEEE Conference on Decision and Control*, 1990.

[BP92]   B. Bamieh and J.B. Pearson. A general framework for linear periodic systems with application to $\mathcal{H}_\infty$ sampled-data control. *IEEE Trans. on Automatic Control*, 37, April 1992.

[BR90]   J.A. Ball and M. Rakowski. Zero-pole structure of nonregular rational matrix functions. *Operator Theory: Advances and Applications*, 47, 1990.

[CB89]   H. Chapellat and S.P. Bhattacharyya. A generalization of Kharitonov's theorem: Robust stability for interval plants. *IEEE Trans. on Automatic Control*, 34, March 1989.

[CD82]   F.M. Callier and C.A. Desoer. *Multivariable Feedback Systems*. Springer-Verlag, 1982.

[CD92]   H. Chapellat and M. Dahleh. Analysis of time-varying control strategies for optimal disturbance rejection and robustness. *IEEE Trans. on Automatic Control*, 37, November 1992.

[CF90]   T. Chen and B.A. Francis. On the $\mathcal{L}_2$ induced norm of a sampled-data system. *Systems and Control Letters*, 15, 1990.

[CF91a]   T. Chen and B.A. Francis. Input-output stability of sampled-data systems. *IEEE Trans. on Automatic Control*, 36, January 1991.

[CF91b]   T. Chen and B.A. Francis. $\mathcal{H}_2$-optimal sampled-data control. *IEEE Trans. on Automatic Control*, 36, April 1991.

[CFN91]   J. Chen, M.K.H. Fan, and C.N. Nett. The structured singular value and stability of uncertain polynomials: A missing link. *Control of Systems with Inexact Dynamic Models, ASME*, 1991.

[Che84]   C.T. Chen. *Linear System Theory and Design*. Holt, Rinehart and Winston, 1984.

[Chv83]   V. Chvatal. *Linear Programming*. Princeton University Press, 1983.

[Con85]   J.B. Conway. *A Course in Functional Analysis*. Springer-Verlag, 1985.

[CP84]   B.C. Chang and J.B. Pearson. Optimal disturbance reduction in linear multivariable systems. *IEEE Trans. on Automatic Control*, 29, October 1984.

[CRR69]   R. Calvin, C. Ray, and V. Rhyne. The design of optimal convolutional filters via linear programming. *IEEE Trans. Geoscience Elec.*, 7, July 1969.

[Dah92]   M.A. Dahleh. BIBO stability robustness in the presence of coprime factor perturbations. *IEEE Trans. on Automatic Control*, 37, March 1992.

[Dah93]   M.A. Dahleh. *Robust Controller Design: A Bounded-Input Bounded-Output Worst-Case Approach*, vol. 53. Academic Press, 1993. In Control and Dynamical Systems, a special issue in *High-performance Systems Techniques and Applications*, ed. C.T. Leondes.

[DBD92]   I.J. Diaz-Bobillo and M.A. Dahleh. State feedback ℓ_1 optimal controllers can be dynamic. *Systems and Control Letters*, 19, August 1992.

[DBD93]   I.J. Diaz-Bobillo and M.A. Dahleh. Minimization of the maximum peak-to-peak gain: The general multiblock problem. *IEEE Trans. on Automatic Control*, 38, October 1993.

[DCB90]   M. Dahleh, H. Chapellat, and S.P Bhattacharyya. Robust stability under structured and unstructured perturbations. *IEEE Trans. on Automatic Control*, 35, October 1990.

[DD90]   M. Dahleh and M.A. Dahleh. Optimal rejection of persistent and bounded disturbances: Continuity properties and adaptation. *IEEE Trans. on Automatic Control*, 35, June 1990.

[DD91]   M. Dahleh and M.A. Dahleh. On slowly time varying systems. *Automatica*, 27, January 1991.

[DDV79]   P.M. Van Dooren, P. Dewilde, and J. Vandewalle. On the determination of the Smith-Macmillan form of a rational matrix from its Laurent expansion. *IEEE Trans. on Circuits and Systems*, 1979.

[DF92]   G.E. Dullerud and B.A. Francis. $\mathcal{L}_1$ analysis and design of sampled-data systems. *IEEE Trans. on Automatic Control*, 37, April 1992.

[DFT92]   J.C. Doyle, B.A. Francis, and A.R. Tannenbaum. *Feedback Control Theory*. Macmillan, 1992.

[DG92]   G.E. Dullerud and K. Glover. Necessary and sufficient conditions for robust stability of SISO sampled-data systems to LTI perturbations. In *Proceedings of the American Control Conference*, June 1992.

[DG93]   G. E. Dullerud and K. Glover. Robust stabilization of sampled-data systems to structured LTI perturbations. *IEEE Trans. on Automatic Control*, 38, October 1993.

[DGKF89]   J.C. Doyle, K. Glover, P. Khargonekar, and B.A. Francis. State-space solutions to standard $\mathcal{H}_2$ and $\mathcal{H}_\infty$ control problems. *IEEE Trans. on Automatic Control*, 34, August 1989.

[DK93]   M.A. Dahleh and M.H. Khammash. Controller design for plants with structured uncertainty. *Automatica*, 29, January 1993.

[DL82]   C.A. Desoer and R.W. Liu. Global parametrization of feedback system with nonlinear plants. *Systems and Control Letters*, 1, January 1982.

[DO88]   M.A. Dahleh and Y. Ohta. A necessary and sufficient condition for robust BIBO stability. *Systems and Control Letters*, 11, 1988.

[DO92]   M.A. Dahleh and Y. Ohta. $\mathcal{L}_1$ sensitivity minimization for plants with commensurate delays. *Mathematics of Control, Signals and Systems*, 5, 1992.

[Doy82]   J.C. Doyle. Analysis of feedback systems with structured uncertainty. *IEE Proceedings*, 129, 1982.

[Doy83]   J.C. Doyle. Synthesis of robust controllers and filters. *Proceedings of the 22th IEEE Conference on Decision and Control*, December 1983.

[Doy84]   J.C. Doyle. *Matrix interpolation theory and optimal control*. UC Berkeley, 1984. Ph.D Thesis.

[Doy85]   J.C. Doyle. Structured uncertainty in control design. *IFAC Workshop on Estimation and Control of Uncertain Systems*, June 17–18 1985.

[DP87a]   M.A. Dahleh and J.B. Pearson. ℓ_1 optimal feedback controllers for MIMO discrete-time systems. *IEEE Trans. on Automatic Control*, 32, April 1987.

[DP87b]   M.A. Dahleh and J.B. Pearson. $\mathcal{L}_1$ optimal feedback compensators for continuous-time systems. *IEEE Trans. on Automatic Control*, 32, October 1987.

[DP88a]   M.A. Dahleh and J.B. Pearson. Optimal rejection of persistent disturbances, robust stability and mixed sensitivity minimization. *IEEE Trans. on Automatic Control*, 33, August 1988.

[DP88b]   M.A. Dahleh and J.B. Pearson. Minimization of a regulated response to a fixed input. *IEEE Trans. on Automatic Control*, 33, October 1988.

[DS74]   C.A. Desoer and J.D. Schulman. Zeros and poles of matrix transfer functions and their dynamical interpretation. *IEEE Trans. Circuits and Systems*, 21, 1974.

[DS81]   J.C. Doyle and G. Stein. Multivariable feedback design: Concepts for a classical/modern synthesis. *IEEE Trans. on Automatic Control*, 26, February 1981.

[DTV93]   M. Dahleh, A. Tesi, and A. Vicino. An overview of extremal properties for robust control of interval plants. *Automatica*, 29, May 1993.

[Dur70]   P.L. Duren. *Theory of $\mathcal{H}_p$ Spaces*. Academic Press, 1970.

[DV75]   C.A. Desoer and M. Vidyasagar. *Feedback Systems: Input-Output Properties*. Academic Press, 1975.

[DV90]   G. Deodhare and M. Vidyasasgar. ℓ_1-optimality of feedback control systems: The SISO discrete-time case. *IEEE Trans. on Automatic Control*, 35, September 1990.

[DVV92]   M.A. Dahleh, P. Voulgaris, and L. Valavani. Optimal and robust controllers for periodic and multi-rate systems. *IEEE Trans. on Automatic Control*, 37, January 1992.

[DWS82]   J.C. Doyle, J.E. Wall, and G. Stein. Performance and robustness analysis for structured uncertainties. *Proceedings of the 21st IEEE Conference on Decision and Control*, December 1982.

[END82]   A. Emami-Naeini and P. Van Dooren. Computation of zeros of linear multivariable systems. *Automatica*, 14, 1982.

[FHZ84]   B.A. Francis, J.W. Helton, and G. Zames. $\mathcal{H}_\infty$-optimal feedback controllers for linear multivariable systems. *IEEE Trans. on Automatic Control*, 29, October 1984.

[FL88]   J.S. Freudenberg and D.P. Looze. *Frequency Domain Properties of Scalar and Multivariable Feedback Systems*. Springer-Verlag, 1988. Lecture Notes in Control and Information Sciences.

[FPW90]   G.F. Franklin, J.D. Powell, and M.L. Workman. *Digital Control of Dynamic Systems*. Addison-Wesley, 1990.

[Fra87]   B.A. Francis. *A Course in* $\mathcal{H}_\infty$ *Control Theory*. Springer-Verlag, 1987.

[FTD91]   M.K.H. Fan, A.L. Tits, and J.C. Doyle. Robustness in the presence of mixed para-
            metric uncertainty and unmodelled dynamics. *IEEE Trans. on Automatic Control*,
            36, January 1991.

[FZ84]   B.A. Francis and G. Zames. On optimal sensitivity theory for SISO feedback systems.
            IEEE Trans. on Automatic Control, 29, January 1984.

[GD88]   K. Glover and J.C. Doyle. State-space formulae for all stabilizing controllers that
            satisfy an $\mathcal{H}_\infty$-norm bound and relations to risk sensitivity. *Systems and Control
            Letters*, 11, August 1988.

[Glo84]   K. Glover. All optimal Hankel-norm approximations of linear multivariable systems
            and their $\mathcal{L}_\infty$-error bounds. *International Journal of Control*, 39, June 1984.

[Glo89]   K. Glover. *A tutorial on Hankel-norm approximation*. Springer-Verlag, 1989. *From
            Data to Model*, ed. .J.C. Willems.

[GLR82]   I. Gohberg, P. Lancaster, and L. Rodman. *Matrix Polynomials*. Academic Press, 1982.

[GM89]   K. Glover and D. McFarlane. Robust stabilization of normalized coprime factor plant
            description with $\mathcal{H}_\infty$ bounded uncertainty. *IEEE Trans. on Automatic Control*, 34,
            August 1989.

[GV83]   G.H. Golub and C.F. Van Loan. *Matrix Compuations*. Johns Hopkins University
            Press, 1983.

[Hof62]   K. Hoffman. *Banach Spaces of Analytic Functions*. Prentice-Hall, 1962.

[IG91]   P.A. Iglesias and K. Glover. State-space approach to discrete-time $\mathcal{H}_\infty$ control. *In-
            ternational Journal of Control*, 54, 1991.

[KA92]   J.P. Keller and B.D.O. Anderson. A new approach to the discretization of continuous-
            time controllers. *IEEE Trans. on Automatic Control*, 37, February 1992.

[Kai80]   T. Kailath. *Linear Systems*. Prentice-Hall, 1980.

[KD92]   M. Khammash and M. Dahleh. Time-varying control and the robust performance of
            systems with structured norm-bounded perturbations. *Automatica*, 28, July 1992.

[Kha78]   V.L. Kharitonov. Asymptotic stability of an equilibrium position of a family of sys-
            tems of linear differential equations. *Differential. Uravnen.*, 14, 1978.

[Kha93]   M.H. Khammash. Necessary and sufficient conditions for the robustness of time-
            varying systems with applications to sampled-data systems. *IEEE Trans. on Auto-
            matic Control*, 38, January 1993.

[KP90]   M. Khammash and J.B. Pearson. Robust disturbance rejection in ℓ_1-optimal control
            systems. *Systems and Control Letters*, 14, 1990.

[KP91]   M. Khammash and J. B. Pearson. Performance robustness of discrete-time systems
            with structured uncertainty. *IEEE Trans. on Automatic Control*, 36, April 1991.

[KP93]   M. Khammash and J.B. Pearson. Analysis and design for robust performance with
            structured uncertainty. *Systems and Control Letters*, 20, March 1993.

[Kre89]   E. Kreyszig. *Introductory Functional Analysis with Applications*. John Wiley, 1989.

[KS72]   H. Kwakernaak and R. Sivan. *Linear Optimal Control Systems*. Wiley-Interscience,
            1972.

[KS82]   P. Khargonekar and E. Sontag. On the relation between stable matrix fraction fac-
           torizations and regulable realizations of linear systems. *IEEE Trans. on Automatic
           Control*, 27, June 1982.

[Kuc79]   V. Kucera. *Discrete Linear Control*. John Wiley, 1979.

[Kwa85]   H. Kwakernaak. Minimax frequency domain performance and robustness optimization
           of linear feedback systems. *IEEE Trans. on Automatic Control*, 30, October 1985.

[LGW89]   D.J.N. Limebeer, M. Green, and D. Walker. Discrete time $\mathcal{H}_\infty$ control. *Proceedings
           of the 28th IEEE Conference on Decision and Control*, 1989.

[Lue69]   D.G. Luenberger. *Optimization by Vector Space Methods*. John Wiley, 1969.

[Lue84]   D.G. Luenberger. *Linear and Nonlinear Programming*. Addison-Wesley, 1984.

[LZD91]   W.M. Lu, K. Zhou, and J.C. Doyle. Stabilization of LFT systems. *Proceedings of the
           30th IEEE Conference on Decision and Control*, 1991.

[Mac89]   J.M. Maciejowski. *Multivariable Feedback Design*. Addison-Wesley, 1989.

[MB82]   M. Milanese and G. Belforte. Estimation theory and uncertainty intervals evaluation
           in the presence of unknown but bounded errors: Linear families of models and
           estimator. *IEEE Trans. on Automatic Control*, 27, April 1982.

[Meg]   A. Megretsky. Necessary and sufficient conditions of stability: A multiloop general-
           ization of the circle criterion. To appear in *IEEE Trans. on Automatic Control*.

[Men89]   M.A. Mendlovitz. A simple solution to the ℓ_1 optimization problem. *Systems and
           Control Letters*, 12, 1989.

[Mey88]   D.G. Meyer. Two properties of ℓ_1 optimal controllers. *IEEE Trans. on Automatic
           Control*, 33, September 1988.

[Mey90a]   D.G. Meyer. On finding the minimum possible peak value of multiinput infinite-
           horizon steering controls. *IEEE Trans. on Automatic Control*, 33, February 1990.

[Mey90b]   D.G. Meyer. A parametrization of stabilizing controllers for multirate sampled-data
           systems. *IEEE Trans. on Automatic Control*, 33, February 1990.

[MK76]   A.G. MacFarlane and N. Karcanias. Poles and zeros of linear multivariable systems:
           A survey of the algebraic, geometric and complex-variable theory. *International
           Journal of Control*, 24–1, 1976.

[MP91]   J.S. McDonald and J.B. Pearson. ℓ_1-optimal control of multivariable systems with
           output norm constraints. *Automatica*, 27, March 1991.

[MZ89]   M. Morari and E. Zafirou. *Robust Process Control*. Prentice-Hall, 1989.

[Neh57]   Z. Nehari. On bounded bilinear forms. *Annals of Math.*, 65, No. 1, 1957.

[Net86]   C.N. Nett. Algebraic aspects of linear control systems stability. *IEEE Trans. on Au-
           tomatic Control*, 31, October 1986.

[Neu62]   L.W. Neustadt. Minimum effort control systems. *SIAM J. Contr.*, 1, 1962.

[NJB84]   C. N. Nett, C. A. Jacobson, and M. J. Balas. A connection between state-space and
           doubly coprime fractional representations. *IEEE Trans. on Automatic Control*, 29,
           September 1984.

[NP92]   K.M. Nagpal and K. Poolla. On ℓ_1 filtering and smoothing. *Proceedings of the 31st
           IEEE Conference on Decision and Control*, December 1992.

[OS70]   A.V. Oppenheim and R.W. Schafer. *Digital Signal Processing*. Prentice-Hall, 1970.

[OZ93]   J.G. Owen and G. Zames. Duality theory of robust disturbance attenuation. *Automat-
           ica*, 29, May 1993.

[PD93]   A Packard and J.C. Doyle. The complex structured singular value. *Automatica, Special Issue on Robust Control*, 29, January 1993.

[PKT⁺92]   K.R. Poolla, P. Khargonekar, A. Tikku, J. Krause, and K. Nagpal. A time-domain approach to model validation. *Proceedings of the American Control Conference*, June 1992.

[Pow82]   S.C. Power. *Hankel operators on Hilbert spaces*. Pitman, 1982.

[QD89]   L. Qui and E.J. Davison. A simple procedure for the exact stability robustness computation of polynomials with affine coefficient perturbations. *Systems and Control Letters*, 13, 1989.

[Rab72]   L. Rabiner. Linear program design of finite impulse response (FIR) digital filters. *IEEE Trans. Audio and Electroacoustics*, 20, October 1972.

[Ran92]   A. Rantzer. Stability conditions for polytopes of polynomials. *IEEE Trans. on Automatic Control*, 37, 1992.

[Red60]   R.M. Redheffer. On a certain linear fractional transformation. *J. Math. Phys.*, 39, 1960.

[RF58]   J.R. Ragazzini and G.F. Franklin. *Sammpled-Data Control Systems*. McGraw-Hill, 1958.

[RM]   A. Rantzer and A. Megretsky. A convex parametrization of robustly stabilizing controllers. To appear in *IEEE Trans. on Automatic Control*.

[Rud73a]   W. Rudin. *Complex Analysis*. McGraw-Hill, 1973.

[Rud73b]   W. Rudin. *Functional Analysis*. McGraw-Hill, 1973.

[Saf82]   M.G. Safonov. Stability margins of diagonally perturbed multivariable feedback systems. *IEE Proceedings*, 129, November 1982.

[SD93]   R. Saligrama and M.A. Dahleh. Examples of extreme cases in ℓ_1 and $\mathcal{H}_\infty$ optimization. *Proceedings of the 32th IEEE Conference on Decision and Control*, Dec. 1993.

[Sar67]   D. Sarason. Generalized interpolation in $\mathcal{H}_\infty$. *Trans. Amer. Math. Soc.*, 127, May 1967.

[SD91]   J.S. Shamma and M.A. Dahleh. Time varying vs. time invariant compensation for rejection of persistent bounded disturbances and robust stability. *IEEE Trans. on Automatic Control*, 36, July 1991.

[Sha93]   J.S. Shamma. Nonlinear state feedback for ℓ_1 optimal control. *Systems and Control Letters*, 21, October 1993.

[Sha94]   J.S. Shamma. Robust stability with time-varying structured uncertainty. *IEEE Trans. on Automatic Control*, 39, April 1994.

[SK91]   N. Sivashankar and P.P Khargonekar. $\mathcal{L}_\infty$-induced norm of sample-data systems. *Proceedings of the 30th IEEE Conference on Decision and Control*, December 1991.

[SS89]   C.B. Schrader and M.K. Sain. Research on system zeros: A survey. *International Journal of Control*, 50, 1989.

[Sta90]   O.J. Staffans. On the four-block model matching problem in ℓ_1. *Helsinki University of Technology*, Espoo, 1990.

[Sta91]   O.J. Staffans. Mixed sensitivity minimization problems with rational ℓ_1-optimal solutions. *Journal of Optimization Theory and Applications*, 70, 1991.

[Sto92]   A.A. Stoorvogel. The discrete time $\mathcal{H}_\infty$ control problem with measurement feedback. *SIAM Journal on Control and Optimization*, 30, January 1992.

[SV83]   M.G. Safonov and M.S. Verma. Multivariable $\mathcal{L}_\infty$-sensitivity optimization and Hankel approximation. *Proceedings of the American Control Conference*, June 1983.

[VDVa]   P. Voulgaris, M.A. Dahleh, and L. Valavani. $\mathcal{H}_\infty$ and $\mathcal{H}_2$ optimal controllers for periodic and multi-rate systems. To appear in *Automatica*.

[VDVb]   P. Voulgaris, M.A. Dahleh, and L. Valavani. ℓ_∞ to ℓ_∞ performance of slowly-varying systems. MIT Technical Report #LIDS-P-2208, September 1993.

[Ver88]   M.S. Verma. Coprime factorizational representations and stability of nonlinear feedback systems. *International Journal of Control*, 48, 1988.

[Vid81]   M. Vidyasagar. *Input-Output Analysis of Large Scale interconnected Systems*. Springer-Verlag, 1981.

[Vid85]   M. Vidyasagar. *Control Systems Synthesis: A Factorization Approach*. MIT Press, 1985.

[Vid86]   M. Vidyasagar. Optimal rejection of persistent bounded disturbances. *IEEE Trans. on Automatic Control*, 31, June 1986.

[Vid91]   M. Vidyasagar. Further results on optimal rejection of persistent bounded disturbances. *IEEE Trans. on Automatic Control*, 36, June 1991.

[Vou93]   P. Voulgaris. ℓ_∞–ℓ_∞ filtering. Proceedings of the American Control Conference, June 1993.

[Wil71]   J. C. Willems. *The Analysis of Feedback Systems*. MIT Press, 1971.

[Wim91]   H. K. Wimmer. Normal forms and symplechic pencils and the discrete-time algebraic Riccati equation. *Linear Algebia and its Applications*, 147, 1991.

[WZ91]   L.Y. Wang and G. Zames. Local-global double algebras for slow $\mathcal{H}_\infty$ adaptation: Part 2: Optimization of stable plants. *IEEE Trans. on Automatic Control*, 36, February 1991.

[YJB76a]   D.C. Youla, H.A. Jabr, and J.J. Bongiorno. Modern Wiener-Hopf design of optimal controllers. Part 1: The single-input case. *IEEE-Trans. A-C*, 21, February 1976.

[YJB76b]   D.C. Youla, H.A. Jabr, and J.J. Bongiorno. Modern Wiener-Hopf design of optimal controllers. Part 2: The multivariable case. *IEEE Trans. on Automatic Control*, 21, June 1976.

[You93]   P.M. Young. Robustness with parametric and dynamic uncertainty. Ph.D Thesis, Caltech, 1993.

[Zam66]   G. Zames. On the input-output stability on nonlinear time-varying feedback systems. Part I: Conditions derived using concepts of loop gain, conicity and positivity. *IEEE Trans. on Automatic Control*, 11, April 1966.

[Zam81]   G. Zames. Feedback and optimal sensitivity: Model reference transformations, multiplicative seminorms, and approximate inverses. *IEEE Trans. on Automatic Control*, 26, April 1981.

[ZF83]   G. Zames and B.A. Francis. A new approach to classical frequency methods: Feedback and minimax sensitivity. *IEEE Trans. on Automatic Control*, 28, May 1983.

[ZK88]   K. Zhou and P. Khargonekar. An algebraic Riccati equation approach to $\mathcal{H}_\infty$ optimization. *Systems and Control Letters*, 11, 1988.

[ZW91]   G. Zames and L.Y. Wang. Local-global double algebras for slow $\mathcal{H}_\infty$ adaptation. Part 1: Inversion and stability. *IEEE Trans. on Automatic Control*, 36, February 1991.

Index